# Lemur Social Systems and Their Ecological Basis

# Lemur Social Systems and Their Ecological Basis

Edited by

Peter M. Kappeler and Jörg U. Ganzhorn
Deutsches Primatenzentrum
Göttingen, Germany

Springer Science+Business Media, LLC

Library of Congress Cataloging-in-Publication Data

Lemur social systems and their ecological basis / edited by Peter M. Kappeler and Jörg U. Ganzhorn.
p. cm.
"Proceedings of a Symposium on Lemur Social Systems and their Ecological Basis, held August 16-21, 1992, in Strasbourg, France"--Copr. p.
Includes bibliographical references and index.
ISBN 978-1-4899-2414-8 ISBN 978-1-4899-2412-4 (eBook)
DOI 10.1007/978-1-4899-2412-4
1. Lemurs--Behavior--Congresses. 2. Lemurs--Ecology--Congresses. 3. Social behavior in animals--Congresses. I. Kappeler, Peter M. II. Ganzhorn, Jörg U. III. Symposium on Lemur Social Systems and their Ecological Basis (1992 : Strasbourg, France)
QL737.P95L45 1993
599.8'1--dc20 93-23384
CIP

Proceedings of a symposium on Lemur Social Systems and Their Ecological Basis, held August 16–21, 1992, in Strasbourg, France

Cover photograph courtesy of Peter M. Kappeler

ISBN 978-1-4899-2414-8


Originally published by Plenum Press, New York in 1993
Softcover reprint of the hardcover 1st edition 1993

## PREFACE

The past decade has seen a steady increase in studies of lemur behavior and ecology. As a result, there is much novel information on newly studied populations, and even newly discovered species, that has not yet been published or summarized. In fact, lemurs have not been the focus of an international symposium since the Prosimian Biology Conference in London in 1972. Moreover, research on lemurs has reached a new quality by addressing general issues in behavioral ecology and evolutionary biology. Although lemurs provide important comparative information on these topics, this aspect of research on lemurs has not been reviewed and compared with similar studies in other primate radiations. Thus, as did many in the field, we felt that the time was ripe to review and synthesize our knowledge of lemur behavioral ecology. Following an initiative by Gerry Doyle, we organized a symposium at the XIVth Congress of the International Primatological Society in Strasbourg, France, where 15 contributions summarized much new information on lemur social systems and their ecological basis.

This volume provides a collection of the papers presented at the Strasbourg symposium (plus two reports from recently completed field projects). Each chapter was peer-reviewed, typically by one "lemurologist" and one other biologist. The first three chapters present novel information from the first long-term field studies of three enigmatic species. Sterling describes the social organization of *Daubentonia madagascariensis*, showing that aye-aye ranging patterns deviate from those of all other nocturnal primates. Colquhoun provides a synopsis of the behavioral ecology of *Eulemur macaco macaco* and describes the response of black lemurs to habitat degradation. Rigamonti reports on the feeding ecology and flexible social organization of *Varecia variegata rubra* on the Masoala peninsula. This section concludes with a report by White *et al.* on group transfer in *Varecia variegata*, which illuminates the mating system of this unusual species.

The two following chapters examine the effects of predation on lemur behavior and population biology. Goodman *et al.* provide a meticulous summary of all known predation events on lemurs and add an impressive quantitative data set on owl predation on a *Microcebus murinus* population. Macedonia summarizes his largely experimental studies of the behavioral responses of two diurnal lemurs, *Lemur catta* and *Varecia variegata*, to predators, and provides a convincing explanation for striking interspecific differences in these lemurs' responses to different types of predators. The evidence provided in both of these chapters puts an end to the myth, that lemurs have been faced with comparatively relaxed predation pressure.

The next four chapters present new insights into the behavioral ecology of the best-known lemur species, *Lemur catta*. Using the help of many volunteers, Jolly *et al.* were able to follow several groups of ringtailed lemurs simultaneously. This unique data set allows fascinating insights into the dynamics of home range use and inter-group interactions among the Berenty makis. Sauther and Sussman describe a different pattern of range use among the *L. catta* at Beza Mahafaly and present their interpretation of ringtailed lemur social organization. Rasamimanana and Rafidinarivo report on the effects of seasonality and reproductive activity on the feeding ecology of *L. catta* females at Berenty, while Sauther discusses behavioral aspects of resource and competition.

The following three chapters compare the effects of resource distribution and other ecological constraints on habitat utilization between pairs of closely related species. Ganzhorn shows how habitat ulitization and food selection of *Lepilemur mustelinus* is constrained by structural habitat variables and by the presence of *Avahi laniger*. Overdorff contrasts patterns of range use in sympatric *Eulemur fulvus* and *E. rubriventer*. Her long-term studies suggest that the number of reproducing females per group may have a profound effect on lemur ranging patterns. Meyers and Wright examine the relationship between food availability and reproductive activity in *Propithecus tattersalli* and *P. diadema*, showing that reproductive schedules in both populations are timed so, that weaning of infants coincides with the season of greatest new leaf availability.

The chapters by Morland and Pereira focus on the responses of lemurs to the challenges of a highly seasonal environment. Morland uses data from *Varecia variegata* to illustrate the behavioral responses of lemurs to annual cycles of temperature change. Pereira emphasizes the importance of environmental regulation of life histories. Using data from wild and captive *L. catta*, he demonstrates that seasonal changes in growth rate and adult body weight represent adaptations to conserve energy. He extends this analysis to formulate a new model that neatly integrates many intriguing aspects of lemur life history and behavior.

The last two chapters address several aspects of lemur social evolution. Kappeler examines the morphological responses of lemurs to sexual selection and explores their potential consequences for social behavior. He shows that lemurs with diverse social systems are apparently characterized by similar intensities of pre- and post-copulatory reproductive competition, and concludes that the limited sexual dimorphism may have facilitated the evolution of female dominance in lemurs. In the final chapter, van Schaik and Kappeler analyze the diversity in lemur social systems and suggest that both activity period and fundamental life history traits are important general determinants of social systems. They identify male-female bonds as the basic building block of many lemur societies and test the hypothesis that the main function of permanent male-female associations is to reduce the risk of infanticide.

We feel that this volume provides a good summary of current research on the behavior and ecology of lemurs. It should be useful for graduate students and researchers whose experiences and interests lie outside this specific area, but who are interested in comparative information from this primate radiation. Students of lemur biology will hopefully find many stimulating ideas for future research.

We have enjoyed organizing the symposium and working on this book and thank the authors for their cooperation and support. The overall scope and quality of this volume were significantly improved by expertise comments from the many reviewers, and we gratefully acknowledge their help. Last, but not least, we thank Jutta Schmid and Stefanie Heiduck, who cheerfully helped with numerous editorial chores, Annie Arsène for translating the French abstracts, and Mary Safford of Plenum Press for her enthusiasm and support.

Peter M. Kappeler
Jörg U. Ganzhorn

Göttingen, im März 1993

**CONTENTS**

# PATTERNS OF RANGE USE AND SOCIAL ORGANIZATION IN AYE-AYES (*DAUBENTONIA MADAGASCARIENSIS*) ON NOSY MANGABE

Eleanor J. Sterling

Department of Anthropology
Yale University
51 Hillhouse, Ave
New Haven, CT 06511
U.S.A.

## ABSTRACT

During a two-year field study on aye-aye (*Daubentonia madagascariensis*) ecology and behavior on the island of Nosy Mangabe, I collected information on mating, spacing, and social behavior in 6 males and 2 females. *Daubentonia* spacing patterns on Nosy Mangabe are different from those of other nocturnal primates studied to date. Male home ranges overlaped greatly, while female home ranges rarely overlapped. Large female home ranges make it difficult for males to defend a single female and her territory. This, coupled with preliminary evidence of asynchronous breeding seasons, would explain why males exhibited polygyny rather than monogamy and did not exhibit long-term defense of resources. However, peculiarities of Nosy Mangabe - small size, lack of predator pressure, limited immigration and emigration opportunities - may affect the social organization of the species on the island. The results of this study indicate that spacing systems do not adequately represent the breadth of diversity in social and mating systems across species and that they should only be used as a first order variable in determining categories of social organization in nocturnal primates.

## INTRODUCTION

The proliferation of studies on nocturnal primates during the past ten years has provided a rich basis for reassessing presumed distinctions between diurnal and nocturnal primate social behavior. Twenty years ago, nocturnal primates were widely held to be a socially homogenous group representing a hypothetical ancestral condition in contrast to the more diversified and derived diurnal primates (Charles-Dominique and Martin, 1970; Martin, 1972; Charles-Dominique, 1977; Eisenberg, 1981). Further field work uncovered more behavioral complexity within a species and more variation between species than anticipated (Charles-Dominique, 1978; MacKinnon and MacKinnon, 1980; Clark, 1985; Bearder, 1986; Harcourt and Nash, 1986; Crompton and Andau, 1987).

As data on the behavior and ecology of nocturnal primates have increased, researchers have categorized them using typologies that facilitate the search for patterns (Charles-Dominique, 1978; Bearder, 1986). Lack of rigor in defining key terms, however, has been a major stumbling block in these and other classifications. In this paper the term social

*Lemur Social Systems and Their Ecological Basis*, Edited by
P.M. Kappeler and J.U. Ganzhorn, Plenum Press, New York, 1993

organization encompasses three elements: social system, mating system and spacing system. While these three categories are not mutually exclusive, it is analytically useful to consider them separately. Social system refers to a "set of rules of conduct for the members of a self-identifying group" (Rowell, 1988). A mating system is an inferred set of rules guiding reproductive interactions between individuals. Spacing system refers to the inferred rules regulating the distribution of individuals over time in a self-identifying group in three-dimensional space. Human observers may not be able to infer the rules in social, spacing or mating systems adequately (Rowell, 1988). Instead, patterns gleaned from observed events and interactions are used to represent the system.

Most diurnal primates live in social groups consisting of individuals that know one another, interact regularly, and spend most of their time nearer to one another than to non-group members. In contrast, many nocturnal primates form a social *network* between animals that recognize one another and interact regularly, but may not spend a significant amount of time in proximity to one another (Richard, 1985). Rowell and Olson (1983; Rowell, 1988) identify two ways that primates coordinate their social interactions: direct communication such as gestures or noises and "monitor/adjust" communication wherein individuals monitor others' movements and readjust their own positions in response. Direct communication in nocturnal networks is often achieved using vocalizations over a distance. Indirect communication is mediated via scent-marking, wherein signals and responses are often separated in time. These spatially or temporally disjunct interactions are often difficult to detect. The consequent lack of information about social relationships in nocturnal primates influences our perception of the patterns in their mating, spacing and social systems.

Prior studies of nocturnal primates discussed social organization with reference to range overlap between same and opposite sex individuals, associations within ranges (including social interactions at night and associations at sleeping sites), dispersal patterns of individuals from natal areas, and mating patterns (Charles-Dominique, 1978; Bearder and Martin, 1979; Bearder, 1986; Harcourt and Nash, 1986). Bearder (1986) has summarized this information in a classification of nocturnal primates (Table 1). Perhaps because spatial data are easier to collect and therefore more readily available than information on social interaction in nocturnal primates, his typology uses dispersion patterns (representing spacing systems) as the first and therefore critical step in grouping species into types. Particularly with species for which there is little information about social interactions, parameters of the spatial system were the sole determinant used to assign a type (e.g., *Mirza coquereli*, type 4).

Spatial parameters such as the availability of individuals with which to interact undoubtedly influence social and mating behavior (Rowell, 1988), but these parameters may not be sufficient to predict the nature of unknown components in a social organization. Resource distribution and defensibility, predation pressure and the intensity and nature of inter-specific competition all may influence dispersion patterns and social and reproductive behavior in mammals (Alexander, 1974; Bradbury and Vehrencamp, 1977; Emlen and Oring, 1977; Wittenberger, 1979; Eisenberg, 1981; Terborgh and Janson, 1986). These factors influence males and females differently (Orians, 1969; Trivers, 1972), resulting in sex-specific social and reproductive behavior (Maynard-Smith, 1977).

Many researchers argue that female behavior is influenced more directly by ecological pressures than male behavior because food availability is a major limiting factor on female fitness. Male behavior focuses more on finding mates and achieving mating success, as predicted from sexual selection theory (Bradbury and Vehrencamp, 1977; Emlen and Oring, 1977; Wrangham, 1980; Fedigan, 1992). There are a number of models predicting male behavior on the basis of female distribution patterns (e.g., Wrangham, 1980; Rubenstein, 1986; Clutton-Brock, 1989; Davies, 1991). So far all of the models applied to primates presuppose that individuals are members of a social group. We do not know if, or how, these models might apply to social networks in nocturnal primates.

I derived and tested predictions from these models using the ecology and behavior of a nocturnal prosimian, *Daubentonia madagascariensis*. Preliminary observations on aye-ayes suggested that females might forage and sleep alone most of the year (Petter, 1977, pers. obs.). Models relating spacing to mating systems in mammalian species in which females are solitary provided a basis for the following predictions (Clutton-Brock,

**Table 1.** Selected parameters of social organization among nocturnal prosimians.

| Type | Range Size and Overlap Patterns[1] M-M | F-F | M-F | M>F | Matriarchy | Mating[2] system | Examples |
|---|---|---|---|---|---|---|---|
| 1 | P | P | P | P | P | 1,2 | Bushbabies, excluding *Galago zanzibaricus* |
| 2 | A | A | P | P | A | 1 | *Perodicticus potto* |
| 3 | V | V | C | V | V | 3,1 | *Tarsius bancanus, T. spectrum, Galago zanzibaricus* |
| 4 | A | A | A | ? | ? | ? | *Microcebus murinus, Mirza coquereli* |

[1]M-M: male-male range overlap; F-F: female-female range overlap; M-F: male-female range overlap; M>F: male range size larger than female. Matriarchies are related adult females with overlapping ranges. A = overlap absent; P = overlap present; C = ranges coincide; V = variable pattern. [2]1 = polygyny; 2 = polygyny, females may attract more than one male; 3 = monogamy. After Bearder (1986).

1989; Davies, 1991). If female range area is defensible, then aye-aye males should exhibit monogamy or polygyny; females should exhibit monogamy or polyandry; and both sexes should defend territories. However, if female range area is not defensible, then males and females should exhibit multi-male/multi-female mating systems (formerly called promiscuity, but see Rowell, 1991; Fedigan, 1992); and males should defend estrous females.

On the basis of male genital morphology (elongated bacculum and specialization of the genitalia), Dixson (1987a,b) predicted that aye-ayes will have either multiple-intromission or prolonged single-intromission copulatory patterns. He argued that these patterns occur more frequently in "multimale or dispersed/non-gregarious" mating systems than they do in monogamous or polygynous systems (but see Dewsbury and Pierce, [1989] for a larger sample resulting in ambiguous conclusions). Dixson pointed out that in many nongregarious prosimians mating is restricted to a short period of estrus during which males compete for females. By prolonging intromission, a male may effectively prevent other males from inseminating the female while increasing his chances for fathering the offspring.

In this paper, I describe ranging patterns and inter-individual encounters in a small population of *Daubentonia* on an island off the east coast of Madagascar. I provide details on the temporal and spatial availability of food resources, female and male dispersion patterns, and components of their mating and social systems.

## STUDY AREA AND METHODS

The study area, Nosy Mangabe Special Reserve, is a 520 ha island located 5 km off the east coast of Madagascar in the Bay of Antongil. The island supports an diverse, evergreen rain forest with a moderately high-canopy (A. Gentry, J. Miller and G. Schatz unpub. data; Gentry, 1988). Average yearly rainfall is 3806 mm (Direction de la Metéorologie, unpubl. data). In 1966, the island was designated a Special Reserve to protect the aye-aye, and nine aye-ayes were translocated there from the nearby mainland (Petter and Peyrieras, 1970).

A total of 496 observation hours derived from 46 dusk-to-dawn follows of aye-ayes were used in this analysis. Between November 1989 and February 1991, I and three guides captured eight animals (6 males (M) and 2 females (F)) and attached radio collars to them. Data for this study focus on two males and two females that were followed for between five and eleven months (F1 = 6 months; F2 = 5 months, M1 = 9 months; M4 = 11months).

The other four radio-collared individuals were each followed for less than one month because their collars stopped functioning.

Data presented here were collected after a period of habituation, which took from 14 to 32 days. Aye-ayes were considered habituated when they could be approached to within two meters when feeding at ground level without being disturbed. Uncollared animals were identified using differences in pelage color, size, characteristic marks on the ear pinnae, or in one case, eye color. Sex was determined visually when possible.

Our methods of estimating night range length developed throughout the study. During the first five months, we measured straight-line distances between successive landmark features along a path and each night's path was superimposed on a 20m grid pattern. But due to the non-linear travel pattern of the animals, we concluded that this method underestimated distances traveled. We subsequently recorded a continuous sample of estimated 20m-movements during the course of a night. Twice a week, we flagged every 20m segment travelled by the animal, where it changed points of direction, and where it stopped. On a subsequent day we retraced the path, taking bearings and distance measurements. These data were used to check the accuracy of our 20m-estimates. Concordance was high between estimated and measured distances (pairwise t-test: P=0.93). Data collected using both methods were mapped using Pathfinder 1.5, a software program developed by M. Winslett.

Home range size was determined using the minimum convex polygon method (Mohr, 1947; Southwood, 1966) because it is simple and commonly used, enhanceing comparisons with other studies (Harris et al., 1990). However, in this technique peripheral points strongly influence range size, possibly causing the range to encompass large areas not used by animals (Harris et al., 1990).

Data on food resource density were collected using the plot sampling method (Cox, 1985). One hundred and seventy 10m x 10m plots were placed randomly along three E-W and two N-S transects. In each plot we measured the availability and degree of decaying dead wood, and the diameter at breast height (DBH) and crown volume of tree species that provided food for aye-ayes. Plots were grouped for analysis into three categories: low = 0-110m above sea level (N=60; mid = 111-220m above sea level (N=56); high = 221-330m above sea level (N=54). Phenological data were collected along a 3km trail once a month for 243 individuals from 56 tree species (Sterling, 1993).

## RESULTS

### Diet and Food Availability

Four main foods accounted for over 90% of *Daubentonia's* diet each month on Nosy Mangabe: (1) seed from *Canarium* spp. (Burseraceae) fruit, (2) cankerous growth of *Intsia bijuga* (Fabaceae), (3) insect larvae (Cerambicidae: Lamiinae, Prioninae; Scarabidae: Dynastinae; Passalidae; Pyralidae: Phycitinae) families and (4) nectar from *Ravenala madagascariensis* (Strelitziaceae) flowers.

Aye-ayes gnawed on the endocarp of the *Canarium* with their anterior teeth and then extracted the cotyledon with their slender middle finger. There are two species of *Canarium* on Nosy Mangabe and both were eaten by aye-ayes. Individual trees are large-crowned and abundant on the island (Table 2). Fruit from one species or the other was available throughout the year (Sterling, 1993).

Aye-ayes removed cankers from a leguminous tree, *Instia bijuga*, and scraped a waxy substance from the underlying cambium. This food resource was patchily distributed and restricted to lower elevations of the island (<170m above sea level, Table 2). We measured availability of this resource once a month. There were periods of scarcity, but the periodicity of these patterns could not be established (Sterling, 1993).

Insect larvae were removed from a variety of sources including fallen dead wood, dead branches on a live tree, living trees, dead and living lianas, and seeds. Fallen dead wood occurred throughout the island in discrete patches of varying sizes (Table 2). Larger larvae, particularly those found in the *Canarium* spp., were available throughout the year (Sterling, in prep).

**Table 2.** Density of major resources used by aye-aye on Nosy Mangabe[1].

| Species | Low (N=60) | Mid (N=56) | High (N=54) |
|---|---|---|---|
| *Canarium* (large leaf) | 29.5 | 13.3 | 0.0 |
| *Canarium* (small leaf) | 0.0 | 1.9 | 9.9 |
| *Intsia bijuga* | 11.7 | 6.2 | 0.0 |
| *Ravenala madagascariensis* | 2.2 | 13.3 | 14.3 |
| Dead wood | 27.8 | 21.8 | 49.9 |

[1]For 170 10m x 10m plots from low, mid, and high elevation, the mean number of stems/ha is presented.

*Ravenala* inflorescenses were few per tree, but trees were often clumped in groups of three to twelve. They were most common at higher elevations on the island (>190m above sea level, Table 2), although this may have been a temporary artifact of forest succession processes and not reflective of a *Ravenala* predilection for higher elevations (G. Schatz, pers. comm). Nectar was present from December through April during this study.

### Ranging Patterns

The shortest male range length overlapped with the longest female range length. However, maximum distances travelled by males were over twice as great as the maximum distances travelled by females (Table 3). Male home range size was 3-6 times greater than female home range size (Table 3). There was considerable overlap between male home ranges, varying from 40 to 75%.

The two females studied had non-contiguous home ranges separated by 250-300m. During all-night follows we never encountered other females within their home ranges. We did occasionally see females outside these home ranges and we frequently encountered an uncollared female between and at the range boundaries of female #1 and female #2, suggesting that this uncollared female's home range was located between them.

**Table 3.** Aye-aye home range sizes (determined by a minimum convex polygon method) and average nightly range length on Nosy Mangabe between November 1989 and April 1991.

| Sex | ID | Mass (kg) | Home Range Area (ha) | Nightly Range Length Mean ± SD (m) | N | Max | Min |
|---|---|---|---|---|---|---|---|
| F | #1 | 2.8 | 39.5 | 1227.2 ± 173 | 8 | 1507.7 | 942.0 |
| F | #2 | 2.7 | 31.7 | 1528.0 ± 237 | 4 | 1832.5 | 1200.0 |
| M | #3 | 3.0 | 126.0 | 1551.8 ± 618 | 28 | 3848.0 | 819.0 |
| M | #4 | 2.6 | 214.6 | 2256.0 ± 1097 | 6 | 4391.5 | 1288.0 |

Male home ranges overlapped the ranges of several females. Male #4's home range overlapped home ranges of at least four females (females #1, #2, #3, #4). Three other collared male home ranges (males #1, #2, #3) each overlapped female #1, female #2 and female #3 home ranges.

Unfortunately, neither of the collared females were accompanied by offspring during this study so we were unable to collect information on dispersal patterns. Therefore it is not known whether or not matriarchies are present.

## Social Interactions

Individuals were never seen to sleep together in the same nest during this study, although adult males have been observed to share a nest on mainland Madagascar (Sterling, unpubl. data). Females were never seen to sleep within 50m of other females. Males slept in nests within 50m of other individuals only during the mating season.

Outside periods of mating activity, social interactions varied considerably from month to month. No incidences of grooming were observed between individuals during this study. Individuals occasionally foraged as a unit, using accompanying calls as they moved from resource to resource. Foraging associations were observed between two adult males, adult and young males, and adult males and females. No more than three individuals were seen foraging together outside periods of mating activity.

Although inter-individual encounters were infrequent, certain patterns emerged. Male #4 consistently avoided interaction with males #2 and #3, either by deviating from a well-established trail to leave a minimum of 15m between himself and the other animal, or by leaving a feeding tree when the other animal arrived. Male #4 and female #1 were seen to forage and travel as a unit on 8 occasions outside the mating season. Judging from his body mass (2.7kg), male #4 was presumably an adult or young adult.

## Mating Behavior

In addition to *ad hoc* observation on uncollared animals, we were able to witness mating activity by both of the telemetered females. In each case, estrus lasted for about three days. Ten days prior to full estrus, as inferred by genital swelling, females increased scent-marking frequency and often visited nests occupied by males, a behavior not seen outside the mating season. During the three days of estrus, females called repetitively, starting well before dusk. Up to six males surrounded the females subsequent to the calls and frequently engaged in agonistic chases and biting. Eventually one male copulated with the female, maintaining hold of her for between 55 and 65min (mean=61.5min, N=6).

After copulating, females traveled quickly 500-600m and then started calling again. Females mated with more than one male during the first day, but only mated with one male on the second and third days of estrus. We were unable to establish whether either female conceived and, if so, which male was the father.

Evidence from this study and from captures of young animals on the mainland (Winn, 1989; Durrell, 1991; K. Glander, pers. comm.) suggests that reproductive activity in *Daubentonia* may take place throughout the year (Sterling, unpubl. data). An individual female, however, does not necessarily cycle year-round. Each female was only observed in estrus once and we were unable to follow any female over a full year to determine whether they cycled again.

# DISCUSSION

## Food Availability and the Spacing System

The two females studied had ranges situated across similar elevation gradients (0-250m). This may reflect efforts to encompass all four major food types, including low-elevation *Intsia* and higher elevation *Ravenala* in a home range. These data are consonant with the idea that resource distribution is an important factor determining female dispersion, but provide no direct evidence that it is the primary determining factor. The fact that individual male home ranges overlapped those of many females may support the prediction that males are distributing themselves to take best advantage of the distribution of females. Similarly, preliminary evidence suggests that *Galago senegalensis* females distribute themselves with reference to vegetative resources and that male distribution is strongly influenced by female dispersion patterns (Bearder and Martin, 1979).

Alternatively, male aye-aye ranges may have exceeded in size and overlapped those of many females because males have greater nutritional requirements than females. However, if *Daubentonia* females can obtain adequate food in 30-40ha then males, which are of similar size, should not need 120-215ha in which to gather food resources.

## Ranging Behavior

Male aye-ayes on Nosy Mangabe had a high degree of range overlap and male ranges also overlapped those of many females. There was no evidence that female ranges overlapped. Sex was easily determined when visual inspection of genitalia was possible and there is no reason to believe that we had an identification bias towards males. Ancrenaz (1991) found a similar spacing system in aye-ayes in mainland forests near Mananara. The combination of a high degree of male home range overlap and the lack of evidence for female home range overlap in aye-ayes is surprising. These patterns are not exhibited by any other nocturnal primate under Bearder's (1986) classification. The only nocturnal primates exhibiting no female home range overlap are *Tarsius bancanus* and *Perodicticus potto*. However, in these species home ranges of males do not overlap either.

*Daubentonia* male home ranges were larger than female home ranges in this study. The primary difference was that males periodically went on extended forays into outlying areas, often covering between 2.2 and 4.4km per night on successive nights. These forays probably exaggerate home range size when using the minimum convex polygon method, and I propose to reanalyze movement data using the grid method to help identify areas within range boundaries that are not used (Reynolds, 1984) and to detect differing intensity of range use (Harris et al., 1990). Anyhow, females had much larger home ranges than diurnal lemurs of similar body size in the same habitat (e.g., *Varecia variegata variegata*: (Morland, 1991), *Eulemur fulvus albifrons*: H. Morland, pers. comm.; pers. obs.).

The exploratory sejourns of aye-aye males are striking. Travel over Nosy Mangabe's steep slopes using both terrestrial and arboreal locomotion requires a great expenditure of energy, suggesting that there must be strong incentives for males to travel long distances. On a number of these extended forays males did encounter estrous females. Elsewhere (Sterling, 1993) I have presented data from captive and wild populations indicating that aye-aye reproduction is not restricted to short breeding seasons typical of most lemuroids (e.g., Richard and Dewar, 1991). Estrus brevity and asynchrony mean that a male aye-aye's ability to detect when the female is in estrus is very important and very difficult. Long forays by males and consequent large home ranges may reflect male efforts to locate estrous females. Long forays during the mating seasons have also been reported for males in *Galago* spp. (Bearder and Martin, 1979).

## Mating System

My results support Dixson's (1987a,b) hypothesis that male aye-ayes monopolize an estrous female for prolonged periods, although it was impossible to tell from a distance whether this was through single or multiple intromissions. Copulations lasting more than one hour were also observed in *Galago crassicaudatus* and *Galago senegalensis* and were thought to represent mate guarding behavior by males (Clark, 1984 in Bearder, 1986).

Female mammals may foster competition between males to widen their choice of mates (Jarman and Southwell, 1986; Fedigan, 1992; Richard, 1992). Female aye-ayes showed signs of advertising estrus to wandering males. Several males chasing females at the peak of estrus has also been seen in other nocturnal primates - *Galago senegalensis* and *Nycticebus coucang* (Bearder and Martin 1979, Elliot and Elliot 1967).

Categories of mating behavior in mammals are decidedly male-biased in perspective and terminology (Rowell, 1991; Fedigan, 1992). I differentiate between male and female mating systems and use non-standard but less biased category titles put forth by Fedigan (1992). Male aye-aye mating behavior is similar to mammals with a "scramble competition polygyny" mating system (Peterson, 1955; Croft, 1981; Ramsay and Stirling, 1986). In scramble competition polygyny in males, females are solitary and males range widely in search of estrous females (Clutton-Brock, 1989). It is interesting to note that most mammals exhibiting high levels of male-male competition for females are sexually dimorphic (Wrangham, 1986), but *Daubentonia* exhibits no sexual dimorphism (see also Richard, 1992).

"Pair bonds" were short on Nosy Mangabe and females mated with more than one male in the same estrus, exhibiting polyandry. Therefore, the mating system exhibited by male and female *Daubentonia* on Nosy Mangabe would fall under a multimale-multifemale mating system (Fedigan, 1992).

**Social System**

This study does not provide enough information to characterize completely the social system of *Daubentonia* on Nosy Mangabe. Animals frequently foraged alone, yet there was a higher male-male and male-female encounter rate at feeding trees than to be expected with purely solitary foragers like *Perodicticus potto*. Affiliative interactions between individual aye-ayes were infrequent, and non-agonistic body contact outside the mating season was absent. This is in contrast to *Galago crassicaudatus* where nightly interactions were frequent and prolonged (Clark, 1985).

Patterns of interaction between individuals in this study may indicate a dominance hierarchy, at least among males. Unfortunately, we cannot conclusively evaluate this hypothesis, as *Daubentonia* may monitor signals such as scent marks that we are unable to sense. We observe only the adjustment and not necessarily the individual reacted to, and so a more exact description of the social system eludes us.

There are possible parallels between aye-aye social systems and those described for *Galago demidovii* (Charles-Dominique, 1977) and *Pongo pygmaeus* (Mitani et al., 1991). These systems are characterized by solitary behavior interspersed with intense competition between males for mates during the mating season. However, researchers distinguish between classes of males (two classes in *P. pygmaeus* and four in *G. demidovii*) in these systems based on differences in territorial defense and ranging patterns. *Daubentonia* males exhibited differential range use on Nosy Mangabe, but the small sample size and difficulties in observing social interactions precludes inferences regarding the reasons for these differences.

In the past, comprehensive treatments of social behavior in primates have overlooked primates living in social networks, i.e., nocturnal prosimians and *Pongo pygmaeus*. As we gather more information on social networks we increasingly recognize distinct patterns of behavior within them, some resembling systems in group-living diurnal primates (Bearder and Martin, 1979; Clark, 1985; Bearder, 1986; Harcourt, 1991). Because nocturnal primates interact in social networks, and all diurnal primates other than *Pongo pygmaeus* are gregarious, we often confuse the distinction between two dichotomies - that of diurnal/nocturnal and that of social group/social network. The juxtaposition of *Pongo pygmaeus* with nocturnal primates may help us eventually to distinguish between behavioral differences based on our ability to monitor communication between individuals at low available light levels and those that are inherent to a system.

Given that nocturnal primates often communicate with olfactory signals that are temporally deferred, that vocalizations sometimes cannot be traced to their source, and that observation conditions are often difficult, how much of the inferred difference between group-living diurnal and nocturnal primates is an artifact of sampling difficulties? It is possible that nocturnal primates communicate the same types of information and develop complex relationships similar to gregarious diurnal primates, but we do not have the sensory capacity to record these interactions. The presumed differences may be variations on a theme, with gregarious diurnal primate interactions merely representing a spatially and temporally compressed version of nocturnal primate interactions.

## ACKNOWLEDGEMENTS

This research was supported by a National Science Foundation Doctoral Dissertation Improvement Grant #8913021 (with A.F. Richard), a Fulbright-Hays Fellowship, the Wenner-Gren Foundation for Anthropological Research, the National Geographic Society, a John F. Enders Fellowship and the Williams Fund. I am grateful to the Government of Madagascar for giving me permission to work in Madagascar. I would like to thank the School of Agronomy of the University of Antananarivo, in particular M. Rakotomanga Pothin, for sponsoring my research. I thank G. Schatz and C. Remington for specimen identification. This work would not have been possible without the assistance of Victor Baba, Barthelamy Damary, or Fortuna Toto, nor the collaboration of Betsy Carlson. I thank M. Remis, N. Bynum, S.R. Beissinger and A.F. Richard for comments on this paper.

## REFERENCES


Alexander, R.D., 1974, The evolution of social behavior, *Ann. Rev. Ecol. System.* 5:325.

Ancrenaz, M., 1991, Contribution a l'étude éco-éthologique du aye-aye (*Daubentonia madagascariensis*). DVM Thesis, Ecole Nationale Veterinaire d'Alfort.

Bearder, S.K, 1986, Lorises, bushbabies, and tarsiers: diverse societies in solitary foragers, *in*: "Primate Societies," B.B. Smuts, D.L.Cheney, R.M. Seyfarth, R.W. Wrangham, and T.T. Struhsaker, eds., University of Chicago Press, Chicago.

Bearder, S.K., Martin, R.D., 1979, The social organization of a nocturnal primate revealed by radio tracking, *in*: "A Handbook on Biotelemetry and Radio Tracking," C.J. Amlaner and D.W. MacDonald, eds., Pergamon Press, Oxford.

Bradbury, J.W., and Vehrencamp, S.L., 1977, Social organization and foraging in emballonurid bats, III: Mating systems, *Behav. Ecol. Sociobiol.* 2:1.

Clark, A.B., 1985, Sociality in a nocturnal "solitary" prosimian: *Galago crassicaudatus*, *Int. J. Primatol.* 6:581.

Charles-Dominique, P., 1977, "Ecology and Behaviour of Nocturnal Prosimians," Columbia University Press, New York.

Charles-Dominique, P., 1978, Solitary and gregarious prosimians: Evolution of social structures in primates, *in*: "Recent Advances in Primatology, Vol. 3: evolution," D.J. Chivers, and K.A. Joysey, eds., Academic Press, London.

Charles-Dominique, P. and Martin, R.D., 1970, Evolution of lorises and lemurs, *Nature* 227:257.

Clutton-Brock, T.H., 1989, Mammalian mating systems, *Proc. R. Soc. Lond.* 236:339.

Cox, G.W., 1985, Laboratory Manual of General Ecology. 5th ed., W. C. Brown Publishers. Dubuque.

Croft, D.B., 1981, Behaviour of red kangaroos (*Macropus rufus*) in Northwestern New South Wales, *Aust. Mamm.* 4:5.

Crompton, R.H., and Andau, P.A., 1987, Ranging, activity rhythms, and sociality in free-ranging *Tarsius bancanus*: a preliminary report, *Int. J. Primat.* 8:43.

Davies, N.B., 1991, Mating systems, *in*: "Behavioural Ecology: An Evolutionary Approach," J.R. Krebs and N.B. Davies, eds., Blackwell, London.

Dewsbury,D.A., and Pierce, J.D. Jr, 1989, Copulatory patterns of primates as viewed in broad mammalian perspective, *Am. J. Primatol.* 17:51.

Dixson, A.F., 1987a, Bacculum length and copulatory behavior in primates, *Am. J. Primat.* 13:51.

Dixson, A.F., 1987b, Observation on the evolution of the genitalia and copulatory behaviour in male primates, *J. Zool. Lond.* 213:423.

Durrell, L., 1991, Notes on the Durrell expedition to Madagascar September-December 1990, *Dodo* 27:9.

Eisenberg, J.F., 1981, "The Mammalian Radiations," University of Chicago Press, Chicago.

Elliot, O., and Elliot, M., 1967, Field notes on the slow loris in Malaya, *J. Mamm.* 48:497.

Emlen, S.T., and Oring, L.W., 1977, Ecology, sexual selection, and the evolution of mating systems, *Science* 197:215.

Fedigan, L.M., 1992, "Primate Paradigms," University of Chicago Press, Chicago.

Gentry, A.H., 1988, Changes in plant community diversity and floristic composition on environmental and geographical gradients, *Ann. Miss. Bot. Garden* 75:1.

Harcourt, C., 1991, Diet and behaviour of a nocturnal lemur, *Avahi laniger*, in the wild, *J. Zool., Lond.* 233:667.

Harcourt, C.S., and Nash, L.T., 1986, Social organization of galagos in Kenyan coastal forests: I: *Galago zanzibaricus*, *Am. J. Primat.* 10:339.

Harris, S., Cresswell, W.J., Forde, P.G., Trewhella, W.J., Wollard, T., Wray, S., 1990, Home range analysis using radio-tracking data - a review of problems and techniques particularly as applied to the study of mammals, *Mamm. Rev.* 20:97.

Jarman, P.J., and Southwell, C.J., 1986, Grouping, associations, and reproductive Strategies in Eastern grey kangaroos, *in*: "Ecological Aspects of Social Evolution: Birds and Mammals," D.I. Rubenstein and R.W.Wrangham, eds., Princeton University Press, Princeton.

MacKinnon, J.R., and MacKinnon, K.S., 1980, The behaviour of wild spectral tarsiers, *Int. J. Primat.* 1:361.

Martin, R.D., 1972, Adaptive radiation and behaviour of the Malagasy lemurs, *Phil. Trans. Roy. Soc. Lond. Ser. B* 264:295.

Maynard-Smith, J., 1977, Parental investment - a prospective analysis, *Anim. Behav.* 25:1.

Mitani, J.C., Grether, G.F., Rodman, P.S., and Priatna, D., 1991, Associations among wild orang-utans: Sociality, passive aggregations or chance? *Anim. Behav.* 42: 33.

Mohr, C.O., 1947, Table of equivalent populations of North American small mammals, *Am. Mid. Nat.* 37:223.

Morland, H.S., 1991, Social organization and ecology of black and white ruffed lemurs (*Varecia variegata variegata*) in lowland rainforest, Nosy Mangabe, Madagascar, Ph.D. Dissertation, Yale University, New Haven.

Orians, G.H., 1969, On the evolution of mating systems in birds and mammals, *Amer. Nat.* 103:589.

Peterson, R.L., 1955, North American Moose, University of Toronto Press, Toronto.

Petter, J-J., 1977, The aye-aye, *in*: "Primate Conservation," Prince Rainier III and G.H. Bourne, eds., Academic Press, New York.

Petter, J-J., and Peyrieras, A., 1970, Nouvelle contribution à l'étude d'un lémurien malgache, le aye-aye (*Daubentonia madagascariensis*, E. Geoffroy), *Mammalia* 34:167.

Ramsay, M.A., and Stirling, I., 1986, On the mating system of polar bears, *Can. J. Zool.* 64:2142.

Reynolds, T.D., 1984, Daily summer movements, activity patterns, and home range of pronghorn, *Northwest Scientist* 58:300.

Richard, A.F., 1985, Primates in Nature, W.H. Freeman & Co., New York.

Richard, A.F., 1992, Aggressive competition between males, female-controlled polygyny and sexual monomorphism in a Malagasy primate, *Propithecus verreauxi*, *J. Hum. Evol.* 22:395.

Richard, A.F., and Dewar, R.E., 1991, Lemur ecology, *Ann. Rev. Ecol. Syst.* 22:145.

Rowell, T.E., 1988, The social system of guenons, compared with baboons, macaques and mangabeys, *in*: "A Primate Radiation: Evolutionary Biology of the African Guenons," A.Gautier-Hion, F. Bourlière, J-P. Gautier, and J. Kingdon, eds., Cambridge University Press, Cambridge.

Rowell, T.E., 1991, What can we say about social structure? *in*: "The Development and Integration of Behavior," P. Bateson, ed., Cambridge University Press, Cambridge.

Rowell, T.E., and Olson, D.K., 1983, Alternative mechanisms of social organization in monkeys, *Behaviour* 86:31.

Rubenstein, D.I., 1986, Ecology and sociality in horses and zebras, *in*: "Ecological Aspects of Social Evolution: Birds and Mammals," D.I. Rubenstein, and R. Wrangham, eds., Princeton University Press, Princeton.

Southwood, T.R.E., 1966, "Ecological Methods: with Particular Reference to the Study of Insect Populations," Chapman and Hall, London.

Sterling, E.J., (1993), Behavioral Ecology of the Aye-aye (*Daubentonia madagascariensis*) on Nosy Mangabe. Ph.D. Dissertation, Yale University, New Haven.

Terborgh, J., and Janson, C.H., 1986, The socioecology of primate groups, *Ann. Rev. Ecol. Syst.* 17:111.

Trivers, R.L., 1972, Parental investment and sexual selection, *in*: "Sexual Selection and the Descent of Man, 1871-1971," B. Campbell, ed., Aldine Press, Chicago.

Winn, R., 1989, The aye-ayes, *Daubentonia madagascariensis*, at the Paris Zoological Garden: Maintenance and preliminary behavioural observations, *Folia Primatol.* 52:109.

Wittenberger, J.F., 1979, The evolution of mating systems in birds and mammals, *in*: "Handbook of Behaviour and Communication," P. Masters and J. Vandenbergh, eds., Plenum Press, New York.

Wrangham, R.W., 1980, An ecological model of female-bonded primate groups, *Behaviour* 75:262.

Wrangham, R.W., 1986, Evolution of social structure, *in*: "Primate Societies," B.B. Smuts, D.L. Cheney, R.M. Seyfarth, R.W. Wrangham, and T.T. Struhsaker, eds., University of Chicago Press, Chicago.

# THE SOCIOECOLOGY OF *EULEMUR MACACO*: A PRELIMINARY REPORT

Ian C. Colquhoun

Primate Evolutionary Biology Program
Department of Anthropology
Washington University
St. Louis, MO 63130
U.S.A.

Present address:
Department of Anthropology
University of Western Ontario
London, Ontario N6A 5C2
Canada

## ABSTRACT

The preliminary results of a 15-month field study of the socioecology of four social groups of *Eulemur macaco macaco* in differing forest habitats at Ambato Massif are described. The four study groups ranged in size from 5 to 14 animals during the project. Mean group size was 9.75. The sex ratio of adult animals approximated 1:1. Home ranges were between 3.5 and 7.0ha, with a mean home range area of roughly 5.0-5.5ha. The population density of black lemurs at Ambato Massif was approximately 200 animals per km$^2$. Larger social groups were not as cohesive as smaller groups and frequently fissioned into sub-groups. The large study group inhabiting a relatively more disturbed patch of forest fissioned continually throughout the dry season, a variation in social organization predicted for such an environment with the use of an ecological modelling approach. Marked seasonal variation in activity and dietary patterns were also noted, although these were shared by all study groups and not related to a particular habitat type as were group size and social organization. *E. m. macaco* can thus be described as a generalist species with a relatively wide econiche.

## INTRODUCTION

The black lemurs (i.e., the nominate subspecies, *Eulemur macaco macaco*, and Sclater's lemur, *E. m. flavifrons*), are virtually unknown members of the Lemuriformes (e.g., Sussman, 1988; Oxnard et al., 1990). Indeed, it has not been until this decade that detailed, long-term field research of this species has commenced. Prior to 1991, the only field data available on the species were the result of short-term surveys (Petter, 1962; Jolly, 1966; Birkel, 1987; Meyers et al., 1989; Andrews, 1990). The present paper is a preliminary report of a 15-month research project on the socioecology of *Eulemur macaco macaco* in differing forest habitats in northwestern Madagascar.

*Lemur Social Systems and Their Ecological Basis*, Edited by
P.M. Kappeler and J.U. Ganzhorn, Plenum Press, New York, 1993

*Eulemur macaco macaco* is found only within the Sambirano phytogeographic domain of northwestern Madagascar. The Sambirano is floristically unique - a region of high annual precipitation (> 2000mm), like the humid forests of the Eastern phytogeographic domain, but with a pronounced (3-4 month) dry season during the austral winter (May - October). *E. m. macaco* is also found on the islets of Nosy Be and Nosy Komba off Madagascar's northwest coast (Tattersall, 1977, 1982; Wolfheim, 1983; IUCN, 1990). Threatened by loss of habitat throughout its small geographic range (Wolfheim, 1983; Richard and Sussman, 1987; IUCN, 1987; Andrews, 1990), the black lemur has shown some adaptability to human-altered environments. The species is known to inhabit both primary rainforest and degraded, secondary forest formations known as "savoka".

While a primary objective of this study was to compare and contrast the social organization of black lemur groups under differing ecological conditions, the limited nature of prior investigations on free-ranging *E. m. macaco* provided little basis for the formulation of specific research questions. This problem was overcome by utilizing the conceptual approach of "strategic", or "ecological", modelling (Tooby and DeVore, 1987; Dunbar, 1989; DeVore, 1990). That is, research questions and predictions concerning the socioecology of this relatively unstudied lemurid were generated from the body of ecological theory that has been developed to account for the behavior of a broad range of animal species. In particular, data on the responses of a diverse range of primate species to human-induced habitat disturbance (Wilson and Wilson, 1975; Poirier, 1977; Tilson, 1977; Bishop et al., 1981; Johns, 1986; Skorupa, 1986; Johns and Skorupa, 1987; Marsh et al., 1987), furnished the comparative background for predictions of the socioecology of *E. m. macaco* in differing forest habitats. It was predicted that compared to black lemurs inhabiting relatively less disturbed habitats, those inhabiting more disturbed habitats could exhibit markedly different behavior patterns. These differences could include: the responses exhibited towards humans; a decrease in the proportions of time spent travelling or feeding, with more time spent resting; a reduction in the degree of social group cohesion and an increased tendency for social groups to break into subgroups; an altered distribution of activities throughout the forest strata; and differing seasonal patterns of home range use. It is not possible to present a full account of each of these topics here; they will be considered in detail elsewhere (Colquhoun, in prep.).

## THE STUDY SITE

The research site was located on the mainland of Madagascar immediately to the east of Nosy Komba, along the narrow peninsula that lies adjacent to the islet of Nosy Faly and forms the southwestern shores of the Baie d'Ambaro. Four groups of *Eulemur macaco macaco* were studied intensively on the northwestern flank of Ambato Massif (13°27' south latitude, 48°34' east longitude; summit=461m), near the fishing village of Antsatsaka, or roughly 30km north of the town of Ambanja. Ambato Massif, covering an area of approximately 700ha, is designated as "fortêt classé" (classified forest) (Andrews, 1990). "Classified forests" are tracts of forest that are recognized as significant because, as with Ambato, they protect important watersheds. Although such areas are nominally protected by law, considerable use of the forest occurs.

### Climate and Vegetation

Monthly rainfall and mean maximum and minimum temperatures ranges recorded between September 1991 and September 1992 are presented in Figure 1. These data typify the Sambirano climate. The austral summer (November - April) saw heavy downpours, often daily, beginning in the late afternoon and frequently continuing on through the night. Rainfall amounts for the austral summer accounted for over three-quarters of the total rainfall recorded over the study period. November and December were the wettest months, together accounting for 42.6% of the total precipitation recorded. By contrast, rainfall amounts through the austral winter (May

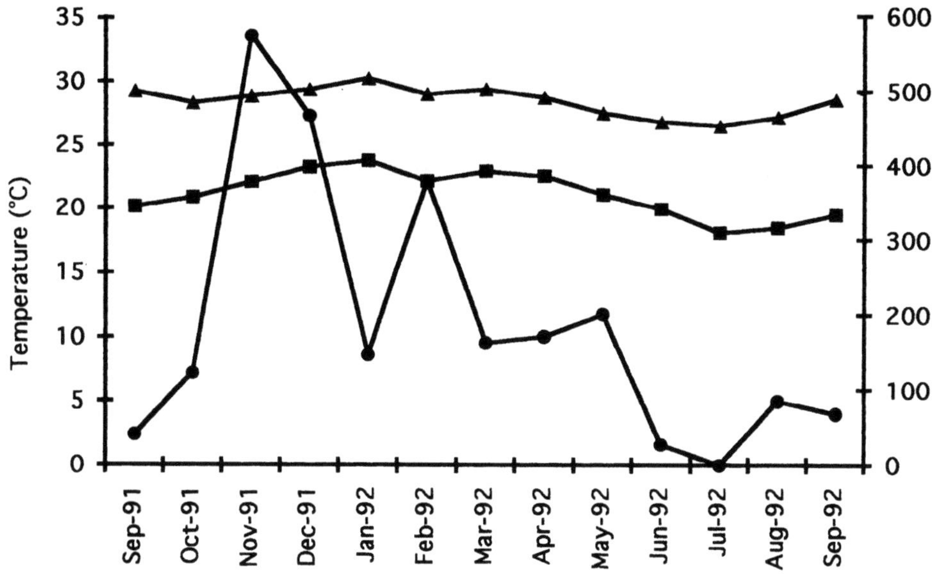


**Figure 1.** Climatic data for the study site at Ambato Massif (circles = total monthly precipitation, squares = monthly mean minimum temperature, triangles = monthly mean maximum temperature).

- October) were substantially reduced, with virtually no rainfall being recorded during July.

Variation in monthly mean temperature maxima and minima was much less marked. January had both the highest mean temperature maximum and minimum, while July had both the lowest mean temperature maximum and minimum. There was less than 4°C variation between the January and July mean temperature maxima; variation between the January and July mean temperature minima was under 6°C.

Because of the marked seasonal variation in precipitation, the vegetation of Ambato Massif can be characterized as seasonally moist semi-deciduous forest. Many of the tree species forming the forest canopy lose their leaves during the dry season. Topography and altitude on Ambato Massif also combine to create a very heterogeneous, botanically complex environment (see also Pollock, 1975). Some areas on the flanks of the massif, as well as the lower reaches of large ravines, were dominated by well-developed secondary forest. In these patches of forest, the canopy reached from 15-20m. Canopy tree species included: *Parkia* sp., *Delonix regia, Trachylobium verrucosum* (Leguminosae); *Uapaca* sp. (Euphorbiaceae); *Ficus* sp. (Moraceae) and *Canarium madagascariensis* (Burseraceae). Fully mature specimens of the latter were rare, as *Canarium* is much sought after for the construction of out-rigger canoes; young, maturing specimens were, however, not uncommon. While undergrowth was present in these forest patches, difficulty of movement on the ground was more due to the often steep angles of the slopes than to thick undergrowth. Epiphytic bird's nest ferns (*Asplenium nidus,* Aspleniaceae) were common, and fern trees (*Filicum decipiens,* Sapindaceae) also occurred, in the moist lower reaches of the large ravines on the massif.

In those patches of forest which showed relatively recent human disturbance, and on the lower slopes of the massif, the canopy was much lower (3-7m) and discontinuous. The endemic "traveller's palm", *Ravenala madagascariensis,* (Strelitziaceae), and the endemic *Chrysalidocarpus* sp. (Palmae), were frequently occurring to abundant in these areas, as was the endemic *Leptolaena* sp. (Sarcolaenaceae). *Pandanus* sp. (Pandanaceae, including subgenus *Vinsonia* section *Acanthostyla* sp.) were more frequent in disturbed areas, particularly near gullies that became streams during the wet season. The endemic palm *Neodypsis* sp. was, however, rare. Interestingly, disturbed forest patches on the slopes of the massif exhibited elfin-like characteristics. For example, *Chrysalidocarpus* palm that occurred on the flanks of the massif or in ravines reached a stature of 10-12m, while apparently mature specimens of the same species on the slopes of the massif could be only 2-4m in height. Very disturbed areas often exhibited vigourous growth of the exotic *Cinnamomum aromaticum* (Lauraceae) in the shrub layer of the forest. Other exotics on the flanks of the massif, and in the study area, included: mango (*Mangifera indica*, Anacardiaceae), cashew (*Anacardium occidentale*, Anacardiaceae), kapok (*Ceiba pentandra*, Bombacaceae), and, jackfruit (*Artocarpus heterophyllus*, Moraceae).

The higher slopes of the massif up to the summit were dominated by thick bamboo forest. Bamboo thickets were also locally abundant at lower altitudes. "Ambato" translates as "place of rocks" and, indeed, huge rock outcrops were a frequent feature of the forest landscape, particularly near stream courses. Woody lianas and the climbing fern *Lygodium* sp. (Schizaeaceae) were also frequently occurring to abundant through much of the forest in the study area; the density of liana growth in some small patches of forest on the flanks of the massif made them virtually impenetrable. Both woody lianas and *Lygodium* were rarer, however, in the low stature forest patches on the slopes of the massif. Black lemur social groups were found in all forested environments on the massif except for the bamboo forest on the upper slopes.

### Fauna

Besides *Eulemur macaco macaco*, the only other lemurid occurring at Ambato Massif that was active during the day was the western subspecies of the gray gentle lemur (*Hapalemur griseus occidentalis*). Several nocturnal lemuriformes also occurred at the study site. The gray-backed sportive lemur (*Lepilemur dorsalis*) was common, as was Coquerel's mouse lemur (*Mirza coquereli*). Much rarer was the greater dwarf lemur (*Cheirogaleus major*). There is circumstantial evidence to suggest the presence of *Phaner furcifer parienti* and *Daubentonia madagascariensis*. In my 15 months at Ambato Massif, I did not record a single *Microcebus rufus*.

Potential mammalian predators at Ambato Massif included the Malagasy ring-tailed mongoose (*Galidia elegans*, Viverridae). This species was sighted fairly regularly. Villagers reported that "fossa" (*Cryptoprocta ferox*) occurred on Ambato Massif, as did "jobady", the introduced lesser oriental civet, (*Viverricula indica*). While neither was sighted at the study site, *Viverricula* was observed to the south of the study site, on Ankify peninsula. I heard no reports, nor had any sightings, of the endemic Malagasy civet (*Fossa fossana*). Other potential lemur predator species included the raptors *Polyboroides radiatus*, *Buteo brachypterus*, *Milvus migrans* and *Accipiter madagascariensis*, and the constrictor *Acrantophis madagascariensis* (Boidae). The wild pig *Potamochoerus porcus* (Suidae) was fairly common at Ambato Massif.

## METHODS OF OBSERVATION

Intensive study of *Eulemur macaco macaco* spanned the period from September 1991 to October 1992 with a two month lead time. A total of 1,620 hours of quantitative observations were conducted on the four study groups. Two of the study groups (BG and FG, Table 1) had home ranges which were predominantly within areas of well-developed, relatively less disturbed, secondary forest. The other two study groups (G4 and PG, Table 1) had home ranges which were primarily in the elfin-like forest on the lower slopes of the massif and patches of secondary forest on the flanks of the massif which exhibited significant human disturbance. Because of the patterns of land use in areas surrounding Ambato Massif, all of the study groups had access to, and made use of, forest edge environments.

Prior to the intensive study period all four study groups were habituated to the point where approach of any of the groups was made with ease and in no way elicited alarm or flight responses among the group members. Individuals often approached to within one metre of the observer without displaying any signs of tension or excitement. Despite this high level of habituation, however, it was frequently not possible to follow a given study group throughout the day because of the topography of the study site and the often difficult, to virtually impassable, nature of the forest structure.

Focal animal observations were recorded on prepared data sheets at 5 minute intervals (Crook and Aldrich-Blake, 1968), and later entered on portable computer. Observation sessions alternated among the four study groups; an effort was made to find each study group at least once a week (although sometimes as little as three days passed between observation sessions on a given group). Focal animals in each group were chosen on a rotating basis in order to acquire approximately equal monthly data totals on group members. In general, one focal animal was followed during each observation session, morning or aftenoon, for as long as contact could be maintained with that animal's group. Morning observations began as soon after first light as contact with a study group was established and continued until contact with that group was lost or until midday, when observations generally switched off to another of the study groups in order to maximize the amount of comparative data. Similarly, afternoon observations were conducted as long as the group could be followed or until about 17:00-17:30 hours, when diminishing light conditions made the consistent identification of individual animals difficult to impossible. Quantitative data were recorded on 12 behavioral categories (foraging, feeding, sitting, scentmarking, autogrooming, allogrooming, resting, huddling, sleeping, playing, travelling and other), as well as distance and identity of the nearest neighbor to the focal animal and data concerning the level of the forest environment in which the behavior occurred. The position of the focal animal in the social group's home range, made in reference to particular landmarks and the marked path system, was also recorded at each 5 minute interval. Supplementary notes commenting on the behavior of the focal animal, and other members of the social group, were kept continuously through each observational session.

## BLACK LEMUR SOCIOECOLOGY

### Group Size and Composition

The striking sexual dichromatism of *Eulemur macaco macaco* (males are black with black ear tufts; females are golden to reddish brown with white ear tufts), meant that the sex ratio of social groups could be readily determined. Even though the dichromatism of infants is not as marked as that of adult animals, the white ear tufts of the females are still visible. Individual male and female animals in the study groups were recognized on the basis of morphological cues. Juveniles were differentiated from adults on the basis of their smaller body size and more gracile build; similarly, animals younger than 1 year exhibited infant-like traits (e.g., a shorter face relative to mature animals) and were the smallest animals in the social groups. Unfortunately, because of the steep topography of the study site, repeated sampling of the local population of black lemurs through the application of standard census transect methods was impossible.

Including data on other black lemur social groups that occasionally ranged into the study area, group size varied between 2 and 15 animals. Outside of the breeding period, there was only one sighting of a solitary male (this was an animal that had disappeared from study group BG only to be sighted again three months later as a lone male). Two different social groups of 2 individuals were seen. One was an adult female and her 4-5 month old female infant that were seen several times during January 1992. The other was an adult male and a juvenile-sized male sighted several times during August and September 1992. The group of 15 animals was also observed intermittantly during January 1992 as they came to feed in fruiting mangos on the forest edge adjacent to the study area. The group was composed of 7 males and 8 females, including an albino male infant.

Table 1 presents data on changes in the size and composition of the four study groups between September 1991 (immediately following that year's birth season) and the end of the study period, when all the infants from the 1992 birth season were from one month to six weeks of age. Permanent changes in study group composition included both male and female transfer, as well as the disappearance of some animals. Two young adult males, one in late September 1991 and the other a month later, joined group G4. An adult male immigrated into group FG in late June 1992. Two animals disappeared from group PG; a juvenile female went missing in April 1992, while an adult (and probably pregnant) female disappeared in July 1992. Similarly, one adult female, one juvenile male and one juvenile female all went missing from group BG over the study period. An adult male emigrated from group BG in May 1992; the emigration only became clear when the male was resighted three months after his apparent disappearance. An infant female disappeared from group BG in mid-September 1991 at approximately two weeks of age. At this age, infants are just beginning to occasionally clamber off their mothers on to surrounding branches; the group BG infant female probably had a fatal fall.

**Table 1.** Changes in the sizes and compositions of the four study groups between September 1991 and September 1992. The number of individuals in each age/sex class is represented. M=male; F=female.

| | September 1991 | | | | | | | September 1992 | | | | | | |
|---|---|---|---|---|---|---|---|---|---|---|---|---|---|---|
| | Adult | | Juvenile | | Infant | | total | Adult | | Juvenile | | Infant | | total |
| Group | M | F | M | F | M | F | | M | F | M | F | M | F | |
| G4 | 1 | 2 | - | 1 | - | 1 | **5** | 3 | 3 | - | 1 | 1 | 2 | **10** |
| BG | 4 | 3 | 2 | 2 | 1 | 2 | **14** | 4 | 3 | 1 | 1 | 1 | 2 | **12** |
| PG | 2 | 3 | 2 | 1 | - | 3 | **11** | 2 | 2 | 2 | 3 | 1 | 1 | **11** |
| FG | 2 | 3 | 1 | 1 | - | 1 | **8** | 4 | 2 | - | 1 | - | 1 | **8** |

During the final month of the study period, group BG fissioned into two daughter groups. The smaller of these was composed initially of only an adult female, her newborn male infant and her year-old son. The latter eventually ended up returning to group BG. Shortly after this, an unknown juvenile-sized female joined the adult female and the male infant. This would appear to be a case of a maturing juvenile/young adult female emigrating from her natal group. They actively avoided contact with group BG, moving rapidly in the opposite direction when they progressed towards their location. The newly formed small group also began to be sighted in areas that they had not even remotely been near prior to the final fission from group BG.

Over the study period the four study groups ranged in size from 5 to 14 animals, with an average group size of 9.9 animals. On average, the study groups contained 2.8 adult males, 2.6 adult females, 1.0 male juvenile, 1.4 female juveniles, 0.5 male infants and 1.6 female infants. The adult sex ratio across the four study groups was 1 male to 0.96 females. Although female juveniles and infants outnumbered males in these age classes 2:1, this is probably best interpreted as a small sample effect. During January 1992 when fruiting mangos (*Mangifera indica*) attracted dozens of groups from outside the study area into the same grove of trees, I could detect no bias in female juveniles and infants. Indeed, contrary to previous reports of strongly male-biased sex ratios in black lemur social groups (Petter, 1962; Jolly, 1966), my limited group counts support the conclusion of Andrews (1990) that there is no significant sex ratio bias.

## Activity Cycle

*Eulemur macaco macaco* social groups at Ambato Massif exhibited cathemeral activity (Tattersall, 1987). That is, in addition to diurnal activity, the animals also exhibited significant amounts of nocturnal activity. First activity was noted around 04:00 hours, well in advance of dawn, when animals were often heard travelling. Outside of the dry season, the period from dawn to between 07:00 to 08:00 hours was generally spent feeding and resting; by approximately 09:00 hours the group settled down to a mid-morning nap. This rest period was sometimes interrupted by a second, briefer, bout of travelling followed by feeding, but by around 10:30 hours the animals would again be resting and/or sleeping. The midday period was characterized by little or no activity from the animals, apart from some social grooming. By approximately 15:00 hours the animals began to stir, with an afternoon bout of foraging and feeding beginning soon thereafter and continuing until well after dark.

Variation in the levels of nocturnal activity appeared to be correlated with the lunar phase (see also Nash, 1986; Erkert, 1989; Meier, 1992). Both calling and group progressions were recorded (i.e., heard) more frequently on nights with bright moonlight than on nights when the moon was new, the sky was overcast or when there was precipitation. That the behavior of *E. m. macaco* corresponds well with Tattersall's (1987) definition of cathemeral activity is underlined by the fact that I even occasionally heard nocturnal intergroup encounter vocalizations that in no way differed from those given during daylight intergroup encounters that I observed. Additionally, after a night of considerable activity, the animals generally slept for longer periods the following day.

The dry season was notable for the marked reduction of activity levels exhibited by the animals compared to the wet season. With first light the animals sought positions in canopy trees, where they could sun themselves following the relatively cool overnight temperatures. It was not uncommon for the animals to already have retreated into a liana tangle (a sleeping site not used in the wet season), and be sleeping by between 07:00 to 08:00 hours. Their level of inactivity was such that they could remain sleeping and/or resting for up to the next 10 hours. Nocturnal activity, however, continued to be recorded through the dry season. The dramatic change in activity cycles seems to be related to the relative absence through the dry season of plant species in fruit, the overwhelmingly preferred food item during the wet season.

## Habitat Utilization and Diet

Regardless of whether they were in well-developed secondary forest with a 15m canopy or in short stature forest with a canopy of just 4m, *E. m. macaco* tended to use the mid to upper levels of the tree canopy. In other words, the higher the animals could be from the ground, the higher they would generally go. Oblique branches of 3-10cm in diameter, and larger, were the substrate most frequently used during progressions. The animals employed strongly hindlimb-dominated locomotion (i.e., a form of "indriid-type leaping" - see Oxnard et al., 1990) when moving between oblique branches, such that the torso maintained a semi-erect posture; this was also the case on those occasions that vertical supports were utilized. When negotiating a gap in the forest canopy and leaping between trees, the animals launched themselves and assumed a stretched-out posture in mid-air. During the downward arc of the leap they would bring their hindlimbs forward to meet the terminal branches of the target tree ahead of their forelimbs.

Animals in those groups in the more disturbed patches of forest were more likely to come to the ground. Females seemed to be less hesitant about leaving the trees than were males. On several occasions I observed animals stride bipedally while scanning ahead of them as they crossed ground that was covered by tall grass. The distances which the animals "walked" were no farther than about 3m.

Fruits dominated the diet through the rainy season, in particular the fruit of *Chrysalidocarpus* sp. palm. *Pandanus* sp. fruit was another staple through much of the year, especially for the groups in the more disturbed forest patches where *Pandanus* was more abundant. Also in the wet season diet, although rarely, were mushrooms and millipedes. Early in the dry season, the nectar from the flowers of *Ravenala madagascariensis* composed a significant portion of the diet. Through the dry season, food items not eaten at any other time of the year were incorporated into the diet; these included leaves of the climbing fern *Lygodium* and the base of the leaves of *Pandanus* (subgenus *Vinsonia* section *Acanthostyla*) sp. Flowers were added into the diet late in the dry season. Of note here are the flowers of *Parkia* sp. (Leguminosae); *E. m. macaco* both fed directly on the flowers (i.e., was a flower predator), and simply licked nectar from the floral heads (i.e., was a potential pollinating agent). Black lemurs also fed on the seed pods that developed from the fertilized flowers and thus acted as seed dispersers of this important forest canopy tree. When feeding on *Parkia* sp., as well as when they engaged in terminal branch feeding in other tree species, a bipedal hanging posture was frequently employed and the food item manipulated with the hands and mouth.

Drinking from tree holes was the primary manner in which water was obtained. If the tree hole were large enough, water was taken directly by mouth; if the tree hole were smaller, lemurs would dip their hands into the water and then lick it from their fingers. On one occasion I observed a male stick his head into a small crevice in a rock outcrop to drink. On another occasion the members of one of the groups licked rain droplets from leaves following a very brief shower near the end of the dry season. At this time, it is only possible to present this very general description of black lemur habitat utilization and diet. A percentage breakdown of the use of the various forest strata and of plant parts in the diet is in preparation.

**Group Movements, Home Ranges, and Population Density**

The home ranges of the study groups ranged in size from approximately 3.5-7.0ha, (averaging approximately 5.0-5.5ha). Each study group's home range overlapped considerably with those of adjacent groups, such that the study area of approximately 20ha encompassed the home ranges of the four study groups, including occasional excursions the groups made from the forest edges of their ranges to nearby cashew trees. The biomass of *E. m. macaco* at Ambato Massif is roughly 5kg per ha and the population density is approximately 200 animals per km$^2$. If the entire area of Ambato Massif were suitable forest for black lemurs, this would mean that a population of roughly 1,400 black lemurs exists on the massif. This is an unrealistically high estimate, however, because *E. m. macaco* does not occur in the extensive bamboo forests carpeting the upper slopes and summit of the massif. With this in mind, it still seems reasonable to estimate a population of around 1000 animals on Ambato Massif, plus whatever lemurs continue to exist in the more degraded forest surrounding the flanks of the massif.

The two smaller study groups (G4 and FG) were very cohesive; during the entire study period there were only a couple of instances when either group apparently fissioned into sub-groups. With the two larger groups (BG and PG), however, fissioning was not an infrequent event. Most notably, group PG, which inhabited relatively more disturbed forest, began to fission more often during the dry season to the point that I was more likely to encounter a sub-group than the entire group. Sub-groups of group PG during the dry season often consisted of only a single animal, even 10 month old infant class animals. Solitary animals could remain alone for 3 or 4 hours at a stretch, often spending much of that time just resting and/or sleeping. Sub-groups of group PG tended to coalesce towards dusk.

When black lemurs moved through the forest, whether as an intact social group or a sub-group, they generally exchanged low, gutteral "contact grunts". If withdrawing from a strange human or potential predator, however, movement was made silently. During the wet season, when the animals were relatively more active, day range path lengths could be several hundred metres. This contrasted sharply with the inactivity of the dry season when day range path lengths could be as little as 50m.

As Sussman (1975) suggested for *Eulemur fulvus rufus*, encounters between black lemur social groups seemed to play little part in defining group boundaries *per se*, given the highly overlapping ranges of the groups. Intergroup encounters seem more to define the zone of range overlap between neighboring groups and likely play a role in promoting the recognition of the social group, both for the members of a group and to neighboring groups. Encounters often occurred in the same general areas, usually along or near progression routes. But in the absence of a neighboring group, a social group would regularly move in the zone of overlap beyond the usual points of intergroup displays. When they occurred, intergroup displays saw much tail swishing and leaping back and forth by the various group members accompanied by agitated staccato grunting and raucous loud calls. Physical contact during encounters was never oberved to be more than a glancing cuff. More usual was for one animal to execute a powerful leap, landing heavily on the same branch as, and very near to, an animal from the opposite group. Such an action could then lead to a cascade of back and forth leaps by members of the two groups. Adult females were usually the animals most involved in the displaying although adult males also participated, if only on the periphery of the encounter. Younger animals generally sat away from the encounter and continued to feed, groom, rest, etc. At other times, however, neighboring groups could be in quite close proximity without there being any sign of intergroup display.

When not in sight of one another, neighboring groups could monitor each other's location and maintain their separation by exchanging the same rasping, raucous loud call given during intergroup encounters. Animals often gave loud calls spontaneously in the late afternoon as they were becoming active. Such calls given by distant groups were responded to in kind.

**Intragroup Behavior**

It was my general impression that there was no overt difference in the nature of social interactions between members of the study groups in the relatively less, and relatively more, disturbed forest patches. In all groups, sitting or resting in contact and allogrooming were the most frequently observed social behaviors. Play behavior during the wet season was often quite vigourous and of extended duration, involving play groups of up to 5 animals from all age/sex classes at a time. During the dry season, however, play was not observed at all. Agonistic encounters were rare and usually consisted of just a lunge or a cuff. If the agonism revolved around access to a food source (e.g., a ripe jack fruit), there were often spats and high-pitched squeaks emitted. Females had priority of access and social dominance over males; for example, females invariably had first access to food items, could displace males from particular resting/sleeping sites and it was usually a particular adult female that led group progressions.

As with social behavior, there was no apparent difference in reproductive behavior among the four study groups. Males exhibited a wide range of reproductive strategies. Several males were observed to engage in "roaming behavior" (Richard, 1974), moving well outside their normal range and following neighboring groups. All adult males appeared to become more vigilant during the breeding period, especially towards roaming males. Male-male aggression also increased at this time; I observed a protracted chase between the two adult males in group FG, while both adult males in group PG ended up bearing serious bite wounds. Prior to engaging in a bout of roaming, the oldest male in group G4 spent several days consorting and guarding all the females in the group. He did not allow the other two males in G4 to even sit with group's juvenile or infant females. Yet, on the afternoon of April 20, 1992 when the oldest male went roaming, one of the other males copulate six times within about 30 minutes with one of the groups's females. On the afternoon of August 23, 1992 (following a gestation of 125 days), she was sighted carrying an infant male. The only other copulation that I observed occurred on April 26 1992 in group PG. Again, it was a resident male that was involved in the mating, despite the fact that several roaming males had been seen near group PG. On August 31 or September 1, 1992, this female gave birth to an infant male (after a gestation of 125-126 days). The youngest female to give birth was in group G4. A juvenile when the study began, she was just 2 years of age when she gave birth to an infant female in late August 1992.

**Reactions to Other Lemuriformes**

*Eulemur macaco macaco* exhibited no agonistic behavior towards *Hapalemur griseus occidentalis*. When the two species encountered one another there was sometimes some mild curiosity, e.g., black lemur individuals remaining curled up in a resting posture while scanning and following the progression of *Hapalemur* through the forest. Generally, however, there was no overt recognition of each others presence. On two occasions I observed a black lemur group first mob and then follow for a short distance a *Lepilemur dorsalis* that had been roused from its day sleeping spot by the progression of the *E. m. macaco* group. There were no observations of interactions between *E. m. macaco* and any of the other lemuriformes found at Ambato massif.

**Anti-Predator Behavior**

Humans represented the most abundant potential predator species of *Eulemur macaco macaco* at Ambato Massif. Black lemurs are hunted with snares that are positioned along bridges constructed to form artificial arboreal pathways across a cleared swath of forest. Blow darts, also used to hunt turtledoves (*Streptopelia picturata*), fruit pigeons (*Treron australis*) and crested ibis (*Lophotibus cristata*), were occasionally employed in the hunting of lemurs, especially towards the end of the wet season when the lemurs have put on a good deal of weight thanks to several months of abundant food resources. I suspect that sling-shots (catapults) were also sometimes used to hunt lemurs. Overall, however, I believe that only a low level of hunting was taking place at Ambato Massif because no snares were found within the study area itself. Moreover, the presence of *Potamochoerus* at Ambato Massif meant there was larger, more economically rewarding, game to pursue.

Given there was some level of hunting pressure at Ambato Massif, lemur groups did not generally mob unfamiliar humans. Rather, individuals would make punctuated huff-grunt vocalizations while swishing their tails back and forth and scanning first at the intruder and then in the opposite direction. Rapid withdrawal, i.e., leaping and/or running through the canopy from the human followed immediately thereafter.

There were four potential avian predators at Ambato Massif. The Madagascar sparrowhawk (*Accipiter madagascariensis*), although too small to take an adult lemur, is large enough to pose a threat to infant animals. On one occasion I observed a Madagascar sparrowhawk swoop within striking distance of a pair of approximately four-month old black lemurs that were playing and grappling on a liana, well away from any adults. Other individuals in the group gave short, sharp alarm hack vocalizations as the bird swooped past the infants, but there were no other responses from the group nor other predation attempts by the bird. The Madagascar buzzard (*Buteo brachypterus*) is another potential predator. Circling buzzards, depending on their altitude, were responded to with alarm hacks and rasping loud calls. Alarm backs were also given in response to circling Black kites (*Milyus migrans*, cf. Sauther, 1989). I never observed a buzzard attempt to prey upon a lemur.

The largest potential avian predator, the Madagascar harrier-hawk (*Polyboroides radiata*), was reacted to most strongly of all. When harrier-hawks were detected soaring just above tree-top level, the response from the lemurs was dramatic. A vocalization similar to the rasping loud call, but with a sharply ascending then descending scream-whistle at the tail end of the call, was invariably given by one or several individuals; alarm hacks could also be given. In response to this loud call/scream-whistle vocalization, animals would dive (often several m) out of the upper branches of trees and down well within the canopy. A predation attempt by a harrier-hawk on an adult female black lemur was observed in the early part of the dry season, when the leguminous tree species *Delonix regia* had already begun losing its leaves. It was between 07:00 and 07:30 hours and the focal female had climbed high up into a *Delonix* to sun herself. I first heard a clattering noise and then saw a harrier hawk fall short of the female and momentarily alight only a metre from her before immediately flying off; the female lemur screamed (literally) and leapt from the branch to the trunk of the tree. The raptor had apparently become partially snagged in the uppermost terminal branches of the tree and was thrown off its mark.

Initially, I assumed the loud call/ scream-whistle vocalization to be a call that specifically communicated the dangerous proximity of a harrier-hawk. Late in the study period, however, I recorded *E. m. macaco* giving the same call in response to tree-top level flight by the megachiropteran fruit bat *Pteropus rufus* on a brightly moonlit night (cf. Sussman, 1975). *Pteropus* has a wing span close to that of *Polyboroides* and under moonlit conditions must project a very similar silouette. The loud call/scream-whistle vocalization, then, would appear to be only an ordinal-level communication that identifies the presence of a very large, potentially predatory, animal immediately overhead. Any relatively large, non-predatory bird, e.g., crested coua (*Coua cirstata*), Madagascar coucal (*Centropus toulou*), greater vasa parrot (*Coracopsis vasa drouhardi*) or green fruit pigeon (*Treron a. australis*) that flew close enough to the lemurs to startle them was responded to with alarm hack vocalizations.

Dogs and presumably the endemic viverrids were responded to with swishing tails and punctuated huff-grunts that could escalate into a full mobbing display with rasping loud calls. Continued presence of a dog would lead to the lemur group retreating from the area. I also had one observation of a prolonged mobbing display by a lemur group towards a 2.5m long boa (*Acrantophis madagascariensis*) that was resting on a low tree branch, roughly 2m off the ground. None of the individuals in the group approached the snake any closer than about 1m. After some 15-20 minutes of mobbing the snake, the group moved from the area.

## DISCUSSION

Traditionally, the black lemurs have been considered to hold a close phylogenetic relationship with the brown lemurs (*Eulemur fulvus*). Yet, a preliminary comparison of the behavior and ecology of *E. m. macaco* at Ambato Massif with the best-studied members of the Lemuridae (*Lemur catta* and *E. fulvus* - Sussman, 1974, 1975), suggests a slightly different picture. Sussman (1974) presented data that showed *E. f. rufus* to be a relatively specialized and arboreally adapted species; this contrasted with the ability exhibited by *L. catta* to exploit a wide range of habitats. In terms of habitat utilization, dietary diversity, home range size and population density, the black lemurs of Ambato Massif are apparently closer to the generalist pattern exhibited by *L. catta* than to the more specialized profile of *E. fulvus*. Indeed, dental data for *E. m. macaco* (Kay et al., 1978) suggested that the species is essentially a facultative frugivore/folivore, able to exploit a wide range of seasonally fluctuating food resources (i.e., wet season - frugivory; dry season - folivory). Such a variable foraging strategy would also allow *E. m. macaco* to persist in secondary and disturbed habitats even if favored seasonal food resources were not available. Successful exploitation of such habitats could prove to be crucial to the long-term survival of the species. The extreme inactivity of *E. m. macaco* at Ambato Massif during the dry season, however, may indicate that the animals are being stressed to their limits of environmental tolerance. Lott (1991) points out that niche breadth should correlate positively with social system plasticity. Thus, the social organizational switch exhibited by group PG from a generally cohesive social group during the wet season to a fission-fusion mode of organization during the dry season provides another piece of evidence that *E. m. macaco* is a generalist species with a relatively broad econiche. Given the wide range of habitats and altitudes found across its geographic range, a broad econiche and generalized ecology would best account for the distribution of the species.

## ACKNOWLEDGEMENTS

This project was conducted under the aupices of the Beza Mahafaly Interuniversity Accord and could not have possibly been completed without the kind cooperation of the Ministere de l'Enseignement Superieur and the Direction des Eaux et Forts. The assistance and interest of Madame Berthe Rakotosamimanana and M. Benjamin Andriamihaja were absolutely invaluable. While at Ambato Massif, I was assisted in the collecting of botanic quadrat data by the ever helpful Patrice Antilahimena. M. Inayat Houssen Adamalykarimdjy and Madame Rossanbay Taibaly kindly allowed us to establish our camp on their property at the foot of Ambato. My wife, Sonia Wolf, provided both material and moral assistance in all phases of the project and to her I owe a special debt of gratitude. Peter Kappeler and Jörg Ganzhorn, with both enthusiasm and generosity, allowed me to submit this contribution despite being in Madagascar during the Strasbourg symposium. And finally, I must thank Dr. Robert Sussman, who first suggested that I work with black lemurs. The research was supported in part by a NSF Dissertation Improvement Award Grant No. BNS-9101520, the National Geographic Society grant number 4496-91, the Boise Fund of the University of Oxford, and a Grant-in-Aid of Research from Sigma Xi, The Scientific Research Society.

## REFERENCES

Andrews, J.R., 1990, A Preliminary Survey of Black Lemurs, *Lemur macaco* in North West Madagascar. "Royal Geographic Society Reports," London.

Birkel, R., 1987, International Studbook for the Black Lemur (*Lemur macaco*), Linnaeus, 1766", St. Louis Zoological Park, St. Louis, Missouri.

Bishop, N., Blaffer Hrdy, S., Teas, J. and Moore, J., 1981, Measures of human influence in habitats of South Asian monkeys, *Int. J. Primatol.* 2:153.

Colquhoun, I.C., In preparation. Ph.D. dissertation, Washington University, St. Louis.

Crook, J.H. and Aldrich-Blake, P., 1968, Ecological and behavioral contrasts between sympatric ground dwelling primates in Ethiopia, *Folia Primatol.* 8:192.

DeVore, I., 1990, Introduction: Current studies on primate socioecology and evolution, *Int. J. Primatol.* 11:1.

Dunbar, R.I.M., 1989, Ecological modelling in an evolutionary context, *Folia Primatol.* 53:235.

Erkert, H.G., 1989, Light requirements of nocturnal primates in captivity: a chronobiological approach, *Zoo. Biol.* 8:179.

IUCN, 1987, "Madagascar: An Environmental Profile," M.D. Jenkins, ed., IUCN, Cambridge.

IUCN, 1990, "Lemurs of Madagascar and the Comoros: The IUCN Red Data Book," Compiled by the World Conservation Monitoring Centre, Cambridge.

Johns, A.D., 1986, Effects of selective logging on the behavioral ecology of West Malaysian primates, *Ecology* 67:684.

Johns, A.D., and Skorupa, J.P., 1987, Responses of rain-forest primates to habitat disturbance: a review, *Int. J. Primatol.* 8:157.

Jolly, A., 1966, "Lemur Behavior: A Madagascar Field Study," University of Chicago Press, Chicago.

Kay, R.F., Sussman, R.W. and Tattersall, I., 1978, Dietary and dental variations in the genus *Lemur*, with comments concerning dietary-dental correlations among Malagasy primates, *Am. J. Phys. Anthropol.* 49:119.

Lott, D.F., 1991, "Intraspecific Variation in the Social Systems of Wild Vertebrates," Cambridge University Press, Cambridge.

Marsh, C.W., Johns, A.D. and Ayres, J.M., 1987, Effects of habitat disturbance on rain forest primates, *in*: "Primate Conservation in the Tropical Rain Forest," C.W. Marsh and R.A. Mittermeier, eds., Alan R. Liss, Inc., New York.

Meier, B., 1992, Rotbauchmaki, *in*: "Lemuren im Zoo," V.Ceska, H.-U. Hoffmann, K.-H. Winkelsträter, eds., Paul Parey, Berlin

Meyers, D.M., Rabarivola, C. and Rumpler, Y., 1989, Distribution and conservation of Sclater's lemur: Implications of a morphological cline. *Primate Conserv.* 10:77.

Nash, L.T., 1986, Influence of moonlight levels on travelling and calling patterns in two sympatric species of *Galago* in Kenya, *in*: "Current Perspectives in Primate Social Dynamics," D.M. Taub and F.A. King, eds., Van Nostrand Reinhold Co, New York.

Oxnard, C.E., Crompton, R.H. and Lieberman, S.S., 1990, "Animal Lifestyles and Anatomies: The Case of the Prosimian Primates," University of Washington Press, Seattle.

Petter, J.J., 1962, Récherches sur l'écologie et l'éthologie des lémuriens Malgaches. *Mem. Mus. Natl. Hist. Nat. (Paris)* 27:1.

Poirier, F.E., 1977, The human influence on genetic and behavioral differentiation among three nonhuman primate populations, *Yrbk. Phys. Anthropol.* 20:234.

Pollock, J.I., 1975, Field observations on *Indri indri*: a preliminary report, *in*: "Lemur Biology," I. Tattersall and R.W. Sussman, eds., Plenum Press, New York.

Richard, A.F., 1974, Patterns of mating in *Propithecus verreauxi verreauxi*, *in*: "Prosimian Biology," R.D. Martin, G.A. Doyle and A.C. Walker, eds., Duckworth, London.

Richard, A.F. and Sussman, R.W., 1987, Framework for primate conservation in Madagascar, *in*: "Primate Conservation in the Tropical Rain Forest, Monographs in Primatology," vol. 9, C.W. Marsh and R.A. Mittermeier, eds., Alan R. Liss, Inc, New York.

Sauther, M.L., 1989, Antipredator behavior in troops of free-ranging *Lemur catta* at Beza Mahafaly Special Reserve, Madagascar, *Int. J. Primatol.* 10:595.

Skorupa, J.P., 1986, Responses of rainforest primates to selective logging in Kibale Forest, Uganda: a summary report, *in*: "Primates: The Road to Self-Sustaining Populations," K. Benirschke, ed., Springer-Verlag, New York.

Sussman, R.W., 1975, A preliminary study of the behavior and ecology of *Lemur fulvus rufus* Audebert 1800, *in*: "Lemur Biology," I. Tattersall and R.W. Sussman, Plenum Press, New York.

Sussman, R.W., 1988, The adaptive array of the lemurs of Madagascar, *Monogr. Syst. Bot. Missouri Bot. Gard.* 25:215.

Tattersall, I., 1977, Distribution of the Malagasy lemurs part I: the lemurs of Northern Madagascar, *Ann. N.Y. Acad. Sci.* 293:160.

Tattersall, I., 1982, "The Primates of Madagascar," Columbia University Press, New York.

Tattersall, I., 1987, Cathemeral activity in primates: a definition, *Folia Primatol.* 49:200.

Tilson, R.L., 1977, Social organization of Simakobu monkeys (*Nasalis concolor*) in Siberut Island, Indonesia, *J. Mammal.* 58:202.

Tooby, J. and DeVore, I., 1987, The reconstruction of hominid behavioral evolution through strategic modeling, *in*: "The Evolution of Human Behavior: Primate Models," W.G. Kinzey, ed., State University of New York Press, Albany.

Wilson, C.C. and Wilson, W.L., 1975, The influence of selective logging on primates and some other animals in East Kalimantan, *Folia Primatol.* 23:245.

Wolfheim, J.H. 1983, "Primates of the World: Distribution, Abundance, and Conservation", University of Washington Press, Seattle.

# HOME RANGE AND DIET IN RED RUFFED LEMURS *(VARECIA VARIEGATA RUBRA)* ON THE MASOALA PENINSULA, MADAGASCAR

Marco M. Rigamonti

Department of Biology 1A-7B
University of Milan
Via Celoria, 26
20133Milan
Italy

## ABSTRACT

Two groups of red ruffed lemurs (*Varecia variegata rubra*) were observed in primary rain forest, on the Masoala peninsula, for 11 months between December 1990 and November 1991. Quantitative observations on ranging behavior, diet, activity, group composition and group dispersion were collected on 8 focal animals. The two study groups, consisting of 5 and 6 members, used home ranges of 23.3 and 25.8ha, respectively. The daily average path length for the main study group was 436m. In the cool-wet season, both groups fragmented into subgroups that used different core areas. The position of core areas within the home range changed from month to month. Red ruffed lemurs were mainly frugivorous (73.9%) and supplemented their diet with leaves (20.9%) and flowers (5.3%). During the study period, focal animals fed on 42 tree species showing a marked preference for only 7 of these species (72.5 % of total feeding observations). Home range size and use were related to the location of large fruit trees.

## INTRODUCTION

The genus *Varecia* comprises one species with two subspecies. The black and white ruffed lemur *(Varecia variegata variegata)* lives in the forests along the eastern coast of Madagascar, while the red ruffed lemur *(V. variegata rubra)* is found east of the Antainambalana river on the Masoala peninsula, where no protected areas currently exist (Fig. 1). These large lemurs (adult weight = 3.6kg, Pereira et al., 1987) were reported to have diurnal-crepuscular habits, to live in small family groups of three or four individuals, and to be mainly frugivorous (Petter et al., 1977). Data obtained on social structure, reproduction, infant care and problems related to the management of captive animals raised the need for detailed information about social organization, territorial behavior, and diet composition of ruffed lemurs in the wild (Kress et al., 1978; Foerg, 1982; Rasmussen, 1985; Brockman et al., 1987; Pereira et al., 1987, 1988; Blanckenhorn, 1990). In recent years, stimulated by these questions, primatologists started field studies to add to the existing information. Results from field studies on social organization, group size and home range size of black and white ruffed lemurs revealed flexibility in different populations and led authors to conflicting conclusions. During a two month study, at Vatoharanana in

*Lemur Social Systems and Their Ecological Basis*, Edited by
P.M. Kappeler and J.U. Ganzhorn, Plenum Press, New York, 1993

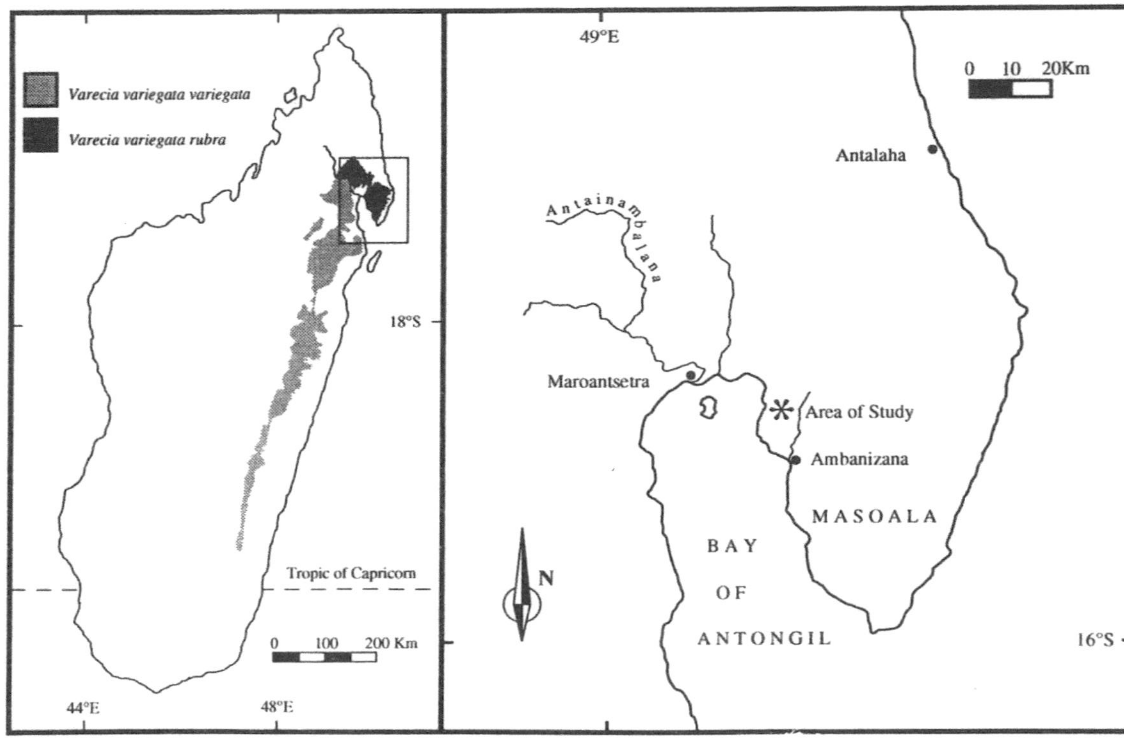


**Figure 1.** Distribution of *Varecia variegata* and location of the study area on the Masoala peninsula.

Ranomafana National Park (south-eastern Madagascar), a pair of ruffed lemurs maintained a territory of 197ha and showed "a type of monogamy that parallels the social system observed for gibbons" (White, 1991). White suggested that the presence of groups larger than a couple was due to the inability of offspring to disperse. In a study conducted in the Special Reserve of Nosy Mangabe during 12 months of an 18-month period, "ruffed lemurs were found to live in dispersed female-based communities with cooperatively defended communal home ranges" (Morland, 1991). The communities observed in this study were composed of between 11 and 16 animals.

Apart from some brief reports (Petter et al., 1977; Constable et al., 1985; Lindsay and Simons, 1986) no surveys had been carried out in the wild on the red ruffed lemur, the rarer of the two subspecies. The objective of this study was to conduct quantitative observations on home range size, home range use and feeding strategies to provide some comparative data for the study of *Varecia*. These data may be useful for defining a conservation strategy for this lemur in its threatened habitat.

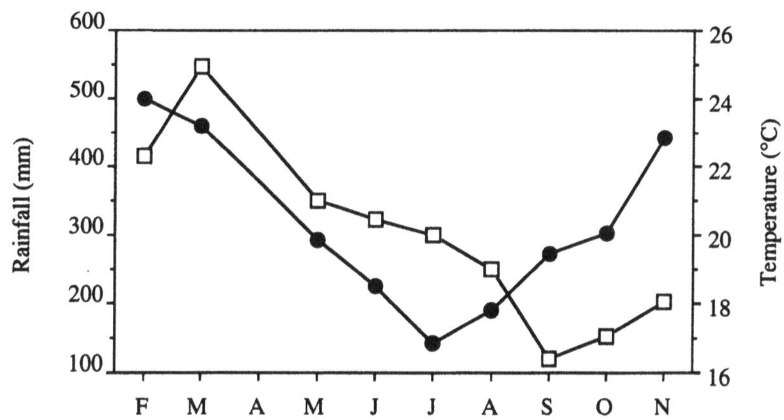


**Figure 2.** Mean monthly rainfall (squares) and temperature (circles) at the study site between February and November 1991. Climatic data are not available for the month of April.

## METHODS

This field study was conducted between December 1990 and November 1991 in rain forest about 5km inland on the western coast of the Masoala peninsula (Fig. 1). The study site is located in primary rain forest between about 450 and 650m above sea level, and it is known to locals as the Ambatonakolahy forest. A 40 x 40m gap in the canopy due to an old "tavy" (slash and burn) was located about 50m from the focal group's home range border. Locals reported that nobody had been working at this tavy plot for at least five years. Other abandoned tavy plots were scattered toward the coast, starting at 600m from the study site. On the west coast of the Masoala peninsula, rainfall exceeds 4200mm per year (Donque, 1972). Rainfall and temperature (maximum and minimum) were recorded daily at the study site and the mean monthly values were calculated (Fig. 2).

Behavioral observations were made during the cool-wet season (lasting from mid-May to mid-August) and the main part of the transitional-dry season (from mid-September to mid-December).

In addition to *Varecia variegata rubra, Eulemur fulvus albifrons, Phaner furcifer, Lepilemur mustelinus, Cheirogaleus major, Microcebus rufus, Avahi laniger laniger, Hapalemur griseus*, and *Daubentonia madagascariensis* also are found on the Masoala peninsula (Petter et al., 1977; Tattersall, 1982). In a dozen nights, sometimes with help from assistants experienced at working with nocturnal lemurs, I spent 2 -3 hours searching the different species in the forest. With the possible exception of *L. mustelinus* and *H. griseus*, which were never observed during the study, all species mentioned above are sympatric with *V. variegata rubra* at Ambatonakolahy.

### Behavioral Observations and Diet

Quantitative observations on ranging behavior, diet, activity, group composition and group dispersion were made from May to November 1991. Data were collected on focal animals and recorded at 3min intervals, using an instantaneous sampling technique (Martin and Bateson, 1986). With this sampling procedure, 539h of observations were collected on a main study group consisting of 5 members (5 focal animals: 2 females and 3 males) and an additional 165h were collected on a second neighboring group of 6 members (3 focal animals: 2 females and 1 male). Observations were made on focal animals between 9 and 13 (mean=10.3) consecutive days per month. All day follows (from dawn to dusk) were completed on 47 of the 72 observation days. All animals were individually recognized by the position and shape of natural markings, especially irregular white patches, but also by body size, scars, and intensity of fur color. With a few exceptions, focal animals were observed following a scheduled sequence; the sleeping tree of the animal to be observed was located on the previous day by field assistants. The dispersion of the group was measured by noting the number of individuals within 50m from the focal animal at 3min-intervals.

All feeding trees were marked, and, on a subsequent day, identified, and measured for height, diameter at breast height (DBH), crown height and crown mean diameter. When identification was uncertain, samples of leaves and eaten item were preserved. The part eaten was noted and classified as shoot, young leaf, mature leaf, unripe fruit, ripe fruit, bud, or open flower. Young leaves were identified by their small size, transparency, and paler, rosy-reddish color in some species. Ripe fruit was identified on the basis of characteristics that varied from species to species. Diagnostic elements were: dimension, external color, pulp hardness, pulp texture, smell and taste (in some cases, fruit was tasted). Advice from field assistant experienced with the species was also considered.

### Animal Habituation

In December 1990 I followed two groups of red ruffed lemurs to habituate them to my presence. Before habituation, when animals saw somebody or heard noises on the ground, they gave alarm calls and stared at the intruders before they fled and continued to vocalize for about half an hour. In January and February 1991 a grid of trails was measured and labelled to estimate home range. This work helped to get the two groups used to the presence of people. After habituation, red ruffed lemurs no longer gave alarm calls when they saw me. In March, the methods used to record data were tested and the categories of the variables to be measured were fixed. In this phase, animals descended toward me to reach a 10m high fruit tree, and didn't even turn their head if I stepped on undergrowth; they were considered habituated.

### Home Range Size and Use

The study area was mapped using a tripod compass, distance tape and a clinometer. A grid of 0.25ha quadrats (50m x 50m) was laid over the map of the study area and a corresponding system of trails was opened and marked every 10m (Fig. 3). The grid quadrat occupied by the focal animal was recorded every 3 min with the same technique used for behavioral data. Data from the focal animals were used to determine the total size of the respective group home range. The core area consisted of all quadrats that a group used in more than 10% of the observations (Kavanagh, 1981).

The 165h of quantitative observations made on the second group were insufficient to determine the size of their total home range precisely. Thus, I estimated the second group's home range borders, also using opportunistic observations, such as sightings during other activities or calls heard while following the main study group. After the grid of trails was labelled in the forest, with practice, it was possible to estimate reliably the locations of calling animals. Estimates were made from relatively elevated positions using map, compass and protractor. Whenever the focal animal crossed the border between two quadrats, time and position were noted. All trees used for feeding, sleeping and long resting periods were marked. Daily paths were traced and measured on the grid map of the study area. A total of 47 complete day ranges were mapped for 8 focal animals.

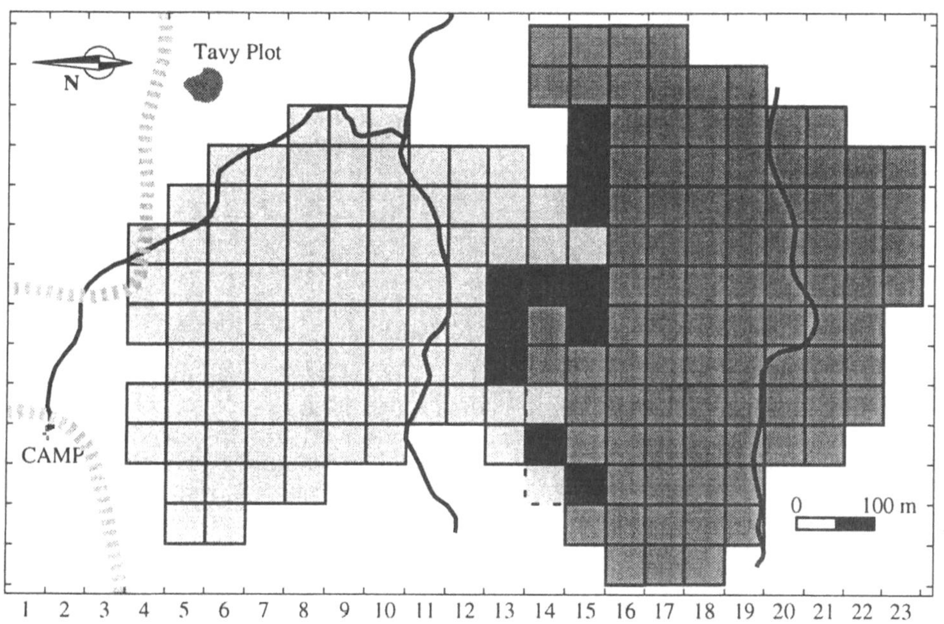


**Figure 3.** Map of the study area. Clear and dark grey quadrats indicate the home ranges of the two studied groups and their overlap is shown in black. Grey dashed lines identify the estimated borders of two neighboring groups that were not studied. Black curbs indicate hill ridges.

## Vegetation Analysis

Species composition and forest structure in the study area were analyzed in order to explore the relationship between red ruffed lemur ranging behavior and environmental factors. Forest structure was measured in 94 sample plots placed randomly in 0.25ha quadrats within the focal group's home range. Each sample plot measured 10m x 10m. Within each sample plot, all trees greater than 10cm DBH were determined, and measured for height, DBH, crown height, and crown diameter. The density (ha$^{-1}$) of each species of tree was calculated as the ratio between the total number of trees of that species in a sample plot and the area of the sample plot itself. An additional 37 sample plots were placed at trees used regularly for sleeping and

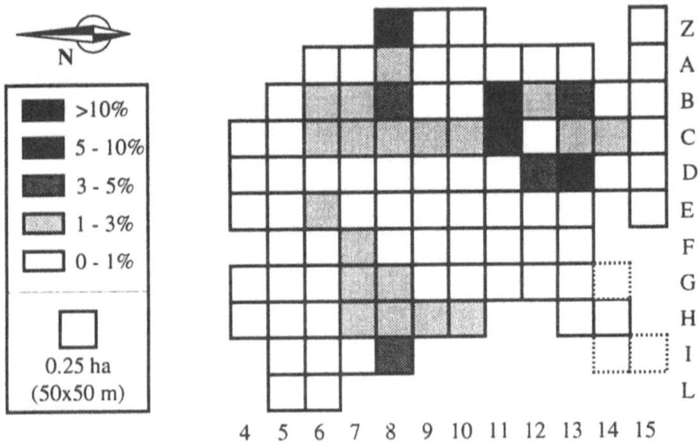


**Figure 4.** Map showing focal group's home range in the Ambatonakolahy forest during the entire study period. Differences in intensity of use are ranked in 5 categories representing different percentages of total observations, and are identified by different patterns. The core area (full black) was considered to consist of all the quadrats that the group used in more than 10% of the observations. This map includes 3 additional quadrats (dashed squares) which were entered only once by a focal animal. No feeding or long resting bouts were observed in these quadrats, only aggressive interactions with a neighboring group.

feeding in order to identify ruffed lemur habitat preferences. These plots represented a total of 5.6 % of the main study group's home range area.

Phenology data were collected once per month from July to November on 511 trees from 72 species, of which 32 species were eaten by red ruffed lemurs. Vegetative and reproductive plant parts were monitored for abundance and stage of ripeness. Data were taken on leaves (young, mature, senescent), flowers (buds, open flowers) and fruit (unripe, ripe, rotting). Abundance was classified in five categories in relation to the portion of the tree crown bearing the considered item: absent, 0-10%, 10-50%, 50-75%, 75-100%.

From two adult trees of each species, I collected three samples of leaves, flowers, and fruit, and noted the local name. Samples were dried and, in some cases, flowers and fruit were photographed and preserved in alcohol. All plant material was identified in the Tsimbazaza Botanical Garden in Antananarivo by Prof. Armand Razanamafy and his assistant.

## RESULTS

### Ranging Behavior

**Home Range Size.** During the study period, the five members of the main focal group used 93 0.25ha quadrats, for a total home range size of 23.3ha (Fig. 4). In addition to these 93 quadrats, Figure 4 shows three other quadrats (dashed lines) that were entered only once by a female on July 12 during the mating season. In these three quadrats, the focal animal was observed to travel and interact aggressively with animals of the neighboring group, but not to feed or forage. A total of 8 of the 93 quadrats entered during the study period accounted for 50.2% of all sightings. For the entire study period, the core area represented only 1.1% of the total home range. On a monthly basis, however, the core area varied between 2.6% and 13.6%

(mean=7.4%, SD=4.0, N=7) of the home range for that month (Fig. 5). On the basis of systematically recorded quantitative observations and opportunistic information, the home range size for the second group was estimated at 25.8ha (103 0.25ha quadrats) (Fig. 3). The overlap between the home ranges of the two groups was 2.5ha (10 0.25ha quadrats), corresponding to 9.3% of the main group's home range. Animals did not use the entire area of each quadrat, sometimes only a small portion of a quadrat was used. Thus, home range size calculated using 0.25ha quadrats is probably overestimated.

**Home Range Use and Territoriality.** Considering only data from the 47 full day follows, daily path length (dpl) varied from 0 to 1279m (mean=436m, SD=290, N=47). Red ruffed lemurs ranged less in the cool-wet season (mean=364, SD=220, N=27) than in the transitional-dry season (mean=550, SD=356, N=20). Each day, animals entered 1.1% to 22.3% of the total number of quadrats in their home range (mean=9.2%, SD=5.5%, N=47). The result dpl = 0 was observed only once. On August 24, a large male, observed since 4:51, was sleeping in the "amotana" feeding tree (F.T.) 136 and started activity at 7:54. The animal spent all day in this tree where it frequently chased flocks of birds and a large group of *Eulemur fulvus albifrons*. At 17:39, it assumed a hunched resting position and did not move until dark.

Red ruffed lemurs exploited only a small portion of their total home range (mean=32.3%) each month. They used this portion of the home range exclusively and intensively for periods of two to three weeks and then the range shifted partially to new areas. Monthly home ranges included between 18% and 70% (mean=45.3%, SD=18.8%) of the quadrats used in the previous month (Fig. 5). Thus, areas that had been actively used would periodically become completely neglected. Monthly core areas regularly corresponded to quadrats containing the most exploited feeding trees for that month. Table 1 shows the relative percentages of feeding observations recorded in each monthly core area quadrat for the transitional-dry season only, because subgroups did not use the same core area (see "Group Dispersion") in the cool-wet season.

Monthly variations in daily path length appeared to be related to temperature (Fig. 6). As temperature increased, the animals ranged further. An exception was observed in July when the lowest temperatures coincided with a peak in daily path length (DPL). Omitting the July data, a correlation coefficient of 0.97 between DPL and temperature was calculated.

**Table 1.** Relative percentages of feeding observations recorded in each monthly core area quadrat for the transitional-dry season.[1]

| % of Feeding Observations in Core Area Quadrats | | Core Area Quadrat | % of Feeding Observations for the Feeding Trees in the Core Area Quadrats | |
|---|---|---|---|---|
| **August** | 68.8 % | Z8 | 64.6 % | Amotana F.T.136 (Ficus lutea) |
| | | C10 | 4.2 % | Nonosay F.T. 86 (*Ficus reflexa*) |
| **September** | 96.0 % | Z8 | 90.8 % | Amotana F.T.136 (*Ficus lutea*) |
| | | B13 | 5.3 % | Vono F.T.155 (*Garcinia sp.1*) |
| **October** | 62.1 % | G7 | 12.5 % | Vono F.T.170 (*Garcinia sp.1*) |
| | | H7 | 5.3 % | Voapaka mena F.T.193 (*Uapaka thouarsii*) |
| | | I8 | 33.2 % | Antevaratra br F.T.190 (*Ocotea sp.2*) |
| | | I8 | 11.1 % | Antevaratra br F.T.203 (*Ocotea sp.2*) |
| **November** | 52.9 % | D13 | 15.3 % | Giodina F.T. 215 (*Asteropeia multiflora*) |
| | | D13 | 33.3 % | Amotana F.T.103 (*Ficus lutea*) |
| | | D13 | 4.3 % | Tavolo ravensara F.T.102 (*Ravensara pervillei*) |

[1]F.T. = Feeding Tree; br = beravina (large leaves).

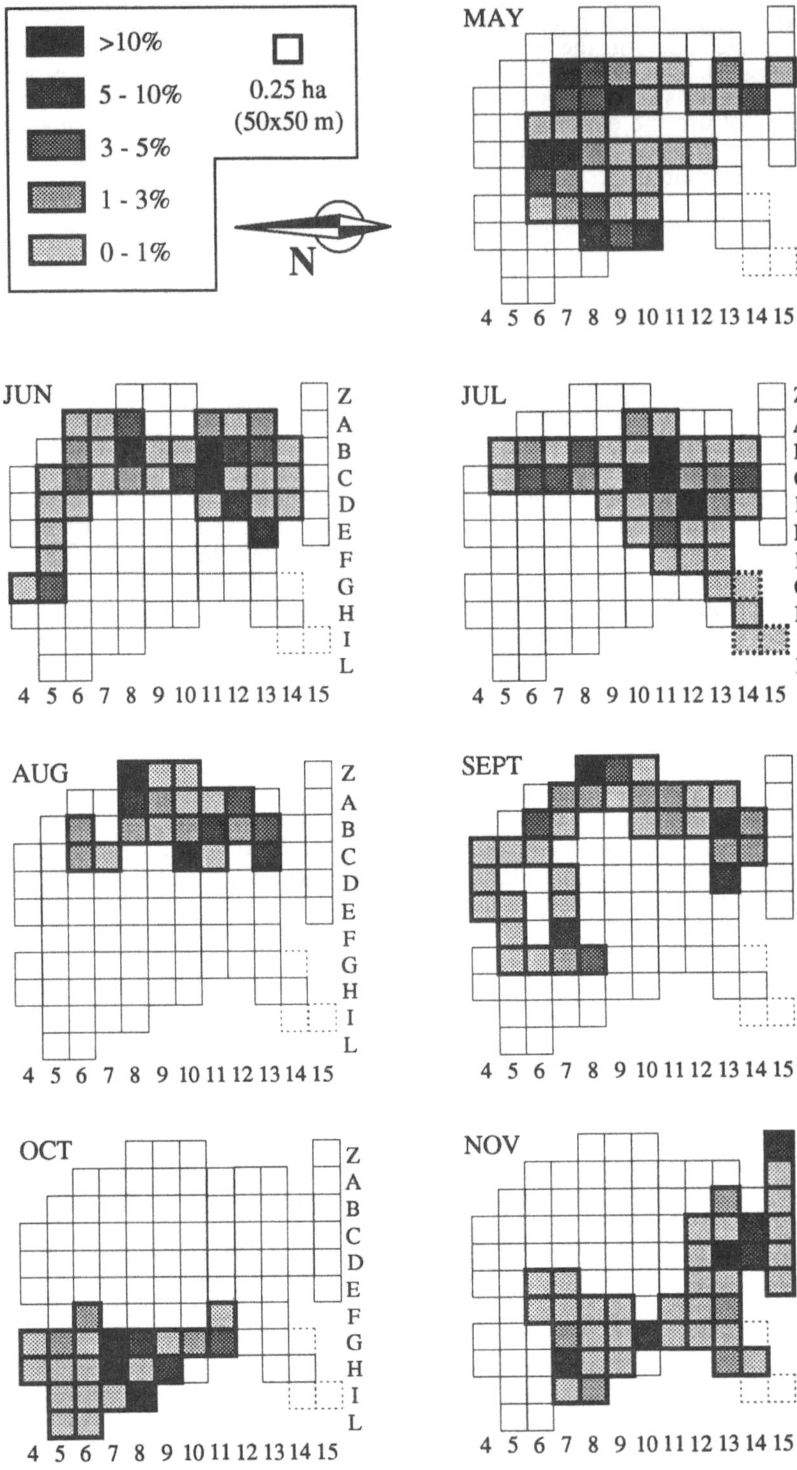


**Figure 5.** Map showing focal group's monthly home ranges in the Ambatonakolahy forest. Open squares were not used (see Fig. 4 for details).

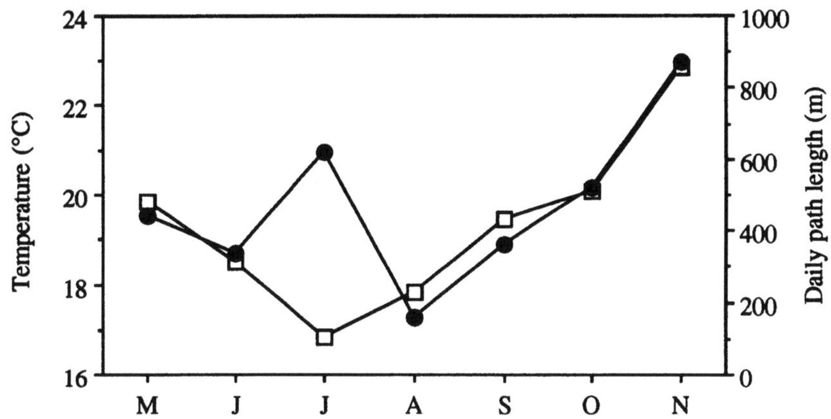


**Figure 6.** Relation between mean daily temperature (squares) and mean daily travel distance (circles) (see text).

Ruffed lemurs answered chorusing vocalizations by animals of neighboring groups promptly. Especially from the end of October to February, a sequence of calls could be heard travelling from group to group throughout the forest. Two other groups were present in the area (Fig. 3), but only loud calls were exchanged between them and the focal groups. Scent marking was always observed at the borders of the home range.

Only two aggressive interactions were observed, both between females and both in the overlap area of the two study groups (D13 and I15-I16 quadrats). During these disputes low snorts and, when in physical contact, low chuckling vocalizations were emitted. Animals scent marked branches very frequently. In the first case, on July 12, the adult female of the main study group covered 270m and reached 4 members of the second group (two males and two females). For 21 min she alternatively chased and escaped from the two females of the second group. While escaping, a female of the second group jumped on a dead branch that cracked, and fell on leafy branches 15m below. On the second occasion, on September 15, a male and the young female of the main group that were feeding on a big *Ficus lutea* were reached by a female of the second group. The two females interacted aggressively, as described in the first case, for 108min with some minutes of rest throughout. Aggressive interactions as those described here have never been observed between members of the same group.

**Group Dispersion.** When foraging during the dry season, the two groups often fragmented into sub-groups consisting of 2-3, or more rarely, one animal. During the mid-day rest period and in the evening, the subgroups usually gathered on the leafy branches of the same tree or in neighboring trees. Focal animal generally slept at a distance of 2-5m from other group members. In the main study group, however, I never saw all of the five animals in the same place at the same time, and the typical number of animals that settled for the night was four. This pattern was very different from the cool-wet season, when group members lived in small subgroups for weeks. In May, the main study group split into three subgroups that used distinct core areas (Fig. 7). In this case, subgroups were formed by two male-female couples and by a single male, but, in the dry season, daily subgroups consisting of two females (probably a mother with her young daughter) and more rarely two males were common. This tendency to form subgroups in the cool-wet season and to stay more cohesive in the transitional-dry months was particularly evident in some periods of each season. In Figure 8, data from May are compared with data from November. In the cool-wet season, as a consequence of this fragmentation, subgroups preferentially used different core areas within the home range. In the transitional-dry season, the core area was the same for all group members, and they were more cohesive.

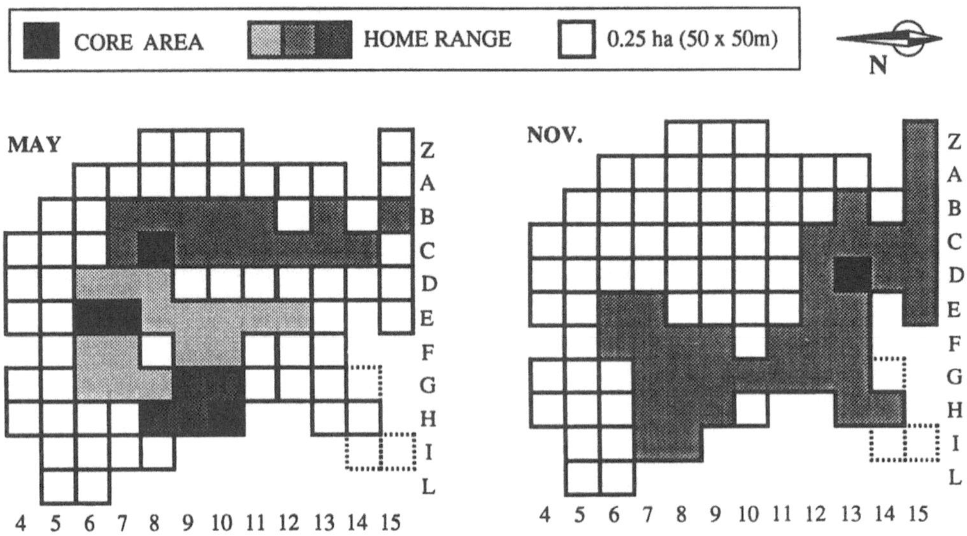


**Figure 7.** The home range of the main study group, showing the distinct core areas of the three subgroups in May, and the core area common to the entire group in November. Subgroups 1 and 2 consisted of 1 male and 1 female each, and subgroup 3 of a single male.

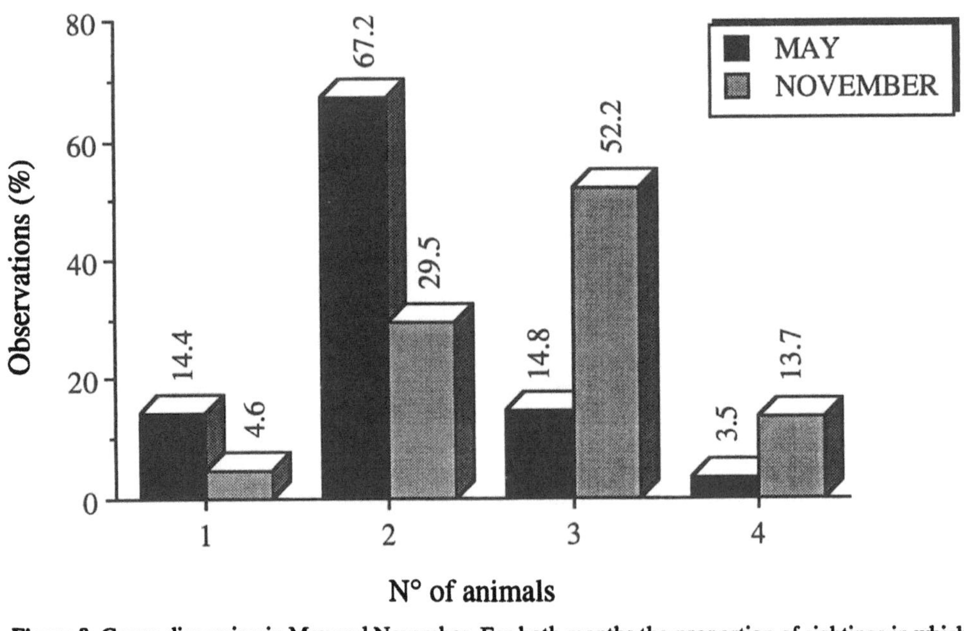


**Figure 8.** Group dispersion in May and November. For both months the proportion of sightings in which the focal animal was solitary (1) or had 1-3 individuals in its vicinity (within 50m) is shown.

**Vegetation Analysis and Diet**

Red ruffed lemurs slept, fed, and rested in large trees with an average DBH of 59.8cm (SD=37.6, N=203). In 84.9% of observations, the focal animal was on a tree with a DBH of between 41 and 80cm. Twelve percent of trees in the forest were of this size. Phenological patterns at the study site changed in relation to climatic variables (Fig. 9). The second half of August was characterized by a decrease in rain fall and by a sudden increase in flower production. During September, October and November, fruit production gradually increased. Formation of new leaves slowed down in September and October, months in which trees were subjected to energetic stress related to fruit growth. This general view does not take the strategies of different tree species into account, however.

During the study period, *Varecia variegata rubra* fed on 42 different tree species. Focal animals were mainly frugivorous (73.9% of feeding observations) and supplemented their diet with leaves (20.9%) and flowers (5.3%) (Table 2). One male was also observed to eat mushrooms once. The dietary importance of each item varied from month to month (Fig. 9). For instance, flowers accounted for between 0% and 40.5% of the diet in different months.

**Table 2.** Composition of diet of *Varecia variegata rubra*.[1]

| Food Category | % of Feeding Observations | Food Subcategory | % of Feeding Observations |
|---|---|---|---|
| Fruit | 73.9 (SD=7.0) | Ripe fruit | 73.9 (SD=7.0) |
| | | Unripe fruit | 0 |
| Flowers | 5.3 (SD=2.2) | Open Flowers | 5.1 (SD=2.3) |
| | | Buds | 0.2 (SD=0.2) |
| Leaves | 20.9 (SD=11.5) | Shoots | 1.7 (SD=1.1) |
| | | Young leaves | 0.9 (SD=0.5) |
| | | Mature leaves | 18.3 (SD=6.8) |

[1]Data from 8 focal animals

Red ruffed lemurs ate fruit from 28 different species during the study period, but only some of these species were preferentially consumed. Two Moraceae, *Ficus lutea* and *Ficus reflexa*, accounted for 58.3% of all observations of fruit eating. Three species (*Ocotea* sp.1, *Garcinia* spp., and *Ficus*) accounted for 77.7% of observations. Phenology data indicated that the two fig tree species provided a rare but consistent source of food. Fruiting *Ficus* were large, low density trees (mean radius = 5.3 and 6.1m, density = 1.1 and 2.1ha$^{-1}$) that carried fruit for many months. The small *Garcinia*, on the contrary, was the most common tree in the study area (mean radius=2.7m, density=47.9ha$^{-1}$) and generally offered only a few fruits per tree. "Antevaratra" (*Ocotea* sp.2) was a rather common tree (density=16.9ha$^{-1}$ that fruited in October. Focal animals never ate unripe fruit during the study. On trees that produced gradually ripening fruit, like *Garcinia* and *Ficus* species, red ruffed lemurs carefully inspected the branches, and smelled for several minutes to find ripe fruit.

The selectivity observed for fruit extended to leaves as well: among the 11 species eaten, three species of Lauraceae (from the genera *Ravensara* and *Ocotea* ) accounted for 88.7% of the records. The preference for leaves of the family Lauraceae has been documented also for *Varecia v. variegata* on Nosy Mangabe (H. Moland, pers. comm.).

**Table 3.** List of species eaten SH=shoots, LY=young leaves, LM=mature leaves, FW=flowers, FR=fruits, ?=not identified.

| | Family | Species | Local name | Months | | | | | | | % total feeding observations |
|---|---|---|---|---|---|---|---|---|---|---|---|
| | | | | May | Jun. | Jul. | Aug. | Sept. | Oct. | Nov. | |
| 1 | Anacardiaceae | *Poupartia chapdieri* | Rahiny | - | - | - | - | ? | - | - | .172 |
| 2 | Asclepiadaceae | *Cryptostegia madagascariensis* | Lombiry | - | - | - | - | - | FR | - | .043 |
| 3 | Burseraceae | *Canarium* sp. | Ramy tsytsiha | FR | FR? | - | - | FW? | FW | - | 1.461 |
| 4 | Burseraceae | *Canarium madagascariensis* | Ramy br | - | - | - | - | - | SH | - | .387 |
| 5 | Combretaneae | *Terminalia* sp. | Mantady br | - | - | - | - | FW | - | FR | .129 |
| 6 | Ebenaceae | *Diospyros* spp. | Hazomainty | - | LM | LM | - | - | FR | - | .215 |
| 7 | Euphorbiaceae | *Croton mongue* | Alampona mr | - | - | - | - | - | FW | FR | 6.360 |
| 8 | Euphorbiaceae | *Croton* sp.1 | Alampona br | - | - | - | FW | - | - | - | .774 |
| 9 | Euphorbiaceae | *Croton* sp.2 | Fotsyavadika | - | - | - | - | - | FW | - | .516 |
| 10 | Euphorbiaceae | *Uapaca Thonarsii* | Voapaka m | - | - | - | - | - | FR | FR | .817 |
| 11 | Euphorbiaceae | *Amyrea Humbertii* | Ampaliala mr | - | - | - | FR | - | - | - | .043 |
| 12 | Flacourtiaceae | *Casearia* sp. | Mampandoa | - | - | - | FR | (FR-LM) | - | - | 1.074 |
| 13 | Guttiferae | *Calophyllum* sp. | Vintano | - | - | - | - | - | - | - | .043 |
| 14 | Guttiferae | *Garcinia* spp. | Vono | FR | FR | FR | FR | FR | FR | FR | 12.334 |
| 15 | Guttiferae | *Symphonia* spp. | Hazinina | - | FR | (FR-FW) | - | - | - | (FR-FW) | 3.782 |
| 16 | Lauraceae | *Ocotea macrocarpa* | Tavolo br | - | LM | LM | - | LM | - | - | 2.836 |
| 17 | Lauraceae | *Ocotea* sp. 1 | Tavolo mr | - | LM? | - | - | LM? | - | FR | .430 |
| 18 | Lauraceae | *Ocotea* sp. 2 | Antaivaratra br | - | - | - | - | - | FR | FR | 6.145 |
| 19 | Lauraceae | *Ravensara Pervillei* | Tavolo raventsara br | LY | (LY-LM-FR) | ? | (SH-LM) | LM | - | FR | 5.329 |
| 20 | Lauraceae | *Ravensara* sp. | Tavolo raventsara mr | LM | FR | (LM-FR) | (LM-FR) | (LM-FR) | - | - | 6.016 |
| 21 | Lauraceae | *Cryptocasya* sp. | Longotra | FR | - | - | - | FR | FR | - | 1.719 |
| 22 | Liliaceae | *Dracaena* sp. | Asimbe' | FR | - | - | - | - | - | - | .215 |
| 23 | Moraceae | *Ficus* sp. 1 | Nononsay | - | FR | FR | FR | FR | FR | FR | 15.900 |
| 24 | Moraceae | *Ficus* sp. 2 | Voara | - | - | - | - | - | - | FR | .387 |
| 25 | Moraceae | *Ficus baroni* | Amotana | - | - | SH | FR | FR | FR | FR | 20.541 |
| 26 | Myrtaceae | *Eugenia* sp. 1 | Ompa mena br | ? | - | - | - | - | - | - | .344 |
| 27 | Myrtaceae | *Eugenia* sp. 2 | Rotro br | FR | - | - | - | FR | - | - | .774 |
| 28 | Myrtaceae | *Eugenia* sp. 3 | Rotro tbr | (FR-LM) | - | - | - | - | - | - | .774 |
| 29 | Nusaceae | *Ravenala madagascariensis* | Ravinala | - | FR? | - | - | - | - | - | .043 |
| 30 | Orchidaceae | *Phayus* sp. | Vatakasaka | - | - | - | - | - | - | - | .430 |
| 31 | Rubiaceae | *Canthium* sp. | Tsifo mr | - | - | - | - | - | FR | FR | .258 |
| 32 | Rubiaceae | *Genipa* sp. | Tsifo br | - | - | FR | - | - | - | - | .043 |
| 33 | Rubiaceae | *Breonia madagascariensis* | Tsiloparimbarika | - | - | - | - | - | FR | - | .086 |
| 34 | Sapindaceae | *Filicium decipiens* | Soretra br | FR | - | - | - | - | FW | - | 2.149 |
| 35 | Sapotaceae | sp. | Nanto antodingana | - | - | - | - | ? | - | - | .301 |
| 36 | Sapotaceae | *Sideroxiylon gerrardii* | Nanto vasy | - | - | - | - | - | - | FR | .172 |
| 37 | Sapotaceae | *Sideroxilon* sp. | Nanto mr | - | FR | FR | - | - | - | - | .688 |
| 38 | Sapotaceae | *Gambeya madagascariensis* | Famelona | FR | FR | FR | FR | - | - | - | 1.805 |
| 39 | Theaceae | *Asteropeia multiflora* | Jody | LM | - | - | - | - | - | FR | 2.450 |
| 40 | Tiliaceae | *Grewia* sp. 1 | Afipotsy | - | - | LM? | - | - | - | - | .215 |
| 41 | Tiliaceae | *Grewia* sp. 2 | Afotramena | - | - | - | - | - | - | FR | .258 |
| 42 | - | Liana spp. | Vaheny | - | FR | (LY-FR) | FR | - | (SH-FR) | FR | 1.547 |

Regarding flowers, two Euphorbiaceae (genus *Croton*) accounted for 68.1% of the records. Red ruffed lemurs ate only whole open flowers, and they were never observed feeding on nectar. Thus, only 7 of 42 food species accounted for 72.5% of the total feeding observations (Table 3).

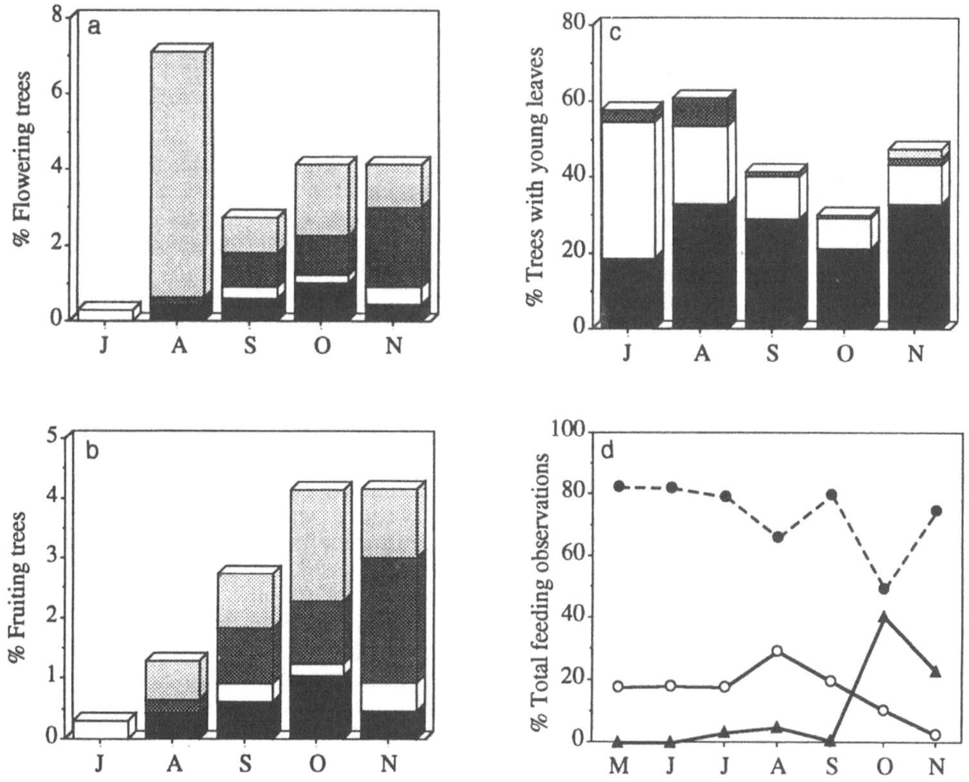


**Figure 9.** Phenological data for the tree community. The stacked column graphs show the monthly variations in percentages of analyzed trees bearing a) flowers b) fruits c) young leaves. Different patterns in each column specify the portion of crown bearing an item: black (0 - 10%), white (10 -50%), grey (50 - 75%), and dotted (75 - 100%). The line graph d) shows seasonal variations of percentages of total feeding observations spent eating leaves (open circles), flowers (triangles) and fruit (filled circles).

## DISCUSSION

Large feeding trees seem to have an influence on ranging patterns of *Varecia v. rubra*. Red ruffed lemurs used their range like "nomads": they intensively exploited a core area with some trees that provided good food resources and after a few weeks, when these trees were exhausted, they changed to a different area. In general (Fig. 4), it is not possible to distinguish an evident core area. The only quadrat used for more than 10% of the time (Z8) contained a big "amotana" tree (*Ficus lutea* F.T. 136, radius=11m) that was the most heavily exploited tree in the home range. Analyzing monthly home ranges in detail (Table 1), shows that these "nomad core areas" corresponded to food patches that were the main food resource for that month. In the cool-wet season, when fruit was scarce, the group split in subgroups, each one exploiting a different core area.

Together with scent marking, the exchange of loud calls between neighboring groups was an efficient way to establish territories and avoid conflict. Only females were involved in aggressive intragroup interactions, which occurred over food resources in the overlap area or during the mating season. These results concur with observations on Nosy Mangabe where ruffed lemurs are organized in communities of animals whose home ranges are cooperatively defended by females. Moreover, subgroups of a few animals use different portions of the common home range in the cool season, as well (Morland, 1991).

In the Ambatonakolahy forest, the main study group of five animals, observed for 539h in 7 months, used an home range of 23.3ha. The study at Vatoharanana (White, 1991) reported that a pair of *V. v. variegata*, observed for 112h over two months, maintained a territory of 197ha. Considering the number of hectares per animal, the area is 20 times larger for the ruffed lemurs at Vatoharanana. It was reported that selective logging occurred in Ranomafana National Park, and that many of the large fruit trees on which ruffed lemurs depended, were removed (Wright and White, 1990). The difference in home range size may partially be explained by the fact that large feeding trees, preferred by ruffed lemurs in both study areas (mean DBH=55.1cm at Vatoharanana and 59.8cm at Ambatonakolahy) are twice as dense at Ambatonakolahy (41cm<DBH<80cm=12.5% in sample quadrats) than at Vatoharanana (41cm<DBH<80cm=6.4% in sample quadrats). However, the wide difference in home range size between these populations of *V. variegata* remains difficult to explain. A census of large feeding trees in the home range of *Varecia* at Vatoharanana, focussing on Moraceae species, could provide data to evaluate this hypothesis.

*Varecia variegata rubra* ate fruit, flowers, and leaves of relatively few species. During the study period, 72.5% of feeding observations were on only 7 tree species out of the 42 species eaten and the 106 available in their range. This high selectivity may explain the fact that the amount of different food categories consumed in different months did not tightly correspond to the availability of each resource produced by the forest. The most evident case is flowers. Red ruffed lemurs had a preference for the flowers of two species of the genus *Croton*. In the Ambatonakolahy forest, most trees flowered in August, but flower consumption was low until October when *Croton* flowered. Diet composition varied on a monthy basis, but red ruffed lemurs remained basically frugivorous. Dependence on fruit was very marked. In fact, even in periods of scarcity of this resource, compared with alternative food categories such as leaves, the importance of fruit was always very high. Besides, as shown in Table 1, core areas regularly corresponded to large fruit trees. The tendency to split in subgroups probably was related to more successful foraging for a few animals. In this case, subgroups' presence could be temporary when food resources are abundant and become permanent where large fruit trees are rarer.

Wright and White (1990), have suggested that, compared to other lemur species, *Varecia* is more susceptible to local extinction when its habitat is disturbed. The forests of the Masoala peninsula are heavily degraded in the east but relatively well conserved in the west. Here the coastal zone and the river valleys are already degraded but very large intact areas still remain in the interior. Especially if on low slopes, these portions of the forest are seriously threatened by growing human population density (Green and Sussman, 1990). From the top of the Ambatonakolahy mountain, on a clear day in mid-December, I counted eleven smoke columns rising from new tavy scattered along the coast. The Masoala peninsula was partially protected in the natural Reserve N° 2. In 1964, however, the reserve was degazetted and timbering allowed (Simons and Lindsay, 1987). Though *Varecia variegata rubra* is listed as endangered in the IUCN Red Data Book (1990), no protected areas currently exist on the Masoala peninsula. The establishment of new areas to preserve the red ruffed lemurs' habitat is urgently needed. For such an operation, it should be taken into account that these animals depend on large trees of selected species, especially of the genus *Ficus*.

## ACKNOWLEDGEMENTS

I thank the Ministère des Affaires Etrangères, Ministère de l'Enseignement Supérieur, Ministère de la Recherche Scientifique et Technologique pour le Développement and Direction des Eaux et Forêt for giving me the authorisation to conduct this study, E. Sterling and H. Morland who provided precious suggestions and assistance from both sides of the equator and commented on this manuscript, V.G. Leone, G. Scarì, C. Vismara and E. Cereda who encouraged this work and were always helpful, A. Razanamafy who kindly determined plant species, the Costo family who solved most logistic problems, and my field assistants Pallotte, Fidaly, Beduny, Tatol, Bonga and Riss who were my eyes, my ears and my shoulders in the forest. A special thanks to F. Rubboli who believed in me.

## REFERENCES

Blanckenhorn, W.U., 1990, A comparative study of tolerance and social organization in captive lemurs, *Folia Primatol.* 55:133.

Brockman, D.K., Willis, M.S., and Karesh, W.B., 1987, Management and husbandry of ruffed lemurs, *Varecia variegata*, at the San Diego Zoo. I. Captive population, San Diego Zoo housing and diet, *Zoo Biol.* 6:341.

Constable, I.D., Mittermeier, R.A., Pollock, J.I., Ratsirarson, J., and Simons, H., 1985, Sightings of aye-ayes and red ruffed lemurs on Nosy Mangabe and the Masoala peninsula, *Primate Conserv.* 5:59.

Donque, G., 1972, The climatology of Madagascar, *in*:"Biogeography and Ecology in Madagascar," Battistini, R., Vindard, R., eds., The Hague.

Foerg, R., 1982, Reproductive behavior in *Varecia variegata, Folia Primatol.* 38:108.

Green, G.M., and Sussman, R.W., 1990, Deforestation history of the eastern rain forest of Madagascar from satellite images, *Science* 248:212.

IUCN, 1990, "Lemurs of Madagascar and the Comoros," IUCN The World Conservation Union, Gland.

Kavanagh, M., 1981, Variable territoriality among tantalus monkeys in Cameroon, *Folia Primatol.* 36:76.

Kress, J.H., Conley, J.M., Eaglen, R.H., and Ibanez, A.E., The behavior of *Lemur variegatus* Kerr 1792, *Zeitschr. Tierpsychol.* 48:87.

Lindsay, N.B.D., and Simons, H.J., 1986, Notes on *Varecia* northern limits ofits range, *Dodo* 23:19.

Martin, P., and Bateson, P., 1986, "Measuring Behaviour: an Introductory Guide," Cambridge University Press, Cambridge.

Morland, H.S., 1991, Preliminary report on the social organization of ruffed lemurs (*Varecia variegata variegata*) in a northeast Madagascar rain forest, *Folia Primatol.* 56:157.

Pereira, M.E., Klepper, A., and Simons, E.L., 1987, Tactics of care for young infants by forest-living ruffed lemurs (*Varecia variegata variegata*): Ground nests, parking, and biparental guarding, *Am. J. Primatol.* 13:129.

Pereira, M.E., Seeligson, M.L., and Macedonia, J.M., 1988, The behavioral repertoire of the black-and-white ruffed lemur, *Varecia variegata variegata* (Primates: Lemuridae), *Folia Primatol.* 51:1.

Petter, J-J., Albignac, R., and Rumpler,Y., 1977, "Mammifères Lémuriens (Primates Prosimiens)," Faune de Madagascar Vol. 44, ORSTOM-CNRS, Paris.

Rasmussen, D.T., 1985, A comparative study of breeding seasonality and litter size in eleven taxa of captive lemurs (*Lemur* and *Varecia* ), *Int. J. Primatol.* 6:501.

Simons, H.J., Lindsay, N.B.D., 1987, Survey work on ruffed lemurs (*Varecia variegata*) and other primates in the northeastern rain forest of Madagascar, *Primate Conserv.* 8:88.

Tattersall, I., 1982, "The Primates of Madagascar," Columbia University Press, New York.

White, F.J., 1991, Social organization, feeding ecology, and reproductive strategy of ruffed lemurs, *Varecia variegata, Primatology Today, Proceedings of the XIII Congress of the IPS, Nagoya and Kyoto 1990*, A. Ehara, T. Kimura, O. Takenaka, M. Iwamoto, eds., Elsevier Sience Publishers, Amsterdam.

Wright, P.C., and White, F.J., 1990, The rare and the specialized: conservation needs of rain forest primates in Madagascar, *Am. J. Phys. Anthropol.* 81:320 (Abstract).

# MALE TRANSFER IN CAPTIVE RUFFED LEMURS, *VARECIA VARIEGATA VARIEGATA*

Frances J. White[1], Elizabeth A. Balko[2], and ElizaBeth A. Fox[1]

[1]Department of Biological Anthropology and Anatomy
Duke University
Durham, NC 27706
U.S.A.

[2]College of Environmental Science and Forestry
SUNY
Syracuse, NY 13210
U.S.A.

## ABSTRACT

Black and white ruffed lemur (*Varecia variegata variegata*) groups have been studied under both semi-free-ranging conditions, at the Duke University Primate Center (DUPC), since 1985 and in the wild, within Ranomafana National Park in southeastern Madagascar, since 1988. In 1990, fences separating two established enclosure groups of *V. variegata* at the DUPC were removed. The lowest ranking males from each group first visited and then transferred from their natal group into the neighboring group. One male transferred into a single-male, single female group and rapidly attained dominance over the older resident male. The male transferring into the larger group (6-8 individuals) rose gradually in rank and mated with both resident adult females (mother and daughter). In *V. variegata*, female mating with multiple males may be an adaptation to reduce the risk of infanticide, but single male matings appear to be necessary for paternal care (guarding). In the wild, males have also been observed visiting between study groups.

## INTRODUCTION

The social organization and mating system of ruffed lemurs, *Varecia variegata*, is highly variable. Studies in captivity and the wild have found that ruffed lemurs can live in groups that range from pairs (Petter et al., 1977; Foerg, 1982; Tattersall, 1982; White, 1991; White et al., 1992) to larger multi-male and multi-female groups (Pereira et al., 1988; Morland, 1990, 1991; White, 1991; White et al., 1992). Although the composition and size of captive groups reflects management decisions, long-term studies of the mating system within larger groups and the tolerances between subsequent evictions from groups indicate that, at least under certain conditions, groups with multiple, related, breeding females may not be stable (White et al., 1992). The ability to disperse from natal groups and the availability of alternative territories may be an important factor influencing the number of breeding males and females in a group (White, 1991, 1992). Observations of captive, free-ranging and wild *Varecia*

*Lemur Social Systems and Their Ecological Basis*, Edited by
P.M. Kappeler and J.U. Ganzhorn, Plenum Press, New York, 1993

for all types of social systems have reported lower levels of cohesion among individuals than seen in other lemur species (Pereira et al., 1988; Morland, 1990, 1991; White, 1991; White et al., 1992).

Ruffed lemurs have a unique reproductive strategy. Unlike other large, diurnal primates, ruffed lemurs do not invest heavily in a single offspring, but have litters of up to four offspring (Petter et al., 1977; Klopfer and Boskoff, 1979; Klopfer and Dugard, 1976). Ruffed lemur infants do not cling to the fur but are first placed in a nest specifically built for this purpose and then are parked out of the nest. When the infants are moved, they are carried in the mouth. When in the nest, the infants are guarded initially by the mother and later the father (Pereira et al., 1987), younger sibs in a captive group (FJW, pers. obs.), or non-maternal adults of unknown relationship (Morland, 1990). It has been suggested that, as the reproductive strategy of ruffed lemurs relies on shareable parental investment (parking and guarding), this species is able to increase the litter size above that more typical singleton of other primates of equivalent body size where litter size may be restricted by the energetic costs of non-shareable parental investment (carrying of infants) (White, 1991).

With this unique reproductive strategy, males play an important role in parental care of offspring. Paternal care is usually expected only when paternal certainty is high (Trivers, 1972). Ruffed lemur groups with more than one adult male, however, present the opportunity for females to mate with more than one male and so to confuse paternity and reduce the likelihood of male care. Until recently, however, little information has been available on the relationship between having more than one mating male in the group and mating strategies within the group.

This paper summarizes new information from studies started in 1987 and 1988 on reproductive strategies following male transfer between semi-free-ranging social groups. Transfer between groups is defined as emigration from the natal group followed by immigration into another group containing breeding individuals of the opposite sex (Pusey and Packer, 1987). Captive data are supplemented with observations of wild groups in the Ranomafana National Park in southeastern Madagascar.

## METHODS

### Captive Studies

Observations of two groups of black and white ruffed lemurs were made at the Duke University Primate Center (DUPC). Observations were conducted between 07:00 and 21:00 for sampling periods of approximately 1 to 4 h with some all-day follows. Five minute focal animal sampling (Altmann, 1974) was evenly divided between all group members within a sampling period to avoid biases due to time of day. Number of observation hours varied between years from 100 to over 500 hours. Behavior patterns and vocalizations were identified following Pereira et al. (1988). The groups were maintained under free-ranging conditions in adjacent natural forest habitat enclosures. One study group was moved in 1985 from another enclosure into the 3.3ha Natural Habitat Enclosure 2 (NHE2) and the second group was released into the 5.8ha Natural Habitat Enclosure 4 (NHE4) in 1989.

The enclosures contained heated shelter boxes and several feeding stations for Purina monkey chow. Fresh fruit was provided and water was continually available. This diet was supplemented with foraging from the natural vegetation.

The NHE2 group consisted of an original adult male and female (BA and CA) who were unrelated and some of the subsequent offspring born in the enclosure (Fig. 1). BA was given a contraceptive implant in 1988 which was removed in 1989 but she has not cycled since the implant was removed. Several other offspring have died, been removed, or evicted from this social group since it was first formed (White et al., 1992). The group in NHE4 consisted of unrelated adult male and adult female (MA and CL) and their offspring (males TH born 1984 and OC born 1985). On March 21 1990, the electric fence separating the two enclosures was disconnected and the chain link fence was opened at both ends.

Data presented are from focal animal sampling and dominance hierarchies are based on unambiguous won (winner gave aggressive or no signals and loser gave only submissive signals) agonistic interactions. Preferences in grooming or nearest neighbors were tested by comparing observed frequencies to expected frequencies based on equal choice of social partners by transferred males using Goodness of Fit G test with Williams correction (Sokal and Rohlf, 1981).

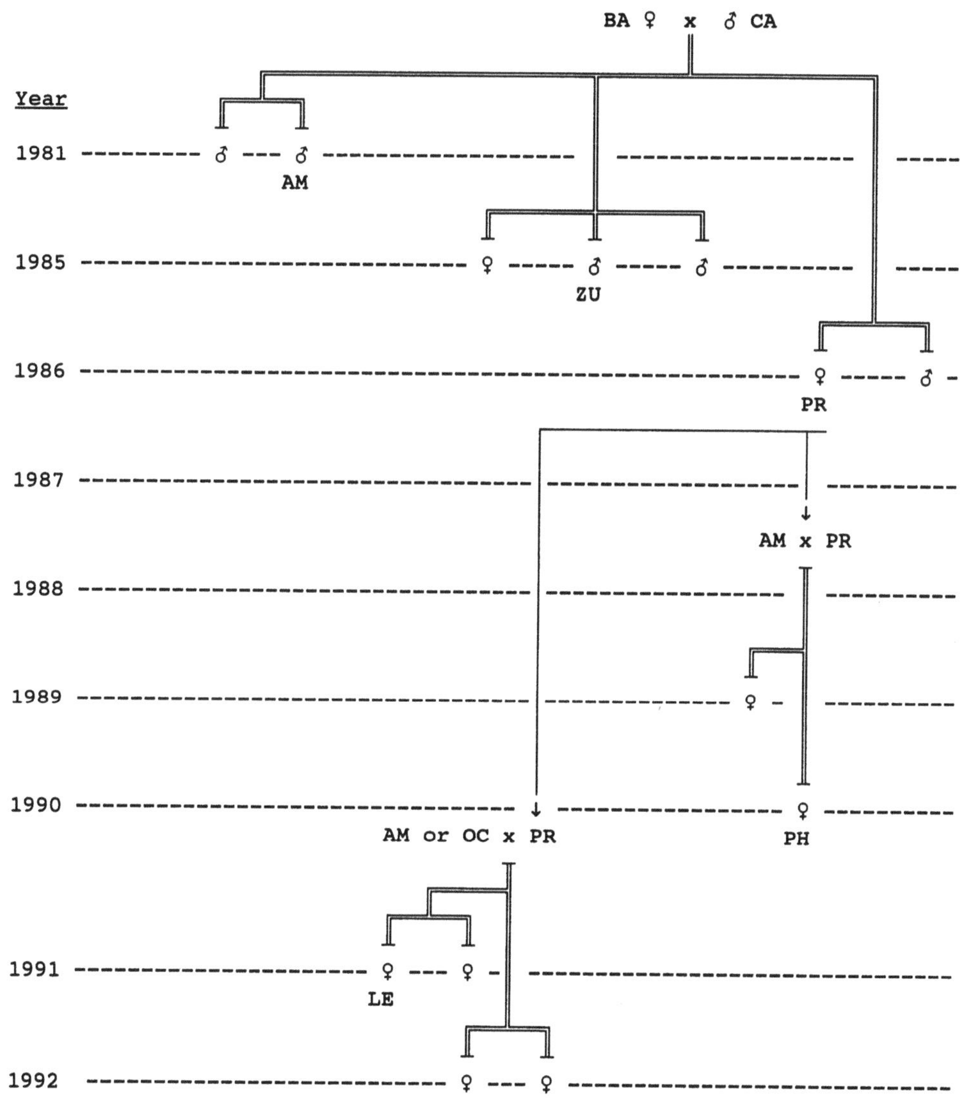


**Figure 1.** Group age and sex composition and relatedness of ruffed lemurs in NHE2 during this study. Only individuals that were in the group during this study are named. Not all litters born to BA and CA are shown. Further details of previous litters are shown in White et al. (1992).

## RESULTS

### Before Tranfer

**NHE2.** In 1989 as PR approached maturity, she outranked her mother BA. However, following an injury to PR when she became tangled in fencing wire mesh, BA was able to temporarily regain her dominant status. PR became dominant again some time after recovering from the accident. PH, PR's daughter, ranked as third female in the group, but died before becoming an adult. AM replaced his father as dominant male in the birth season of 1990 and mated with PR in 1988, and 1989. ZU remained the lowest ranking animal in the group (Fig. 2).

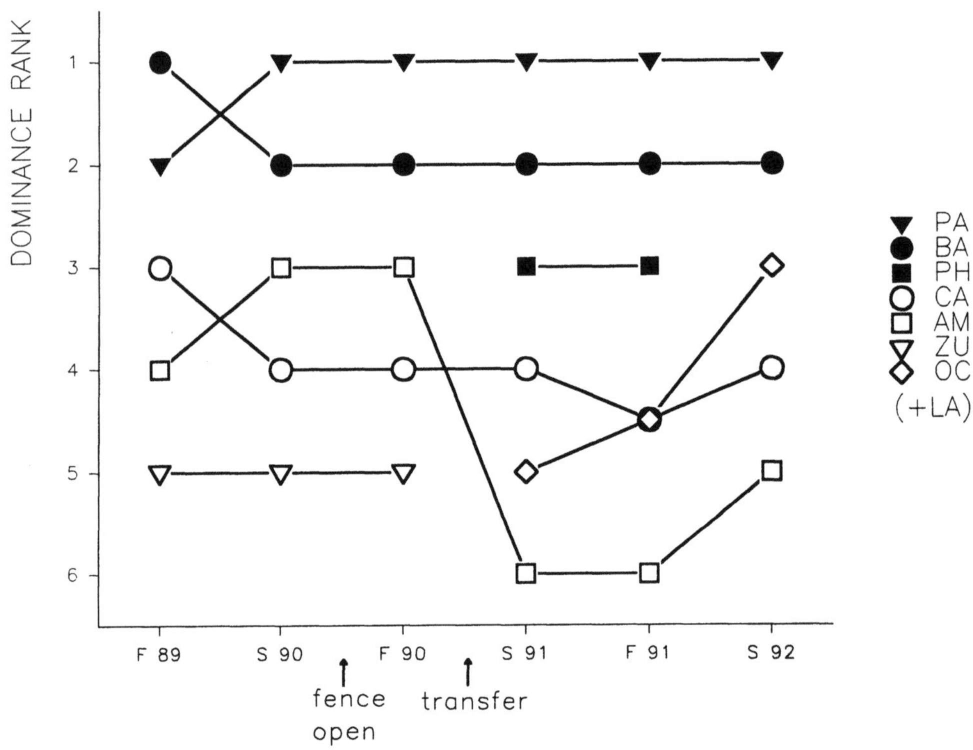


**Figure 2.** Changes in the dominance hierarchy in NHE2 based on decided agonistic interactions. Dominance ranks are based on at least 50 observation hours per season. Age and sex of individuals are given in Figure 1. Key: F=Fall, S=Spring.

**NHE4.** Following the release of the group to semi-free-ranging CL, as the only female, was the top ranking animal. OC ranked above both his father, MA and his older brother TH. TH was removed from the enclosure in 1989 due to frost-bite. The rankings in this small group were stable (Fig.3).

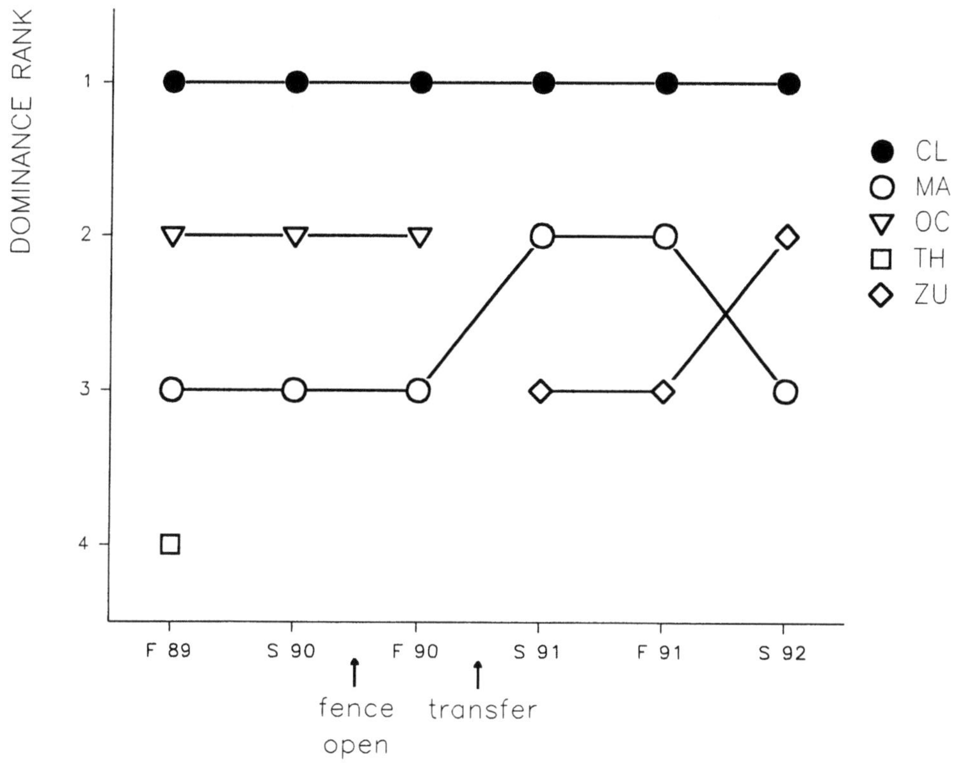


**Figure 3.** Changes in the dominance hierarchy in NHE4 based on decided agonistic interactions. Age and sex of individuals are given in text. Key: F = Fall, S = Spring.

## Behaviors Following Opening of Fence

Two sections of the fence separating NHE2 and NHE4 were opened at approximately 13:30 on March 21, 1990. The first discovery of change in the enclosure fencing occurred when CL lead OC and MA towards the west fence. OC and MA then ran at the fence and engaged in a chase-jump away territorial defense display directed at the NHE2 males (CA, AM, and ZU) that responded with a similar display at them from the opposite site of the fence until 15:54. AM then entered NHE4 through the open west gate. OC chased AM back into NHE2 but stopped abruptly at the boundary line marked by the previous location of the closed gate. OC then entered NHE2 and ran along the NHE2 side of the fence before climbing over the fence back into NHE4. The NHE2 group watched but did not interact with OC. ZU entered NHE4 at 16:05 and slowly walked the NHE4 side of the fence. ZU investigated areas frequently scent marked by the NHE4 group but did not overmark. The NHE4 group did not approach ZU. After a roar-shriek chorus by the NHE4 group, ZU left NHE4 at 16:08. ZU gave a growl-snort on his return to NHE2. Both groups returned to their usual sleep areas by 18:00. On the following day, ZU entered NHE4 ahead of PR and AM by the east gate. Both ZU and PR stopped and sniffed at the previous closed gate line. These three explored approximately $20m^2$ of NHE4 while the NHE4 group was at the opposite end of NHE4. Only ZU scent-marked (chest) in NHE4 in this area not frequently used by the NHE4 group. After a roar-shriek chorus by the NHE4 group at 09:27, ZU and AM returned to their enclosure and abrupt roared. PR remained exploring in NHE4, returning to NHE2 at 09:32. On March 24 at 16:40, OC chased ZU back into NHE2. ZU had entered approximately half-way into NHE4 when first detected. AM and OC then engaged in a jump-away and chase displays along the west boundary. There was little interaction between the groups after this, as feeding was moved to the furthest chow hoppers from the fence line.

### Male Transfer Between Groups

On December 16, 1990 OC was observed approaching the group in NHE2. He was chased away by AM. The dominant female in NHE2, PR, was in estrus at this time. On December 21, 1990 OC was again observed in NHE2 and the youngest male from NHE2, ZU, was observed in NHE4. The adult male in NHE4, MA, was observed to have injuries of a bloodied ear and lost fur, presumably from fighting, and was observed chasing ZU. The transfers of OC and ZU took place on December 30, 1990 and there were many aggressive interactions between resident and immigrant males (DUPC records). Mating occurred between BA and CA on February 11 and between BA and OC on February 12, 1991.

### After Transfer

Following transfer of OC into the NHE2 group, AM became the lowest ranking male in NHE2 and did not rise in rank again until 1992 (Fig. 2). OC initially affiliated mostly with CA (test of equal probability of grooming with all group members $G=14.68$, $P<0.05$; test of equal probability of having other group members as nearest neighbor $G=16.21$, $P<0.05$; Table 1). In the subsequent breeding and birth season, OC groomed with AM and other group members but maintained affiliation in nearest neighbor with CA (test of equal probability of grooming with all group members $G=2.51$, NS; test of equal probability of having other group members as nearest neighbor $G=9.79$, $P<0.05$; Table 1).

**Table 1.** Affiliative interactions and nearest neighbors (NN) of CA and ZU after transfer

**a) OC**

| Interaction Partners | | Spring 1991 % of Grooming | Spring 1991 % as NN | Fall 1991 - Spring 1992 % of Grooming | Fall 1991 - Spring 1992 % as NN |
|---|---|---|---|---|---|
| Adult Females | PR | 10 | 18.7 | 12.5 | 28.1 |
| | BA | 0 | 9.7 | 12.5 | 14.0 |
| | PH | 20 | 19.4 | - | - |
| Adult Males | CA | 70 | 31.3 | 25.0 | 31.0 |
| | AM | 0 | 20.9 | 50.0 | 26.1 |
| | (n) | 10 | 134 | 8 | 133 |

**b) ZU**

| Interaction Partners | | Spring 1991 % of Grooming | Spring 1991 % as NN | Fall 1991 - Spring 1992 % of Grooming | Fall 1991 - Spring 1992 % as NN |
|---|---|---|---|---|---|
| Adult Females | CL | 100 | 73.9 | 50 | 57.0 |
| Adult Male | MA | 0 | 26.1 | 50 | 43.0 |
| | (n) | 3 | 92 | 22 | 79 |

ZU transferred into NHE4 as the subordinate of the two males (Fig. 2). He affiliated only with the adult female of this group (test of equal probability of grooming with all group members $G=3.56$, $P<0.05$; test of equal probability of having other group members as nearest neighbor $G=10.71$, $P<0.05$; Table 1) and fought extensively with the other male, MA. This situation has changed as in 1992 ZU groomed equally with CL and MA but retained a nearest neighbor affiliation with the female CL (test of equal probability of grooming with all group members $G=0.00$, NS; test of equal probability of having other group members as nearest neighbor $G=9.40$, $P<0.05$; Table 1).

After the transfer, OC spent a large amount of time near CA and groomed CA more than any other individual. Later in 1991-1992, OC also began to affiliate with AM. ZU originally only affiliated with the adult female and was involved in more major disputes with MA. In 1992 ZU began to affiliate with MA. OC mated first with BA, the lower ranking of the two females, but in 1992 started showing sexual advances, that are tolerated, towards PR. BA also mated with CA, but there were no offspring.

**Paternal Care**

Females build their nests away from the group and remain separate from the group for the first 2 to 3 weeks *post partum*. Guarding of the nest and of the infants once they leave the nest by the male parent and by both male and female siblings has been observed at DUPC. When guarding, the individual remains alert and vigilant near the nest and aggressively attacks any ruffed lemur, other lemur species, and occasionally observers who approach the infants. As a dramatic and obvious behavior pattern, guarding could be scored as present or absent. Until 1991, guarding was observed during all birth seasons.

In the 1989 breeding season, after repeated unsuccessful attempts to reach the unrelated male TH in the neighboring enclosure, PR mated with AM. In the subsequent 1990 birth season, both AM and ZU participated equally in guarding PR and offspring PH. PR gave birth to twin infants in both 1991 and 1992. In the 1990 breeding seasons, PR was observed mating with AM while OC was near the group. It is possible, but cannot be confirmed, that PR also mated with OC during this estrous period. In the 1991 birth season, none of the males present in the group (AM, CA, OC), including the presumed father (AM) participated in infant guarding, and only one infant (female LE) survived. In the 1991 breeding season, PR mated with OC and AM and, in the subsequent birth season, no males participated in any guarding of the infants. Both infants disappeared from a nest in a pine tree 13 days after birth.

## DISCUSSION

Although there were two related females of breeding age in the NHE2 group, neither female transferred into the neighboring group when the enclosures were open. From the observations of individual behaviors, it is clear that most interest and exploration of the neighboring groups and territorial defense of the home group was done by the males. Transfer was preceded by a long period of male visiting to the neighboring enclosure followed by simultaneous transfer of OC and ZU. After the transfer, there was affiliation among the males. OC's transfer, although initially marked with minor disputes, has been relatively unresisted by the resident males and the affiliation among all the males is noteworthy.

During studies at three sites of Ranomafana National Park in southwestern Madagascar males were observed to visit neighboring groups at each of the three sites. These males split off from their home groups for periods of one to two days. The males appeared to be ranging between adjacent groups before returning to their home groups. Responses of neighboring group members to intruders included chasing and engaging in roar-shriek choruses with the intruder. The males did not join the groups that they visited, but these visits may be preludes to possible later transfer. This interpretation, however, can only be examined with further long-term study of the wild groups.

*Varecia* is not unusual among primates in showing male transfer. It is, however, interesting that there was no evidence of female transfer in this pair-bonded group, despite the presence of two adult females and observations of evictions of mothers by mature daughters in this group (White et al., 1992). While this may be partly explained by the lack of successful breeding by the mother, it may also be that the presence of more than one unrelated male in the group (CA for BA and OC for PR) removed the potential competition among the females for the male's care of offspring.

It is interesting that it was ZU and not the higher ranking AM that transferred from the natal group. After rising in rank above his father (CA), AM mated with his full-sib (PR). During her estrous periods, however, PR clearly preferred the unrelated males, especially TH, in the neighboring enclosure. After transfer, PR continued to mate with AM even though the unrelated OC was present. PR may have continued to mate with AM to avoid potential future infanticide of the infants. Evidence of infanticide in ruffed lemurs comes from observations of a *V. variegata rubra* group where the dominant female was in a enclosure with two related males (father and son) both of whom were unrelated to her. The son outranked the father at the time of estrus. The female was confined for the duration of the estrus with the lower-ranking male in a cage within the enclosure. After the birth of a litter of three, the higher-ranking male attacked the infants through the wire mesh and severely injured one infant, resulting in amputation of one forelimb, before being removed from the group (DUPC records).

By mating with both males, PR confused the paternity of the subsequent offspring and we observed no infanticidal attempts by either of the mating males. This strategy, however, involved mating with the less preferred, related male AM as well as the preferred male (OC). An important consequence of mating with more than one male, however, appears to be that the female loses the male's guarding of the infants. The lack of guarding following the multiple matings was dramatic and an unusual situation. It is interesting that infants that were not guarded had a low chance of survival. One set of infants disappeared from a nest and are suspected to have been taken by a raptor. Ruffed lemurs usually attack and mob both model raptors and local North Carolina raptors that enter the enclosure (Macedonia, 1993). The absence of this response from any male may have contributed to the disappearance of the infants from the nest. If male care is important for infant survival, but multiple matings confuse paternity and result in no male care, infanticide avoidance through multiple matings may not be an effective long-term strategy. Multiple mating may, therefore, be a consequence of the process of male transfer and instability in male rank.

Given the variation in ruffed lemur social organization and mating system, there may be confusion between visits to neighboring groups and what may be a wider community network. The situation is complicated by seasonal variation in cohesion in captive groups (White, 1991; White et al., 1992). It is important to make clear, however, that reports of the variation in social organization of ruffed lemurs is real and not simply a consequence of incomplete observations or short term sampling. Ruffed lemurs introduced to the island of Nosy Mangabe, Madagascar, clearly live in larger, more fluid networks of individuals (Morland, 1990, 1991). Similarly, one pair at Vatoharanana (Ranomafana NP) has now been stable and defending the same territory from 1988 to 1992. This monogamous pair successfully produced offspring in 1992 (C. Hemingway, pers. comm.). Larger groups have also been observed in the RNP, but these are clearly group-based social systems rather than networks. This range of variation in social organization is truly remarkable. The causes of this variation will be an important aspect of future research on this species.

## ACKNOWLEDGEMENTS

We thank Director Kenneth Glander, Scientific Director Elwyn Simons, and the DUPC staff for making the study at DUPC possible. FJW's research at DUPC on ruffed lemurs has benefited greatly from the enthusiasm of many undergraduate students, including Christine Kratt, Leigh Ertel, Leslie DeBardeleben, Elizabeth Dirga, Laurin Hayworth, Kristin Calvert, Pat Gaynor, Rhonda Washington, Christy

Karavanich, Daniel Buck, Laura Johnson, Loretta Ishida, and Melissa Vernon. We thank Kathryn Davis and Claire Hemingway for sharing unpublished data on DUPC groups. Permission to undertake research in Madagascar was given by M. Raymond Rakotoandriana and M. George Rakotoanrivo of the Department of Water and Forests of Madagascar, M. Berthe Rakotsamimanana and M. Benjamin Andriamihaga of the Ministry of Higher Education. These studies could not have been undertaken without the generous help of Patricia Wright, Deborah Overdorff, Michael Winslett, Peter Kappeler, and Claire Hemingway. This is DUPC publication number 554.

## REFERENCES

Altmann, J., 1974, Observational study of behavior; sampling methods, *Behaviour* 49:227.

Foerg, R., 1982, Reproductive behavior in *Varecia variegata*, *Folia Primatol.* 38:108.

Klopfer, P. H. and Boskoff, K., 1979, Maternal behavior in prosimians, *in*: "The Study of Prosimians," G.A. Doyle and R. D. Martin, eds., Academic Press, New York.

Klopfer, P. H. and Dugard, J., 1976, Patterns of maternal care in lemurs III: *Lemur variegatus*, *Z. Tierpsychol.* 40:210.

Macedonia, J.M., 1993, Adaptation and phylogenetic constraints in the antipredator behavior of ringtailed and ruffed lemurs, *in*: "Lemur Social Systems and Their Ecological Basis," P.M. Kappeler and J.U. Ganzhorn, eds., Plenum Press, New York.

Morland, H. S., 1990, Parental behavior and infant development in ruffed lemurs (*Varecia variegata*) in a northeast Madagascar rain forest, *Am. J. Primatol.* 20:253.

Morland, H. S., 1991, Preliminary report on the social organization of ruffed lemurs (*Varecia variegata variegata*) in a northeast Madagascar rain forest, *Folia Primatol.* 56:157.

Pereira, M. E., Klepper, A. Simons, E. L., 1987, Tactics of care for young infants by forest-living ruffed lemurs (*Varecia variegata variegata*): ground nests, parking, and biparental guarding, *Am. J. Primatol.* 13:129.

Pereira, M. E., Seligson, M. A. and Macedonia, J. M., 1988, The behavioral repertoire of the black-and-white ruffed lemur, *Varecia variegata variegata. Folia Primatol.* 51:1.

Petter, J. J., Albignac, R., and Rumpler, Y., 1977, Faune de Madagascar, Vol. 44, Mammifères Lémuriens, ORSTOM, CNRS, Paris.

Pusey, A. E. and Packer, C., 1987, Dispersal and Philopatry, *in*: "Primates Societies," B. B. Smuts, D. L. Cheney, R. M. Seyfarth, R. W. Wrangham, and T. T. Struhsaker, eds., University of Chicago Press, Chicago.

Sokal, R. R. and Rohlf, F. J., 1981, Biometry 2nd edn. Freeman, New York.

Struhsaker, T. T., 1969, Correlates of ecology and social organization among African cercopithecines, *Folia Primatol.* 11:80.

Tattersall, I., 1982, "The Primates of Madagascar", Columbia University Press, New York.

Trivers, R. L., 1972, Parental investment and sexual selection *in*: "Sexual Selection and the Descent of Man, 1871-1971," B. Campbell, ed., Aldine, Chicago.

White, F. J., 1991, Social organization, feeding ecology and reproductive strategy of ruffed lemurs, *Varecia variegata*, *in*: "Primatology Today: Proceedings of the XIII Congress of the International Primatological Society, Nagoya and Kyoto 18-24 July 1990," A. Ehara, T. Kimura, O. Takenaka and M. Iwamoto, eds., Elsevier Science Publishers, Amsterdam.

White, F. J., Burton, A., Buchholz, S. and Glander, K. E., 1992, Social organization of free-ranging black and white ruffed lemurs, *Varecia variegata variegata*: mother-adult daughter relationship, *Am. J. Primatol.* (in press).

# A REVIEW OF PREDATION ON LEMURS: IMPLICATIONS FOR THE EVOLUTION OF SOCIAL BEHAVIOR IN SMALL, NOCTURNAL PRIMATES

Steven M. Goodman[1], Sheila O'Connor[2] and Olivier Langrand[2]

[1]Field Museum of Natural History
Roosevelt Road at Lake Shore Drive
Chicago, IL 60605
U.S.A.

[2]World Wide Fund for Nature
B.P. 738
Antananarivo 101
Madagascar

## ABSTRACT

It has been postulated that predation has been an important selective force in molding social behavior in mammals. However, observations of predators taking primates are rare and most cases concern relatively large diurnal primates. For the lemurs of Madagascar little quantified information is available, and it has generally been assumed that predation by carnivores and raptors is rare. Contrary to expectations there is a considerable amount of data on the topic, derived from several different sources, which is presented herein. The most detailed information on lemur predation is for *Microcebus murinus*. The population dynamics of this species is reviewed in light of heavy predation pressure from two owls (*Tyto alba* and *Asio madagascariensis*), particularly the implications of predation on social behavior and life-history traits of this small nocturnal primate.

## INTRODUCTION

Considerable attention has been given to the importance of predation as a selective force in molding social behavior in primates (Alexander, 1974; van Schaik, 1983; van Schaik and van Hooff, 1983; van Schaik et al., 1983; Terborgh, 1983; Dunbar, 1988; Boesch, 1991). However, observations of non-human predators taking primates are rare, and thus a substantive portion of the data is based on anecdotal or inferred information (Cheney and Wrangham, 1987). Moreover, most documented cases of predation and the theoretical implications concern diurnal primates, which are relatively large compared to Malagasy prosimians. The types of predators taking larger diurnal primates are generally different from those feeding on smaller nocturnal primates, which implies important differences in the selective pressures on prey species.

Techniques are available to quantify rates of predation. For example, systematic sampling of carnivore's scats (Hoppe-Dominik, 1984; Emmons, 1987) or, in the case of raptors, regurgitated pellets and prey remains (Izor, 1985; Skorupa, 1989; Struhsaker and Leakey, 1990; Boshoff et al., 1991). These methods allow prey profiles to be constructed which include: species, number, frequency and age structure. Also,

*Lemur Social Systems and Their Ecological Basis*, Edited by
P.M. Kappeler and J.U. Ganzhorn, Plenum Press, New York, 1993

by sampling the food remains of potential predators, broad-based predator/prey relationships can be derived. Such data, combined with information on densities of predator and prey, can be used to estimate predation rates.

The diurnal lemurs of Madagascar have been the subject of behavioral and ecological field studies since the 1960's, however few predatory events on these animals have been reported. Laboratory and field studies have demonstrated that the relatively large, diurnal and social lemurs react to the presence of potential predators such as carnivores, birds of prey, and snakes (Macedonia and Polak, 1989; Sauther, 1989; Macedonia, 1990; Macedonia and Yount, 1991; Bayart and Anthouard, 1992). In contrast, the social behavior and life history of wild nocturnal species (Charles-Dominique et al., 1980), including their responses to predators, remains poorly documented.

A review of non-human predation on lemurs is presented. Contrary to expectations, a considerable amount of information, from the perspectives of predators and prey, is available. These data come from three sources: a literature review of Malagasy vertebrates; a survey of zoologists, primarily primatologists; and analyses of pellets of raptors and scats of carnivores. Details about localities are only given for those sites not mentioned in Nicoll and Langrand (1989). The discussion looks in more depth at the population dynamics of *Microcebus* spp., in light of heavy predation pressure from owls, concerning specifically the implications of high levels of predation on the social behavior of small nocturnal lemurs.

## SYSTEMATIC SURVEY OF PREDATION ON LEMURS

### Family *Cheirogaleidae*

***Microcebus murinus* (Grey mouse lemur).** Rand (1935) found that a pair of Henst's goshawks (*Accipiter henstii*), collected near Tabiky, northeast of Tulear, each had a portion of *Microcebus murinus* in their "gullet". This hawk tends to be diurnal and the lemur nocturnal, thus the *Microcebus* may have been taken at dawn or dusk. At Ankarafantsika, Barre et al. (1988) reported that an unidentified large raptor was attracted to the alarm call of a *M. murinus*.

Richard (1978) watched a colubrid snake, *Ithycyphys miniatus*, take an adult grey mouse lemur at Ampijoroa. Attention was drawn to the snake by a group of *Propithecus verreauxi* giving a bark alarm call. At Beza Mahafaly, Randrianarivo (1979) reported a boa (*Sanzinia madagascariensis*) predating on *M. murinus*.

Langrand (1990) identified the remains of *M. murinus* in barn owl (*Tyto alba*) pellets obtained at Analabe. Barn owl pellets from Beza Mahafaly were collected between November 1990 and November 1991 (Goodman et al., submitted). A minimum of 58 individual *M. murinus* was identified within the sample, making up 5.7% of the total individuals and 22.9% of the total biomass taken. When these samples are analyzed by month, there are clear seasonal differences in the number of *Microcebus* taken by barn owls (Fig. 1). A collection of barn owl pellets was recovered below a nest site at Lac Tsimanampetsotsa (S. Goodman, unpubl. data). The minimum number of vertebrates identified in the remains was 81 individuals, of which 10 (12.3%) were *M. murinus*.

Three collections of Madagascar long-eared owl (*Asio madagascariensis*) food remains from Beza Mahafaly were analyzed by Goodman et al. (in press). This material included a sample from 2 November 1990 in which *M. murinus* constituted 8.3% of the individuals and 7.8% of the biomass; a March-April 1990 sample representing 26.3% and 24.9% (respectively); and a 16 June 1990 sample including 25.0% and 40.3% (respectively) - the overall average in the three samples was 20.5% and 21.2% (respectively).

Russell (1977) observed a carnivore, *Cryptoprocta ferox*, clawing at the opening of a tree cavity, about 2.5 m above the ground, which held a *M. murinus*. Near Morondava, J.-J. Petter (in litt.) observed two separate incidences of the carnivore, *Mungotictis decemlineata*, extracting *M. murinus* from day shelters. In one case, two or three of these carnivores worked in unison to enlarge the entrance to the tree cavity, and then captured at least four mouse lemurs (Petter et al., 1977).

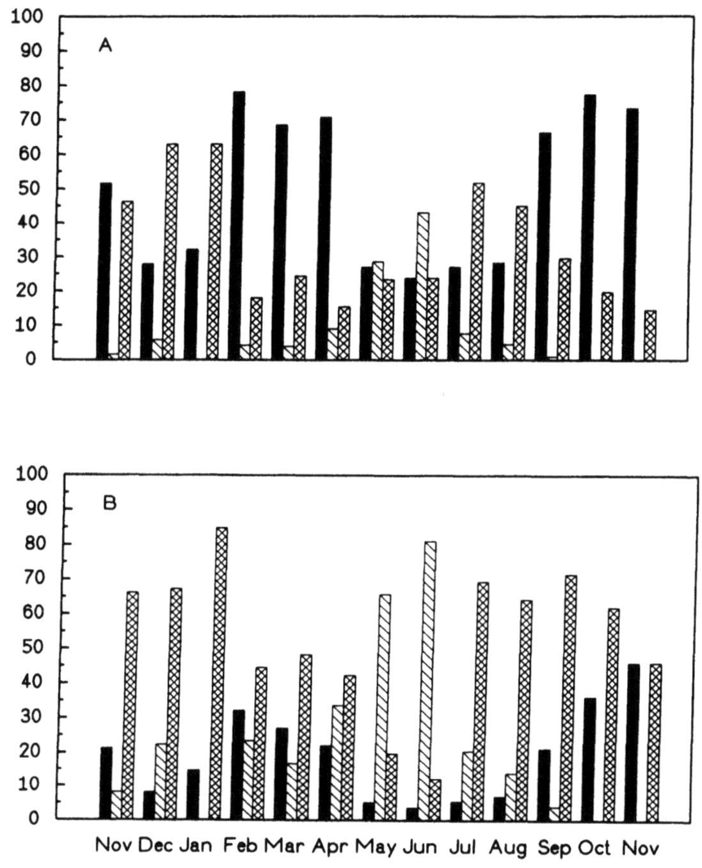


**Figure 1.** Percent composition of prey in the diet of barn owls near the Beza Mahafaly Reserve by A) total number of individuals and B) biomass. Black bars = reptiles and amphibians, hatched bars = lemurs, crossed bars = rodents.

***Microcebus rufus* (Brown mouse lemur).** Remains of this lemur have been identified from barn owl pellets collected at Andasibe (Goodman et al., submitted). *M. rufus* made up 1.1 % of the total individuals and 2.5% of the biomass. At this same locality, Ganzhorn et al. (1985) observed an unidentified owl swoop down on a *M. rufus* as it was moving in a vine tangle. L. Durrell (in litt.) observed at Manantantely, approximately 15km west of Fort Dauphin, a hooked-billed vanga (*Vanga curvirostris*) carrying off a young *M. rufus*, which it probably obtained from a nest. Three days earlier the nest contained two infants, each weighing 10g, and after the incident the nest was empty.

*M. rufus* bones have been found in two separate Madagascar long-eared owl pellets obtained in the Bezavona Forest, near Fort Dauphin: in a 27 to 30 December 1989 collection, *M. rufus* comprised 12% of the total individuals and 8.2% of the biomass, and in a 17 September to 13 October 1990 sample, 3 % and 1.8%, respectively (Goodman et al., 1991, in press). Martin (1973) found *M. rufus* inhabiting lowland montane forest in the general vicinity of the Bezavona Forest and *M. murinus* in the littoral forest a few km away. The remains recovered from the Bezavona Forest are presumed to be of *M. rufus*, but it is possible that some are of *M. murinus*.

In secondary forest at Ranomafana National Park, E. Rajeriarison (pers. comm.) scared a Madagascar harrier hawk (*Polyboroides radiatus*) perched in a tree. As the bird flew away, it released an infant *M. rufus* that had a crushed head, but was still alive. The hawk made several passes overhead as the lemur was being handled. A ball nest, typical of *Microcebus*, was located in a tree about 8m above the spot where the lemur was found. Petter et al. (1977) reported that when the carnivore, *Galidia elegans*, finds an active *Microcebus* nest, they are likely to raid it and feed upon the occupants. At Ambalafary, Manongarivo, a village dog was observed taking a mouse lemur at dusk in secondary forest (A. P. Raselimanana, pers. comm.).

***Mirza coquereli* (Coquerel's dwarf lemur).** Near Morondava, J.-J. Petter (in litt.) observed a raptor, probably a Madagascar buzzard (*Buteo brachypterus*), catch a *M. coquereli*, fly off with it, and then release the animal about 50m above the ground. The lemur landed on a bush, partially breaking its fall. Its belly had been torn open and the intestines exposed. Thereafter five or six different *Mirza* were trapped that had scars on the belly, presumably from predation attempts by raptors. Since the majority of the large hawks on Madagascar are diurnal (Langrand, 1990) and *Mirza* is nocturnal (Petter et al., 1977) two interpretations are suggested: either the lemurs are also active at dawn or dusk, or they are removed from their day nests.

***Cheirogaleus medius* (Fat-tailed dwarf lemur).** Petter (1988) suggested that near Morondava this nocturnal lemur is vulnerable to predation at night by snakes and owls, and during the day while sleeping by the Madagascar harrier hawk and carnivores.

***Cheirogaleus major* (Greater dwarf lemur).** In 1991, three separate incidents of predation on this lemur were recorded in Ranomafana National Park. *Galidia elegans* was observed killing a young male (with abdominal testes and weighing 375g). The face of the animal was severely damaged, with the maxilla fractured and eyeballs removed (M. Dagosto and A. Mehrenlander, pers. comm.). The second case, similar to the first in terms of damage, was an adult female (weighing 570g) found alive in the jaws of a *G. elegans* (C. Hemingway, in litt.). The third case concerned a Madagascar buzzard found resting in a tree and clutching a dead *C. major*. Attention was drawn to the bird by a group of brown lemurs (*Eulemur fulvus rufus*) alarm-calling on the ground below the raptor. When the bird was approached, it flew off, leaving the lemur behind (M. Fidi, pers. comm.). The stiff and cold animal was examined; it was an adult female and weighed 310g (C. Hemingway, in litt.). The lemur had dirt in its mouth and the feet were clenched. The dorsal fur was plucked from the shoulders down to the upper portion of the legs. The distal portion of the right limb was broken off at the humerus and portions of the right shoulder and clavicle were eaten away. It is not clear if the *Buteo* killed or scavenged the lemur. *Cryptoprocta* is also suspected to feed on *Cheirogaleus* spp. (Albignac, 1973).

***Phaner furcifer* (Fork-marked lemur).** About 50km north of Morondava, C. Cassel (pers. comm.) observed a Madagascar buzzard capture a fork-marked lemur. The raptor flew a short distance with the talons of one leg planted in the lemur's nape and the other in the upper back, and then landed on a trail in the forest. The observer scared the hawk away and the lemur, which did not have any noticeable external injuries, moved off quickly into a tree cavity. Charles-Dominique and Petter (1980) reported that in the Marosalaza Forest, 50km north of Morondava, a Madagascar cuckoo falcon (*Aviceda madagascariensis*) caught a *Phaner* at twilight. After being carried a short distance, the *Phaner* broke loose from the raptor's grip and fell some 20m into vegetation.

**Family Megaladapidae**

***Lepilemur leucopus* (White-footed sportive lemur).** Bone remains of an adult *L. leucopus* were found in a barn owl pellet collected along the cliffs of the Ehazoara Valley, near Beza Mahafaly (S. M. Goodman, unpubl.).

***Lepilemur mustelinus* and *L. septentrionalis* (Weasel sportive and Northern sportive lemurs).** Ratsirarson (1986) reported that boas (*Sanzinia madagascariensis*) wait in ambush outside the shelters of these *Lepilemur* spp., and that snakes may also enter day shelters of these lemur species to feed upon them. Russell (1977) noted that *L. mustelinus* has a specific alarm call in response to boas. In Ankarana, J. Wilson (in litt.) reported that *L. septentrionalis* reacted strongly by giving alarm calls to large birds flying overhead.

***Lepilemur ruficaudatus* (Red-tailed sportive lemur).** Near Morondava, N. Rakotoarison (pers. comm.) observed a Madagascar harrier hawk extract a *L. ruficaudatus* from its day shelter. At this same locality, R. Albignac (pers. comm.) observed a similar, although unsuccessful, act of predation.

**Family Lemuridae**

***Lemur catta* (Ring-tailed lemur).** At Beza Mahafaly, Ratsirarson (1985) observed a Madagascar harrier hawk take a young *L. catta*. Immediately after the attack, a nearby troop of ring-tailed lemurs (the predated individual was not a member) started to give alarm calls while watching the hawk carry away the struggling lemur. Bones of *L. catta* have been found at Lac Tsimanampetsotsa below a roost site of this hawk, consisting of a minimum of four individuals ranging from a young juvenile (having the ends of long bones slightly ossified) to adults (S. M. Goodman, unpubl. data). This lemur exhibits a number of anti-predator behaviors towards the Madagascar buzzard, Madagascar harrier hawk and the black kite (*Milvus migrans*) (Jolly, 1966; Budnitz and Dainis, 1975; Sauther, 1989; Macedonia, 1990).

There are a number of predation records from Berenty. L. Durrell (in litt.) observed an attack by an unidentified raptor on a young *L. catta*. The lemur was hit by the bird, fell to the ground, and when recovered alive by the observer was found to have puncture wounds to the head. A black kite was observed swooping at an infant *L. catta* being carried on the back of a female (C. Harcourt, in litt.). Koyama (1992) reported finding an infant ring-tailed lemur lying on the ground with a small hole in its head and apparent paralysis on the left side of the body. The researcher was attracted to the site by a troop of *L. catta* giving alarm calls, and postulated the predator may have been France's sparrowhawk (*Accipiter francesii*). However, on the basis of a photograph (Koyama, 1992, plate 1D) of the lemur, the raptor responsible for the attack was almost certainly larger than *A. francesii*. A *L. catta* was noted tracking an *Acrantophis* as it moved along the ground and in low branches (S. O'Connor, unpubl. data).

At Beza Mahafaly, Ratsirarson (1985) observed an incident when *L. catta* upon seeing a *Cryptoprocta* quickly climbed a tree and started alarm-calling, which continued until the carnivore left the area. Sussman (1977) noted that this lemur vocalizes at the initial sight of *Cryptoprocta* and then remains silent, closely watching the animal as it moves away. A *Viverricula indica* was observed predating on a young ring-tailed lemur (M. Enafa, pers. comm.). The carnivore was noted lying flush to the ground near a group of *L. catta* and waving its tail, which is similar in color and pattern to that of the ring-tailed lemur. The troop was apparently attracted to the tail movement and when they came down out of a nearby tree to investigate, the carnivore immediately attacked the young lemur by grabbing its muzzle (P. Rakotomanga, pers. comm.). P. Rakotomanga found bones and fur of *L. catta* at the entrance of a feral cat's shelter. *L. catta* has different alarm calls towards snakes, carnivores, and aerial predators (Macedonia and Polak, 1989; Sauther, 1989; S. O'Connor, unpubl. data).

***Eulemur coronatus* (Crowned lemur).** Wilson et al. (1989) found grey fur and broken bones in the faeces of *Cryptoprocta ferox* that were "apparently" the remains of *E. coronatus*. The presence of *Cryptoprocta* provokes alarm-calling in this lemur, which may also be predated upon by crocodiles, *Crocodylus niloticus* (J. Wilson, in litt.). M. Hadani (pers. comm.) observed in Montagne d'Ambre National Park a *Cryptoprocta* attempt to catch an adult crowned lemur in a tree. Crowned lemurs give alarm calls at the approach of the Madagascar harrier hawk and other relatively large birds (Wilson et al., 1989; Langrand, 1990).

***Eulemur macaco* (Black lemur).** In mid-May 1992, along the Ambato Massif, north of Ambanja, I. Colquhoun (in litt.) heard a troop of *E. m. macaco* giving alarm barks and loud calls. The still vocalizing troop was located and two infants were found huddled together in dense forest understory. Soon thereafter, a Madagascar harrier hawk swooped into the area and landed for an instant on a branch near an adult female, which screamed and leapt to a nearby tree. At Nosy Be, a group of *E. m. macaco* was noted to vocalize when a raptor passed overhead (D. Meyers, in litt.). Bayart and Anthouard (1992) reported that captive *E. macaco* respond to the presence of snakes.

***Eulemur mongoz* (Mongoose lemur).** Andriatsarafara (1988) reported that *E. mongoz* react with great excitement when *Buteo brachypterus* fly overhead. This researcher also suggested that snakes are predators of young lemurs.

***Eulemur rubriventer* (Red-bellied lemur).** At Ranomafana National Park, D. Overdorff (in litt.) observed red-bellied lemurs sleeping together in a tree about 5m away from a boa. The snake was discovered by two lemurs that maintained an unusual posture as they cautiously prodded and sniffed it for five minutes.

***Eulemur fulvus* (Brown lemur).** There are several recent reports of attacks on *E. f. rufus* at Ranomafana National Park. A male *Cryptoprocta* ferox was trapped and held in captivity (G. Naylor, pers. comm.). Scats recovered from the animal contained fur of *E. f. fulvus* (S. Goodman, unpubl. data). J. M. Radiva (pers. comm.) heard in primary forest a troop of *E. f. rufus* giving alarm calls nearby. He observed a large unidentified raptor swooping in and hiting an adult lemur, immediately killing it. The raptor quickly opened the lemur's cranium and consumed the brain and most of the flesh. D. Overdorff (in litt.) observed a large raptor perched about 25m from a group of *E. f. rufus*, which gave alarm calls and moved quickly to within 1m of the ground. The group remained motionless for about 30min. The bird was identified as *Eutriorchis astur*, but was likely to have been *Accipiter henstii*. At other localities on Madagascar this lemur is known to react to the presence of *Polyboroides* (Harrington, 1975; Sussman, 1975; Langrand, 1990) and *A. henstii* (Thompson and Evans, 1991).

On Nosy Mangabe, A. Peyrieras (pers. comm.) observed an *Acrantophis madagascariensis* swallowing an adult *E. f. albifrons*. There are also reports that this lemur reacts to the presence of *Sanzinia madagascariensis* (Andriatsarafara, 1988; E. Sterling, in litt.). Bayart and Anthouard (1992) noted that captive *E. f. mayottensis* respond to the presence of snakes.

***Varecia variegata* (Ruffed lemur).** Barden et al. (1991) noted that visual contact by the ruffed lemur with the Madagascar harrier hawk or Henst's goshawk provokes alarm-calling. At Nosy Mangabe, Morland (1991) observed an encounter between a male *V. v. variegata* and a raptor that was tentatively identified as an *Accipiter*. The ruffed lemur was feeding in a tree when the raptor flew overhead, the lemur responded by giving roaring-calls, and then moved out of the tree. The raptor landed in an adjacent tree and the lemur rushed towards it, causing the bird to fly away. At the same locality, E. Sterling (in litt.) also reported that *V. variegata* gives alarm calls when raptors pass overhead. Near Maroantsetra, A. Peyrieras (pers. comm.) noted a large hawk, perhaps *Buteo brachypterus*, dive at a group of *V. v. rubra* with infants. This presumed act of predation was unsuccessful and was terminated when two adult lemurs chased the hawk away as it landed nearby.

***Hapalemur griseus* (Grey gentle lemur).** *H. griseus* responds to the presence of the Madagascar harrier hawk (Langrand, 1990). A calcaneus recovered from a pellet of the Madagascar long-eared owl in the Bezavona Forest belongs to either a *H. griseus* or *Avahi laniger* (Goodman et al., in press). Since both of these lemurs occur in the Bezavona Forest, and the cuboid facet and the tuberosity are missing, it is difficult to identify the species. However, the well developed sustentaculum and groove of the calcaneus indicates it was an adult (I. Tattersall, in litt.).

R. Rakoto (pers. comm.) observed a *Sanzinia madagascariensis* consuming a *H. griseus* on the outskirts of the Ranomafana village (Ifanadiana) in an area of extensive bamboo. He came upon the snake about 3m high in a tree, and it had already consumed the head of the adult *Hapalemur*.

***Hapalemur aureus* (Golden bamboo lemur).** J. Ratsimbazafy (pers. comm.) reported that at Ranomafana National Park, golden bamboo lemurs respond to the presence of large diurnal raptors. He observed one incident when a large raptor flew over a group of lemurs feeding in bamboo, they gave quiet "whimpering calls" and descended to the ground.

**Family Indridae**

***Avahi laniger* (Woolly lemur).** A calcaneus of *Avahi* or *Hapalemur griseus* was recovered from a Madagascar long-eared owl pellet (see *H. griseus*, above). At noon on 4 December 1992 in relatively undisturbed rain forest at 1150m elevation in the Ranomafana National Park, (E. Balko, pers. comm.) observed a Henst's Goshawk swoop down through the forest and hit an *Avahi* resting in a tree. The raptor then descended to the ground carrying the animal in its talons, and started to feed by holding the *Avahi*'s body with its talons and tearing apart the head with its beak. The hawk was chased off the prey after a few minutes, and most of the head had already been removed and consumed. Small bone fragments and tufts of hair were spread out around the site. The *Avahi* was a juvenile male that weighed 0.5kg (without the head).

***Indri indri* (Indri).** At Périnet, R. Mittermeier (in litt.) observed a troop of *Indri* with young respond with an unusual call to a *Polyboroides* flying overhead and perching nearby.

***Propithecus diadema* (Diademed sifaka).** At Ranomafana National Park, a study troop of *P. d. edwardsi* consisted of an old adult female, adult male, young adult female and young adult male. One morning the young adult male was observed uttering distress calls, and nearby clumps of black and white hair and portions of a lemur's digestive tract were found. All members of the troop, except the older adult male, were seen later that same day, and the former individual was not observed again (M. Dagasto, J. Powzyk and N. Vasey, in litt.). Simona (1988) reported that at Andapa the principal predators on *P. d. candidus* are *Cryptoprocta*, *Acrantophis* and large raptors; however, specific details were not provided.

*P. diadema* also reacts strongly to the presence of large birds of prey, such as *Polyboroides* (Langrand, 1990). At Ranomafana National Park C. Hemingway (pers. comm.) has observed this lemur feeding in the canopy of relatively open trees, during which time a raptor passed overhead. The lemurs upon noticing the bird immediately gave alarm calls, quickly climbed or jumped down to areas of denser vegetation, and moved closer together.

***Propithecus tattersalli* (Tattersall's sifaka).** At Daraina, N. Vasey (in litt.) reported that *P. tattersalli* gives an alarm call in response to raptors flying overhead, and that an individual of this sifaka intensively watched a snake slither by, but did not call.

***Propithecus verreauxi* (Verreaux's sifaka).** Rasoanindrainy (1985a, 1985b) studied the population dynamics of *P. v. verreauxi* in gallery forest along the Onilahy River, southwestern Madagascar, and concluded that the main cause of infant mortality (individuals up to one year old) was *Polyboroides* predation. A group of Verreaux's sifaka was observed alarm-calling at a Madagascar harrier hawk circling overhead (J. M. Rasoanindrainy, pers. comm.). It quickly descended towards the group and captured an adult female that was nursing a small infant. The hawk laboriously flew off with the lemur struggling in its talons, and when several meters off the ground, the infant released its grip from the mother and fell. This lemur is known to give high intensity alarm calls in response to this hawk (Jolly, 1966; Sussman, 1977; Richard, 1978).

About 80km north of Morondava, C. Cassel (pers. comm.) observed a Madagascar buzzard in the forest dismantling an adult male *P. v. verreauxi*. The lemur had a severe wound to the nape, its cranium had been opened, and the brain partially eaten. Near Morondava, fur of this species was found in *Cryptoprocta* faeces (N. Rakotoarison, pers. comm.), and near Bongolava, remains of *P. v. deckeni* have been found and attributed to *Cryptoprocta* predation (Albignac, 1973). Along the Onilahy River, Rasoanindrainy (1985b, pers. comm.) followed the alarm calls of a group of *P. v. verreauxi*, and found them on the ground around a corpse of a dead adult sifaka. The animal had been eviscerated and the heart and liver eaten. At Berenty, O'Connor (unpubl. data) recovered a severely wounded adult male *P. v. verreauxi*. The animal had been mauled by a carnivore and had punctures in the upper chest cavity and in the groin and upper thighs. The rest of the group were sitting above him vocalizing (contact calls) while he remained inert on the ground. Rabemazava (1990) reported that *Cryptoprocta* is a rare predator on *P. v. coronatus*, and that large birds of prey (e.g., *Buteo brachypterus*) occasionally take young lemurs, but no details were provided.

The only documented case of predation on *P. v. verreauxii* recorded by Richard et al. (1991) during their five year study at Beza Mahafaly, was a young male found eviscerated (P. Rakotomanga, in litt.). The thoracic cavity of the lemur had been completely devoured and other portions of the body were scattered around the site; local guides suggested that the predator was most likely *Cryptoprocta ferox*. Also bones of *P. verreauxi* have been found outside of *Cryptoprocta* dens (Albignac in Richard, 1978).

## DISCUSSION

On the basis of this review, it is clear that a wide range of animals feed on lemurs (Table 1). All of the extant lemur genera, with the exception of *Daubentonia* and *Allocebus* are included. More quantified data are needed, but it appears that predation on some taxa (e.g., *Microcebus murinus*) is frequent enough to be a potentially important selection pressure in the evolution of lemur behavior or life history traits. This general topic, for relatively large diurnal primates has been the subject of considerable discussion (e.g., van Schaik, 1983; Cheney and Wrangham, 1987; Boesch, 1991), while comparatively smaller nocturnal (Charles-Dominique, 1977; Wright, 1989) and diurnal primates (Mitchell et al., 1991) have received little attention. Our information on *Microcebus* predation is the most complete for any of the lemur genera, and below we explore the implications of this pressure on the behavior and ecology of this species.

### Historical aspects

It is important to put predator/prey interactions on the island into a historical perspective. Mostly through human-induced changes, the ecosystems of Madagascar have been tremendously altered in the past 2,000 years, the minimum period humans have been on the island (MacPhee and Burney, 1991). For much of Madagascar, little is known about the original biota before human colonization. For example, the extent and type of forest found on most of the high plateau before human-induced perturbations is not certain (Battistini and Vérin, 1972; cf. MacPhee et al., 1985). However, it is noteworthy that numerous species of lemurs and other vertebrates, including potentially important lemur predators, have gone extinct on the island in the past few millennia (Dewar, 1984; Richard and Dewar, 1991). Little is known about the original recent fauna of Madagascar. Thus, when we look at modern interactions between predators and lemurs, important pieces of information are lacking. For example, material of a large extinct eagle (*Aquila sp.*) has been identified from surface deposits in Mitoho Cave, near Lake Tsimanampetsotsa (Goodman, in press). *Aquila* no longer occur on Madagascar. It is unknown when this eagle disappeared, nor its former distribution on the island. However, based on size comparison to extant African and Eurasian *Aquila*, the Lac Tsimanampetsotsa bird would have been capable of taking adult lemurs the size of *Propithecus verreauxi* (3-4kg). Thus, it is plausible that the strong stereotypic response of several lemurs (e.g., *P. verreauxi* and *Lemur catta*) to the presence of birds of prey (Sauther, 1989) evolved as a result of predation by the extinct *Aquila*, and that this defensive behavior is reinforced by occasional acts of predation by extant raptors.

**Table 1.** Summary of different types of animals predating on lemurs.[1]

| Species of Lemur | Types of Predators | | | |
|---|---|---|---|---|
| | Diurnal Raptors | Nocturnal Raptors | Carnivores | Snakes |
| *Microcebus murinus* | x | x | x | x |
| *Microcebus rufus* | x | x | x | |
| *Mirza coquereli* | x | | | |
| *Cheirogaleus medius* | ? | ? | ? | ? |
| *Cheirogaleus major* | ? | | x | |
| *Phaner furcifer* | x | | | |
| *Lepilemur leucopus* | | x | | |
| *Lepilemur mustelinus* | ? | | x | x |
| *L. septentrionalis* | ? | | x | x |
| *Lepilemur ruficaudatus* | x | | | |
| *Lemur catta* | x | | x | |
| *Eulemur coronatus*[2] | ? | | x | |
| *Eulemur macaco* | ? | | | ? |
| *Eulemur mongoz* | ? | | | ? |
| *Eulemur rubriventer* | | | | ? |
| *Eulemur fulvus* | x | | x | x |
| *Varecia variegata* | ? | | | |
| *Hapalemur griseus* | ? | x[3] | | x |
| *Hapalemur aureus* | ? | | | |
| *Avahi laniger* | x | x[3] | | |
| *Indri indri* | ? | | | |
| *Propithecus diadema* | ? | | x | ? |
| *Propithecus tattersalli* | ? | | | |
| *Propithecus verreauxi* | x | | x | ? |

[1] x = documented case of predation, ? = inferred case of predation. See systematic section of text for further details and references.

[2] *Crocodylus niloticus* is also a presumed predator on this species.

[3] a calcaneus recovered from a Madagascar long-eared owl pellet was of either *Hapalemur griseus* or *Avahi laniger*.

**The dynamics of *Microcebus murinus* predation at Beza Mahafaly**

The habitat of the Beza Mahafaly Reserve is typical of sub-arid thorn scrub of the southern domain (Nicoll and Langrand, 1989). The area is composed of two natural forest types: riverine gallery forest and spiny forest. The gallery forest along the seasonally dry Sakamena River is dominated by *Tamarindus indica*. The area is characterized by a distinct dry season, highly irregular annual amounts of precipitation, and a rainy season generally between December and February.

Information on *M. murinus* predation in the reserve is based on analysis of regurgitated barn owl and the Madagascar long-eared owl prey remains. Pellets of the cosmopolitan barn owl were collected on 23 occasions between November 1990 and November 1991 at a gallery forest site (Goodman et al., submitted). In most cases, two samples were collected per month, and are presumed to be the food remains of one barn owl pair. A minimum of 58 *M. murinus* were identified from the pellet material. On three occasions in 1990, pellets of the endemic Madagascar long-eared owl were collected at a roost close to the barn owl site, and these included a minimum of 16 individuals of *M. murinus* (Goodman et al., in press). It is likely that these two owl species are hunting in the same area of gallery forest.

No information on the territory size of the barn owl and Madagascar long-eared owl on the island is available. Densities of barn owls can be high. Wilson et al. (1986) found 49 breeding pairs on a 300ha plot in Mali, and Steyn (1982) located five nests within a 100m stretch in Zimbabwe. Little else is known about the territory size of African barn owls. This species' territory size is directly related to prey abundance (Cramp, 1985). In areas of central Europe, when food is plentiful, the territory size varies from about 0.4 to 0.6km$^2$ (summarized in Cramp, 1985). Thus, assuming that there are parallels in population densities throughout portions of this species' range and using the more conservative figure of 0.6km$^2$ (60ha), then approximately 9 or 10 pairs of barn owls might occupy the 5.8km$^2$ (580ha) Beza Mahafaly Reserve. Further, assuming that the number of *M. murinus* taken by the pair of barn owls represented in our data is characteristic of the Beza Mahafaly area population at large, then approximately 522-580 *M. murinus* would be taken annually by the 9-10 pairs of barn owls occurring in the reserve.

Some information is available about the densities of *M. murinus* living in dry areas of Madagascar. Near Berenty, *M. murinus* densities have been estimated to be 360 individuals per km$^2$ (Charles-Dominique and Hladik, 1971) and in the Marosalaza Forest about 400 individuals per km$^2$ (Hladik et al., 1980). However, densities can be exceptionally high; L. Durrell (in litt.) found about 60 active *M. murinus* nests in a 5.6ha area of degraded forest near Manantantely. If the estimated density of *M. murinus* at Beza Mahafaly is similar to that found at Berenty and Marosalaza, 360-400 individuals per km$^2$ (100ha), then Beza Mahafaly would hold a base population of about 2100-2300 *M. murinus*. Using these estimates, each year at Beza Mahafaly between 522-580 *M. murinus* would be taken from a population of approximately 2100-2300 individuals, an estimated predation rate of approximately 25%. If these estimates are accurate, then this rate is the highest known for any primate species (Cheney and Wrangham, 1987).

Although a relatively large amount of data has been gathered on life-history traits of captive *M. murinus*, little is known about the natural history of wild animals and most of the information was gathered in littoral forests near Fort Dauphin (Martin, 1972a, 1973). This lemur reaches sexual maturity between the age of 8 to 12 months, breeding is between September and March, the average gestation period is 63 days, the average litter size is 2-3, the weaning stage is reached 40 days after birth, and some females may produce two litters per year (Martin, 1972a; Petter et al., 1977; Tattersall, 1982). In captivity, females have been recorded to start their first estrous cycle at 135 days old (Perret, 1990). Lemurs show little geographic variation in the seasonal pattern of breeding, regardless of different climatic regimes (Martin, 1972b), and the reproductive cycle of *M. murinus* is regulated by photoperiod (Petter-Rousseaux, 1975). Since Fort Dauphin and Beza Mahafaly are at approximately the same latitude, it is reasonable to presume that the reproductive cycles of *M. murinus* at these localities are similar. Thus, using the Fort Dauphin data, if breeding is generally between September and March, then weaning would take place between

December and June, and it is in this latter period that yearly population cycles at Beza Mahafaly should reach their maximum level. *M. murinus* consumption by the barn owl is greatest in the April to July samples (Fig. 1). The age of *M. murinus* taken during these four months is not heavily biased towards sub-adults (Table 2). The only other month that sub-adults were recovered from the samples was December 1990, when two individuals were identified. Of the minimum total of 58 *M. murinus* identified, 13 are sub-adults and 45 are adults. Presumably these numbers reflect preferential selection of adults over sub-adults, rather than low reproductive success during the late 1990 and 1991 breeding season.

On the basis of the above life history traits, the potential for rapid population growth in *M. murinus* is considerable. For example, if we assume an average base population of 2100-2300 *M. murinus* in the reserve and a rate of two young successfully weaned per year, then theoretically the high point of the yearly population cycle would increase to 4200-4600 individuals and with three young weaned per year to 5350-5750 individuals. Thus, what at first appears to be an intolerably high level of predation on a primate for it to maintain relatively stable population levels, is not as drastic in light of the high reproductive potential of *M. murinus*.

**Table 2.** Age structure of *Microcebus murinus* taken by *Tyto alba* at Beza Mahafaly.

| Month | Age-Class Sub-Adult | Adult |
|---|---|---|
| Nov 1990 | 0 | 2 |
| Dec 1990 | 2 | 6 |
| Jan 1991 | 0 | 0 |
| Feb 1991 | 0 | 2 |
| Mar 1991 | 0 | 3 |
| Apr 1991 | 1 | 6 |
| May 1991 | 5 | 12 |
| Jun 1991 | 3 | 6 |
| Jul 1991 | 2 | 4 |
| Aug 1991 | 0 | 3 |
| Sep 1991 | 0 | 1 |
| Oct 1991 | 0 | 0 |
| Nov 1991 | 0 | 0 |
| Total | 13 | 45 |

Quantified information on the food habits of the Madagascar long-eared owl is also available from the Beza Mahafaly area. On the basis of three pellet samples collected at different times of the year, *M. murinus* makes up 20.5% (range 8.3-26.3%) of the total individuals and 21.2% (range 7.8-40.3%) of the total biomass consumed by this owl (Goodman et al., in press). A minimum of 16 *M. murinus* are represented in these pellets. Our seasonal samples of this owl's food remains are insufficient to calculate its annual estimated predation rate on *M. murinus*. However, given the number of *M. murinus* taken by *Tyto* and *Asio* at Beza Mahafaly, the estimated annual predation rate of about 25% is conservative. Further, this estimate does not include predation by any other raptors, carnivores or snakes.

Given the intensity of predation by these two species of nocturnal owls, this lemur might well have behavioral adaptations to reduce its vulnerability to predation. Two basic patterns of predator avoidance are generally distinguished in primates: vigilance and concealment (van Schaik and van Hooff, 1983; Cheney and Wrangham, 1987).

The barn owl can detect, locate and catch prey in total darkness (Payne, 1971; Knudsen, 1981). These acoustic abilities are not known to be paralleled by any other genus of owl (Schwartzkopff, 1963). The other predatory birds found in the reserve are more visually orientated hunters. The total bird of prey community known from Beza Mahafaly includes eight diurnal resident hawks, five nocturnal resident raptors, and one diurnal boreal winter migrant hawk (Langrand, 1990; S. Goodman, unpubl. data).

The natural forest around Beza Mahafaly, including the gallery forest, is relatively open. *M. murinus* is a nocturnal and arboreal quadruped that travels and forages in all strata of the forest, although more commonly in trees and shrubs between one and five meters. Thus, when moving through the forest they would be accessible to predators. This primate is also known to descend to the ground to feed (Martin, 1973). To be effective, vigilance in a non-gregarious foraging species such as *M. murinus* (Pagès-Feuillade, 1988) might consume a considerable amount of time and thus significantly reduce that available for other activities such as foraging. Further, since *Tyto* often uses a sit and wait strategy to locate prey and when hunting on the wing special modifications to the flight feathers makes it extremely difficult to hear (Bunn et al., 1982), vigilance might not be an effective defense against this owl.

Social and visually oriented diurnal lemurs give vocalizations when a member of a troop observes an approaching raptor, thus these act as anti-predator signals (Sauther, 1989; Macedonia, 1990). Petter et al. (1977) described an alarm call for *M. murinus*, but the context in which it is used is unclear. More importantly, since *M. murinus* forages solitarily, the utterance of such a vocalization would only serve to attract more attention to the primate and increase the risk of predation. There is evidence that the general nocturnal activity of *M. rufus* is reduced on moonlit nights (Wright, 1989). This strategy might decrease vulnerability to visually oriented nocturnal predators, but would presumably be of little use against the barn owl, which is both visually and acoustically oriented. Thus, because of their small size and non-gregarious foraging habits, *M. murinus* appears to be highly vulnerable to predation by *Tyto alba* and perhaps to a lesser extent by other nocturnal foraging raptors.

The strategy of concealment might be a more effective way for *M. murinus* to avoid nocturnal predators. Presumably this lemur is most vulnerable to nocturnal predators when moving across or feeding in crowns of trees and shrubs or on the ground. When away from their leaf nests or tree hollows, this species should minimize time that it is in open areas and should maximize periods in dense vegetation such as lianas and vine tangles. Near Andasibe, Ganzhorn et al. (1985) observed an unidentified owl swoop at a *M. rufus*; apparently the lemur "was not aware of the attack (if it was one) but was protected by vegetation which was too dense for the bird to fly through." This observation illustrates the possible importance of concealment for *Microcebus* as a strategy to avoid predation.

Mouse lemurs are also faced with predation pressure from diurnal animals when resting in leaf-nests or in tree cavities. The genus *Polyboroides*, to which the Madagascar harrier hawk belongs, has unique foraging abilities. These hawks have exceptionally long legs which can bend behind the vertical by 70° and from side to side by 30° at the tibiotarsus-tarsometatarsus joint (Burton, 1978). This movement allows them to probe into narrow tree cavities, behind tree bark, and places most other raptors would find inaccessible (Thurow and Black, 1981). Further, throughout the range of *Microcebus*, there are a variety of diurnal and nocturnal carnivores and snakes (e.g., *Acrantophis*) that would pose a constant threat to both active and sleeping individuals. Given the diversity of mouse lemur predators and their different methods of hunting, a variety of strategies would presumably be needed to substantially reduce risk or rates of predation. Too little is known about this species to generalize about behavioral adaptations. However, their reproductive potential may well compensate for the high predation rate.

How prevalent is predation on small nocturnal primates in other parts of the world? One of the better comparisons that can be made is between mouse lemurs and bushbabies (*Galago* spp.). Both of these genera are prosimians and may be closely related (Bearder, 1987). Bushbabies, in general, share many behavioral and anatomical traits with mouse lemurs. These include being nocturnal, foraging solitarily, building spherical leaf nests or resting in tree cavities, using mainly quadruped locomotion (although they are able to make horizontal leaps), and having relatively high reproductive rates (Bearder, 1987). There appears to be no published quantitative food habit study of a raptor species for which a significant portion of its diet is composed of *Galago*. Bearder and Martin (in Cheney and Wrangham, 1987) reported estimated predation rates for *G. senegalensis* in the Transvaal as less then 15% of the total population per year. On the basis of incidental information, it is clear that an array of African birds of prey eat bushbabies. These range from diurnal eagles to several species of owls (Kingdon, 1974; Steyn, 1980, 1982; Fry et al., 1988; Msuya, in press); snakes (Jones, 1969; Struhsaker, 1970), and carnivores (Charles-Dominique, 1971; Kingdon, 1974). Thus, it appears that the types and diversity of predators taking *Galago* and *Microcebus* are similar. Both of these prosimians appear to use parallel strategies to counteract predation, these include spending significant portions of active periods within the protection of dense vegetation and having relatively high reproductive rates.

In conclusion, when attempting to interpret the evolutionary implications imposed on the behavior and ecology of primates by predators, it is important to separate diurnal and nocturnal primates. Long-term studies are needed on the natural history and population dynamics of *Microcebus* and other nocturnal primates, and such work should coincide with detailed analyses of the density and food habits of locally sympatric predators (Isbell, 1990) as well as investigations of *Microcebus* anti-predator behavior.

## ACKNOWLEDGEMENTS

For providing unpublished data, information from theses, and other details about lemur predation we are grateful to R. Albignac, J. Andrews, V. Barre, C. Cassel, I. Colquhoun, M. Dagosto, P. Daniels, L. Durrell, M. Ebono, M. Enafa, M. Fidi, J. Ganzhorn, L. Gould, M. Gunther, M. Hadani, D. Halleux, C. Harcourt, C. Hemingway, K. Howell, A. Jolly, A. Merenlander, D. Meyers, R. Mittermeier, H. Simons Morland, D. Overdorff, G. Naylor, J.-J. Petter, A. Peyrieras, M. Pidgeon, J. Powzyk, J.-M. Radiva, E. Rajeriarison, R. Rakoto, N. Rakotoarison, P. Rakotomanga, A. P. Raselimanana, J. M. Rasoanindrainy, J. Ratsimbazafy, J. Ratsirarson, C. Raxworthy, D. Reid, A. Richard, Y. Rumpler, M. Sauther, E. Sterling, R. Sussman, I. Tattersall, N. Vasey, L. Wilmé, J. Wilson, and P. Wright. For comments on an earlier version of this paper we thank S. Boinski, J. Ganzhorn and P. Kappeler. We are indebted to J. Ganzhorn for presenting this paper at the IPS on our behalf.

## REFERENCES

Albignac, R., 1973, "Faune de Madagascar. No. 36. Mammifères Carnivores," ORSTOM/CNRS, Paris.

Alexander, R. D., 1974, The evolution of social behavior, *Ann. Rev. Ecol. System.* 5:325.

Andriatsarafara, R., 1988, Etude écoéthologique de deux lémuriens sympatriques de la forêt sèche caducifoliee d'Ampijoroa: *Lemur fulvus fulvus* E. Geoffroy, 1766 et *Lemur mongoz* Linné. 1766. Thèse Doct. Trois. Cycle, Université de Madagascar.

Battistini, R. and Vérin, P., 1972, Man and the environment in Madagascar: past problems and problems of today, *in*: "Biogeography and Ecology of Madagascar," R. Battistini and G. Richard-Vindard, eds., Junk, The Hague.

Barden, T. L., Evans, M. I., Raxworthy, C. J., Razafimahaimodison, J.-C., and Wilson, A., 1991, The mammals of Ambatovaky Special Reserve, *in*: "A Survey of Ambatovaky Special Reserve, Madagascar," P. M. Thompson, and M. I. Evans, eds., Madagascar Environmental Research Group, London.

Barre, V., Lebec, A., Petter, J.-J., and Albignac, R., 1988, Etude du Microcèbe par radiotracking dans la forêt de l'Ankarafantsika, *in*: "L'Equilibre des Ecosystèmes forestiers à Madagascar: Actes d'un Séminaire International," L. Rakotovao, V. Barre, and J. Sayer, eds., IUCN, Gland.

Bayart, F. and Anthouard, M., 1992, Responses to a live snake by *Lemur macaco macaco* and *Lemur fulvus mayottensis* in captivity, *Folia Primatol.* 58:41.

Bearder, S. K., 1987, Lories, bushbabies, and tarsiers: diverse societies in solitary foragers, *in*: "Primate Societies," B. B. Smuts, D. L. Cheney, R. M. Seyfarth, R. W. Wrangham, and T.T. Struhsaker, eds., University of Chicago Press, Chicago.

Boesch, C., 1991, The effects of leopard predation on grouping patterns in forest chimpanzees, *Behaviour* 117:220.

Boshoff, A. F., Palmer, N. G., Avery, G., Davies, R. A. G., and Jarvis, M. J. F., 1991, Biogeographical and topographical variation in the prey of the black eagle in the Cape Province, South Africa, *Ostrich* 62:59.

Budnitz, N. and Dainis, K., 1975, *Lemur catta*: ecology and behavior, *in*: "Lemur Biology," I. Tattersall, and R. W. Sussman, eds., Plenum Press, New York.

Bunn, D. S., Warburton, A. B., and Wilson, R. D. S., 1982, "The Barn Owl," T. and A. D. Poyser, Calton.

Burton, P. J. K., 1978, The intertarsal joint of the harrier-hawks *Polyboroides* spp. and the crane hawk *Geranospiza caerulescens*, *Ibis* 120:171.

Charles-Dominique, P., 1971, Eco-éthologie des prosimiens du Gabon, *Biol. Gabonica* 7:121.

Charles-Dominique, P., 1977, "Ecology and Behavior of Nocturnal Primates: Prosimians of Equatorial West Africa," Academic Press, New York.

Charles-Dominique, P., Cooper, H. M., Hladik, A., Hladik, C. M., Pages, E., Pariente, G. F., Petter-Rousseaux, A., Petter, J.-J., and Schilling, A., 1980, "Nocturnal Malagasy Primates: Ecology, Physiology, and Behavior," Academic Press, New York.

Charles-Dominique, P. and Hladik, C. M., 1971, Le lepilemur du sud de Madagascar: écologie, alimentation et vie sociale, *Terre et Vie* 25:3.

Charles-Dominique, P. and Petter, J.-J., 1980, Ecology and social life of *Phaner furcifer*, *in*: "Nocturnal Malagasy Primates: Ecology, Physiology, and Behavior," P. Charles-Dominque, H. M. Cooper, A. Hladik, C. M. Hladik, E. Pages, G. F. Pariente, A. Petter-Rousseaux, J.-J. Petter, and A. Schilling, eds., Academic Press, New York.

Cheney, D. L., and Wrangham, R. W., 1987, Predation, *in*: "Primate Societies," B. B. Smuts, D. L. Cheney, R. M. Seyfarth, R. W. Wrangham, and T.T. Struhsaker, eds., University of Chicago Press, Chicago.

Cramp, S., ed., 1985, "The birds of the western Palearctic," Vol. IV, Oxford University Press, Oxford.

Dewar, R. E., 1984, Extinctions in Madagascar: the loss of the subfossil fauna, *in*: "Quaternary Extinctions: a Prehistoric Revolution," P. S. Martin, and R. G. Klein, eds., University of Arizona Press, Tucson.

Dunbar, R., 1988, "Primate Social Systems," Cornell University Press, New York.

Emmons, L. H., 1987, Comparative feeding ecology of felids in a neotropical rainforest, *Behav. Ecol. Sociobiol.* 20:271.

Fry, C. H., Keith, S., and Urban, E. K., 1988, "The Birds of Africa," Vol. III, Academic Press, London.

Ganzhorn, J. U., Abraham, J. P., and Razanahoera-Rakotomalala, M., 1985, Some aspects of the natural history and food selection of *Avahi laniger*, *Primates* 26:452.

Goodman, S. M., in press, The enigma of anti-predator behavior in lemurs: evidence of a large extinct eagle on Madagascar. *Int. J. Primatol.*

Goodman, S. M., Creighton, G. K., and Raxworthy, C., 1991, The food habits of the Madagascar long-eared owl *Asio madagascariensis* in southeastern Madagascar, *Bonn. zool. Beitr.* 42:21.

Goodman, S. M., Langrand, O., and Raxworthy, C. J., in press, Food habits of the Madagascar long-eared owl (*Asio madagascariensis*) in two habitats in southern Madagascar, *Ostrich*.

Goodman, S. M., Langrand, O., and Raxworthy, C. J., submitted, A comparative study of the food habits of the barn owl (*Tyto alba*) at three sites on Madagascar, *Ostrich*.

Harrington, J. E., 1975, Field observations of social behaviour of *Lemur fulvus fulvus* E. Geoffroy 1812, *in*: "Lemur Biology," I. Tattersall, and R. W. Sussman, eds., Plenum Press, New York.

Hladik, C. M., Charles-Dominique, P., and Petter, J.-J., 1980, Feeding strategies of five nocturnal prosimians in the dry forest of the west coast of Madagascar, *in*: "Nocturnal Malagasy Primates: Ecology, Physiology, and Behavior," P. Charles-Dominque, H. M. Cooper, A. Hladik, C. M. Hladik, E. Pages, G. F. Pariente, A. Petter-Rousseaux, J.-J. Petter, and A. Schilling, eds., Academic Press, New York.

Hoppe-Dominik, B., 1984, Etude du spectre des proies de la panthère, *Panthera pardus*, dans le Parc National de Taï en Côte d'Ivoire, *Mammalia* 48:477.

Isbell, L. A., 1990, Sudden short-term increase in mortality of vervet monkeys (*Cercopithecus aethiops*) due to leopard predation in Amboseli National Park, Kenya, *Am. J. Primatol.* 21:41.

Izor, R. J., 1985, Sloths and other mammalian prey of the harpy eagle, *in*: "The Evolution and Ecology of Armadillos, Sloths, and Vermilinguas," G. G. Montgomery, ed., Smithsonian Institution Press, Washington, D.C.

Jolly, A., 1966, "Lemur Behavior: a Madagascar Field Study," University of Chicago Press, Chicago.

Jones, C., 1969, Notes on ecological relationship of four Lorisids in Rio Muni, West Africa, *Folia Primatol.* 11:255.

Kingdon, J., 1974, "East African Mammals: An Atlas of Evolution in Africa," Vol. 1, University of Chicago Press, Chicago.

Knudsen, E. I., 1981, The hearing of the barn owl, *Sci. Amer.* 245:113.
Koyama, N., 1992, Some demographic data of ring-tailed lemurs (*Lemur catta*) at Berenty, Madagascar, *in*: "Social Structure of Madagascar Higher Vertebrates in Relation to Their Adaptive Radiation," S. Yamagishi, ed., Osaka City University, Osaka.
Langrand, O., 1990, "Guide to the Birds of Madagascar," Yale University Press, New Haven.
Macedonia, J. M., 1990, What is communicated in the antipredator calls of lemurs: evidence from playback experiments with ringtailed and ruffed lemurs, *Ethology* 86:177.
Macedonia, J. M. and Polak, J. F., 1989, Visual assessment of avian threat in semi-captive ringtailed lemurs (*Lemur catta*), *Behaviour* 111:291.
Macedonia, J. M. and Yount, P. L., 1991, Auditory assessment of avian predator threat in semi-captive ringtailed lemurs (*Lemur catta*), *Primates* 32:169.
MacPhee, R. D. E. and Burney, D. A., 1991, Dating of modified femora of extinct dwarf hippopotamus from southern Madagascar: implications for constraining human colonization and vertebrate extinction events, *J. Archaeol. Sci.* 18:695.
MacPhee, R. D. E., Burney, D. A., and Wells, N. A., 1985, Early holocene chronology and environment of Ampasabazimba, a Malagasy subfossil lemur site, *Int. J. Primatol.* 5:463.
Martin, R. D., 1972a, A preliminary field-study of the lesser mouse lemur (*Microcebus murinus*, J. F. Miller, 1777), *Zeitschr. Tierpsy.* 9:43.
Martin, R. D., 1972b, Adaptive radiation and behavior of the Malagasy lemurs, *Phil. Trans. Roy. Soc. Lond.* (Series B) 264:295.
Martin, R. D., 1973, A review of the behaviour and ecology of the lesser mouse lemur (*Microcebus murinus* J. F. Miller, 1777), *in*: "The Comparative Ecology and Behaviour of Primates," R. P. Michael, and J. H. Crook, eds., Academic Press, London.
Mitchell, C. L., Boinski, S., and van Schaik, C. P., 1991, Competitive regimes and female bonding in two species of squirrel monkeys (*Saimiri oerstedi* and *S. sciureus*), *Behav. Ecol. Sociobiol.* 28:55.
Msuya, C. A., in press, Feeding habits of crowned eagles *Stephanoaetus coronatus* in Kiwengoma Forest Reserve, Matumbi Hills, Tanzania, Proc. 8th Pan-African Ornith. Cong.
Nicoll, M. E. and Langrand, O., 1989, "Madagascar: Revue de la Conservation et des Aires Protégées," World Wide Fund for Nature, Gland.
Pagès-Feuillade, E., 1988, Modalités de l'occupation de l'espace et relations interindividuelles chez un prosimien nocturne malgache (*Microcebus murinus*), *Folia Primatol.* 50:204.
Payne, R., 1971, Acoustic location of prey by barn owls (*Tyto alba*), *J. Exper. Biol.* 54:535.
Perret, M., 1990, Influence of social factors on sex ratio at birth, maternal investment and young survival in a prosimian primate, *Behav. Ecol. Sociobiol.* 27:447.
Petter, J.-J., 1988, Contribution à l'étude du *Cheirogaleus medius* dans le forêt de Morondava, *in*: "L'Equilibre des Ecosystèmes Forestiers à Madagascar: Actes d'un Séminaire International," L. Rakotovao, V. Barre, and J. Sayer, eds., IUCN, Gland.
Petter, J.-J., Albignac, R., and Rumpler, Y., 1977, "Faune de Madagascar. No. 44. Mammifères Lémuriens (Primates Prosimiens)," ORSTOM/CNRS, Paris.
Petter-Rousseaux, A., 1975, Activité sexuelle de *Microcebus murinus* (Miller 1777) soumis à des régimes photopériodiques expérimentaux, *Ann. Biol. anim. Bioch. Biophys.* 15:503.
Rabemazava, M., 1990, Statut et distribution de Propithèque couronné (*Propithecus verreauxi coronatus*) et proposition de mesures de protection, Rapport de Stage de l'Ecole pour la Formation de Spécialistes de la Faune, Garoua, Cameroun.
Rand, A. L., 1935, On the habits of some Madagascar mammals, *J. Mammal.* 16:89.
Randrianarivo, R., 1979, Essai d'inventaire des lémuriens de la future réserve de Beza Mahafaly, Mémoire de fin d'études, Etablissement d'Enseignement Supérieur des Sciences Agronomiques, Antananarivo.
Rasoanindrainy, J. M., 1985a, [Population dynamics of *Propithecus verreauxii*,] Nauchnye Dokl. Vyssh. *Shk. Biol. Nauki* 1985 4:43 (in Russian).
Rasoanindrainy, J. M., 1985b, [Comparative biology of *Propithecus verreauxi verreauxi* and *Lemur catta*, and perspectives for their protection,] Ph.D. thesis (Biology), University of Moscow, Lomonossov (in Russian).
Ratsirarson, J., 1985, Contribution a l'étude comparative de l'éco- éthologie de *Lemur catta* dans deux habitats differents de la Reserve Speciale de Beza-Mahafaly, Memoire de Fin d'Etudes, Université de Madagascar.
Ratsirarson, J., 1986, Contribution à l'étude comparée de l'éco- éthologie de deux espèces de lémuriens: *Lepilemur mustelinus* (I. Geoffroy 1850) et *Lepilemur septentrionalis* (Rumpler et Albignac 1975), Thèse Doct. Etat, Université Louis Pasteur de Strasbourg.

Richard, A. F., 1978, "Behavioral Variation: Case Study of a Malagasy Lemur," Bucknell University Press, Lewisburg.
Richard, A. F. and Dewar, R. E., 1991, Lemur ecology, *Ann. Rev. Ecol. System.* 22:145.
Richard, A. F., Rakotomanga, P., and Schwartz, M., 1991, Demography of *Propithecus verreauxi* at Beza Mahafaly, Madagascar: sex ratio, survival, and fertility 1984-1988, *Am. J. Phys. Anthrop.* 84:307.
Russell, R. J., 1977, The behavior, ecology and environmental physiology of a nocturnal primate, *Lepilemur mustelinus* (Strepsirhini, Lemuriformes, Lepilemuridae), Ph.D Thesis, Duke University.
Sauther, M. L., 1989, Antipredator behavior in troops of free- ranging *Lemur catta* at Beza Mahafaly Special Reserve, Madagascar, *Inter. J. Primatol.* 10:595.
van Schaik, C. P., 1983, Why are diurnal primates living in groups? *Behaviour* 87:120.
van Schaik, C.P. and van Hooff, J.A.R.A.M., 1983, On the ultimate causes of primate social systems, *Behaviour* 85:91.
van Schaik, C. P., van Noordwijk, M.A., Warsono, B., and Sutriono, E., 1983, Party size and early detection of predators in Sumatran forest primates, *Primates* 24:211.
Schwartzkopff, J., 1963, Morphological and physiological properties of the auditory system in birds, Proc. XIII Int. Ornith. Congress.
Simona, A., 1988, Investigation sur l'habitat et régime alimentaire de *Propithecus diadema candidus* (Simpona) dans la réserve naturelle intégrale no. XII. Marojejy Andapa - Madagascar, Rapport de Stage de l'Ecole pour la Formation de Spécialistes de la Faune, Garoua, Cameroun.
Simons Morland, H., 1991, Social organization and ecology of black and white ruffed lemurs (*Varecia variegata variegata*) in lowland rain forest, Nosy Mangabe, Madagascar, Ph.D. thesis, Yale University.
Skorupa, J. P., 1989, Crowned eagles *Stephanoaetus coronatus* in rainforest: observations on breeding chronology and diet at a nest in Uganda, *Ibis* 131:294.
Steyn, P., 1980, Observations on the prey and breeding success of Wahlberg's eagle, *Ostrich* 51:56.
Steyn, P., 1982, "Birds of Prey of Southern Africa: Their Identification and Life Histories," David Philip, Cape Town.
Struhsaker, T. T., 1970, Notes on *Galagoides demidovii* in Cameroon, *Mammalia* 34:207.
Struhsaker, T. T. and Leakey, M., 1990, Prey selectivity by crowned hawk-eagles on monkeys in the Kibale Forest, Uganda, *Behav. Ecol. Sociobiol.* 26:435.
Sussman, R. W., 1975, A preliminary study of the behavior and ecology of *Lemur fulvus rufus* Audebert 1800, *in*: "Lemur Biology," I. Tattersall, and R. W. Sussman, eds., Plenum Press, New York.
Sussman, R. W., 1977, Feeding behaviour of *Lemur catta* and *Lemur fulvus*. *in*: "Primate Ecology: Studies of Feeding and Ranging Behaviour in Lemurs, Monkeys, and Apes," T. H. Clutton-Brock, ed., Academic Press, London.
Tattersall, I., 1982, "The Primates of Madagascar," Columbia University Press, New York.
Terborgh, J., 1983, "Five New World Primates: A Study in Comparative Ecology," Princeton University Press, Princeton.
Thompson, P. M. and Evans, M. I., 1991, The birds of Ambatovaky, *in*: "A Survey of Ambatovaky Special Reserve, Madagascar", P. M. Thompson, and M. I. Evans, eds., Madagascar Environmental Research Group, London.
Thurow, T. L. and Black, H. L., 1981, Ecology and behaviour of the gymnogene, *Ostrich* 52:25.
Wilson, J. M., Stewart, P. D., Ramangason, G.-S., Denning, A. M., and Hutchings, M.S., 1989, Ecology and conservation of the crowned lemur, *Lemur coronatus*, at Ankarana, N. Madagascar, *Folia Primatol.* 52:1.
Wilson, R. T., Wilson, M. P., and Durkin, J. W., 1986, Breeding biology of the barn owl *Tyto alba* in central Mali, *Ibis* 128:81.
Wright, P. C., 1989, The nocturnal primate niche in the new world, *J. Human Evol.* 18:635.

# ADAPTATION AND PHYLOGENETIC CONSTRAINTS IN THE ANTIPREDATOR BEHAVIOR OF RINGTAILED AND RUFFED LEMURS

Joseph M. Macedonia

Department of Biological Anthropology and Anatomy
Duke University
Durham, NC

Present address: Animal Communication Lab,
University of California, Davis, CA 95616-8761
U.S.A.

## ABSTRACT

The antipredator responses of forest-living ringtailed (*Lemur catta*) and ruffed (*Varecia variegata*) lemurs were documented over a three-year period at the Duke University Primate Center (DUPC, Durham, NC, USA). Vocal and nonvocal responses to naturally-occurring and simulated predators are described, and their functions are considered with respect to species-specific differences in body size, ecology, and reproductive biology. Nonvocal responses of the two lemur species differed most conspicuously in propensity of predator-directed aggression: whereas ringtailed lemurs generally evaded predators, ruffed lemurs were likely to confront or attack them. Interspecific variation in vocal responses to predators included differences in call diversity, stimulus specificity, and function. Ringtailed lemur antipredator behavior (including large group size) is viewed as an adaptation to nontrivial levels of predator pressure that stem from being a relatively small-bodied, semi-terrestrial primate living in an open habitat. In contrast, the highly aggressive antipredator behavior of the ruffed lemur is seen in part as an effect of a somewhat larger body size, but also as a constraint of producing sessile offspring that do not cling to the mother. Thus, in contrast to ringtailed lemurs, ruffed lemurs with infants cannot flee predators without risking their reproductive success.

## INTRODUCTION

### Effects of Predation on Primate Behavior

Avoiding predation has been argued to have had a profound impact on the evolution of social organization in primates (Alexander, 1974; van Schaik, 1983; van Schaik and van Hooff, 1983; van Schaik et al., 1983; Terborgh, 1983; Struhsaker and Leakey, 1990). Yet, the evidence available for predation on primates remains limited (see Cheney and Wrangham, 1987, for a review). Due to the swiftness with which predation typically occurs, the probability of observing a kill by chance is extremely low (Terborgh and Janson, 1986). Direct evidence of

*Lemur Social Systems and Their Ecological Basis*, Edited by
P.M. Kappeler and J.U. Ganzhorn, Plenum Press, New York, 1993

predation is available in other forms, however, such as bones, teeth, and hair that can be matched to a prey species (Struhsaker and Leakey, 1990; Goodman et al., 1993). The veracity of predation on prey populations can be assessed indirectly as well. Just as astronomers and physicists identify the presence of forces that cannot be observed directly, but which are inferred from the behavior of the bodies they influence, stereotyped antipredator behavior serves as indirect yet compelling evidence that predation pressure has had a significant influence during a prey species' recent evolutionary past (Cheney and Wrangham, 1987).

Large group size in diurnal primates is one factor that has been argued to afford increased protection from predation due to there being "more eyes and ears" available for detecting the presence of predators (see Terborgh and Janson, 1986; Cheney and Wrangham, 1987, for reviews). For example, van Schaik et al. (1983) demonstrated in long-tailed macaques (*Macaca fascicularis*) that larger groups detected approaching observers at significantly greater distances than did smaller groups (though see Isbell, 1990). In another study, van Schaik and van Noordwijk (1988) found that long-tailed macaques lived in larger groups on Sumatra, where large felid predators include the tiger (*Panthera tigris*) and the arboreal clouded leopard (*Neofelis nebulosa*), than on the offshore island of Simeulue, where no felids occur. Thus, modulating group size appears to be one way that primates cope with the threat of predation.

Many species of birds and mammals, including primates, emit vocalizations in response to predators. In general, antipredator vocalizations (or, 'alarm calls') fall into one of three functional categories (see Klump and Shalter, 1984; Hasson, 1991 for reviews). *Alerting/Warning Calls* function to inform conspecifics of a predator's presence while minimizing their detection by the predator. *Mobbing Calls* are emitted during predator harassment. These vocalizations are predator-directed, repetitive, and solicit aggregation of the prey near the predator. *Perception Advertisement Calls* also are predator-directed, but unlike mobbing calls they do not function in predator harassment. These calls neither are repetitive in the absence of stimulation from the predator, nor do prey congregate near the predator during call emission. Rather, they serve to inform predators that they have been detected by the intended prey, thus causing some predators to abort the hunt.

Most primate species will mob mammalian predators, and some larger primates escalate to the point of direct combat with these carnivores. Baboons, for example, are known to kill domestic dogs, mob and chase cheetahs and jackals, and even fight to the death with leopards and lions (see Cheney and Wrangham, 1987). Primates characteristically mob perched raptors, though harassing truly large raptors usually is restricted to adult males in most monkeys. Eason (1989), for example, observed a cautious, yet determined, adult male red howler (*Alouatta seniculus*) physically harassing a Harpy eagle (*Harpia harpyja*), the largest living raptor. Even 6-8kg adult male howlers are not immune to harpy eagle predation, however (Sherman, 1991).

Among the five primate species described in an interaction with a crowned hawk-eagle (*Stephanoetus coronatus*), only adult males of the two largest species (white-nosed guenon: *Cercopithecus nictitans*; gray-cheeked mangabey: *Cercocebus albigena*) participated in chasing the eagle (Gautier-Hion and Tutin, 1989). As in the previous example, there is evidence that such behavior increases the vulnerability of these individuals to eagle predation (Struhsaker and Leakey, 1990).

Snake mobbing has been documented for several primate species, where individuals congregate around a snake and emit aggressive vocalizations (Heymann, 1987; Bartecki and Heymann, 1987; van Schaik and Mitrasetia, 1990). In addition, individuals sometimes will drop or throw branches at snakes (Chapman, 1986; van Schaik and van Noordwijk, 1989), and in one case an adult male used a branch to club a snake repeatedly (Boinski, 1988).

A few species employ antipredator tactics that are unusual for primates. When confronted by a snake in trees, pottos (*Perodicticus potto*) have been observed to drop to the ground, run a short distance and then freeze as if feigning death (Charles-Dominique, 1977). Although well-known for birds, 'predator distraction displays' (Armstrong, 1949) also seem to be very rare in primates. Hall (1965) proposed, however, that a predator distraction display best characterized the behavior of adult male patas monkeys (*Erythrocebus patas*) when dealing with certain predators. In sum, primates are not defenseless against predators, and they employ a variety of techniques to avoid predation.

**Evidence for Predation on Malagasy Lemurs**

With land-hunting eagles (cf. fish eagles) being absent from Madagascar, and there being no truly large carnivores (such as leopards) on the island, predation on lemurs often is considered to have been negligible during the evolution of these primates (van Schaik and van Hooff, 1983). While it is true that there are only a few diurnal raptor species and a single type of carnivore (*Cryptoprocta ferox*) that are large enough to take adult lemurids as prey, predation on infants alone could have significant demographic effects on lemur populations (Sauther, 1989). Given that some lemur species produce only a single infant per year, and that sickness and lethal injuries take their toll of young, the loss of a progeny merely once or twice in a female's lifetime could dramatically reduce her reproductive success. Likewise, predator diversity is not necessarily a good indicator of predator pressure. Despite the fact that the leopard is the only East African carnivore whose preferred prey seems to include vervets, leopards have been responsible for more deaths of healthy vervets in Amboseli (during the years for which predation has been documented there by primatologists) than any other single cause (Cheney and Wrangham, 1987; Isbell, 1990). The compilation of information presented by Goodman et al. (1993), including the recent discovery of a large extinct Malagasy eagle (Goodman, in press), throws new light on the issue of predation on lemurs. It likewise helps to resolve the seemingly curious existence of formalized antipredator behavior in these primates - a phenomenon whose evolution would defy explanation had predation pressure actually been as trivial as has generally been claimed.

Below I describe the antipredator behavior of semi-captive, forest-living ringtailed (*Lemur catta*) and ruffed (*Varecia variegata*) lemurs. The responses of these lemurs to naturally-occurring and simulated predators are compared to those of free-ranging conspecifics, and species-specific antipredator tactics are viewed in terms of body size, ecology, and reproductive biology. By comparing these prosimians with anthropoids of similar physical and ecological characteristics, some insight into the evolutionary forces underlying primate antipredator behavior may be gained.

## MATERIALS AND METHODS

Antipredator behavior was documented in a group of ringtailed lemurs (*Lemur catta*, Lc1 Group) and a group of black-and-white ruffed lemurs (*Varecia variegata variegata*, Vv1 Group) between May 1986 and July 1989 at the Duke University Primate Center (DUPC, Durham, NC). These lemur groups lived year-round in a 3.5ha natural habitat enclosure (NHE-2) composed of mixed pine/hardwood forest surrounded by a mildly electrified fence. The groups were provisioned daily with monkey chow and twice per week with mixed cut fruit, but the lemurs spent much time foraging on local flora. All lemurs in the NHE's wore individually-identifiable collars and tags, excluding young infants who were identified by patterns of tail shaves. Adult (>3 yrs of age) ringtailed lemurs weighed roughly 2kg; adult ruffed lemurs weighed approximately 3.5kg (Kappeler, 1991). (For details of the study site, lemur group histories, and recording/analysis equipment, see Macedonia, 1990).

Observations of responses to naturally-occurring predators were recorded opportunistically. Raptors (e.g., red-tailed hawk: *Buteo jamaicensis*; red-shouldered hawk: *Buteo lineatus*; great-horned owl: *Bubo virginianus*), carnivores (gray fox: *Urocyon cinereoargenteus*; raccoon: *Procyon lotor*; weasle: *Mustela* sp.), venomous snakes (copperhead: *Agkistrodon contortrix*; cottonmouth: *Agkistrodon piscivorus*), and a nonvenomous, semi-arboreal constrictor that could be dangerous to young infants (black rat snake: *Elaphe obsoleta*) inhabited or frequented the enclosures. By June 1992, a number of lemurs had been killed by these predators: 1 young adult female ringtailed lemur by a gray fox, 5 infant ringtailed lemurs by a great-horned owl, 1 infant and 1 young adult male ringtailed lemur by a copperhead, and 2 ruffed lemur newborns by a weasle. Whereas no lemur at the DUPC had been killed by a diurnal raptor, red-tailed hawks attack small mammals in the natural habitat enclosures (pers. obs.) and are capable of taking prey up to the size and weight of adult *L. catta* (e.g., jackrabbits, *Lepus californicus*; pers. obs.).

Because observations of responses to naturally-occurring predators were relatively rare (see Macedonia and Polak, 1989; Macedonia and Evans, in press), predator models were presented to Lc1 and Vv1 Groups to augment acquisition of opportunistic data. To observe responses to perched raptors, a museum specimen of a perched red-shouldered hawk (Fig. 1a) or great-horned owl (*Bubo viginianus*: Fig. 1b) was placed along one of the paths routinely taken by the lemurs. Four presentations were conducted, one for each lemur species with each specimen. To document responses to attacking raptors, on two occasions a plywood raptor shape was hung on a guide wire and flown at the lemurs in a simulated 'stoop' (Fig. 1c).

Two presentations of a large carnivore (27kg dog; Fig. 1d) were conducted. For the 'low urgency' presentation, the dog was walked slowly toward the lemurs from a distance of roughly 50m. For the 'high urgency' presentation, the (tethered) dog was permitted to ambush the lemurs from behind a blind (see Pereira and Macedonia, 1991, for details).

**Figure 1.** (a) Perched specimen of red-shouldered hawk (*Buteo lineatus*); (b) perched specimen of great-horned owl (*Bubo virginianus*); (c) raptor model on test apparatus; (d) dog used as mammalian predator.

**Table 1.** Ringtailed and Ruffed Lemur Vocalizations Emitted in the Predator Context.

**a. Ringtailed Lemur**

| **Vocalization** | **Context and/or Function** |
|---|---|
| Gulp | general context group alert call (Fig. 2a) |
| Rasp | conspecific-directed raptor alarm call (Fig. 2b-d) |
| Shriek | predator-directed antiraptor call (Fig. 2f-h) |
| Chirp | elicits, and may mediate, rapid group relocation (Fig. 2i) |
| Plosive Bark | high-intensity threat vocalization (Fig. 2j) |
| Click | 'location marker' in response to low/moderate arousal disturbances (Fig. 2k) |
| Closed-Mouth Click Series | mammalian disturbance; also during rapid locomotion, particularly arboreal; location marker in response to moderate/high arousal disturbances (Fig. 2k) |
| Open-Mouth Click Series | location marker in response to high arousal disturbances; may aid in synchronization of Yaps (Fig. 2k) |
| Yap | mammalian predator 'mobbing call' (Fig. 2k) |
| Howl | individual/group 'advertisement call', emitted in series' by males (Fig. 2l) |

**b. Ruffed Lemur**

| **Vocalization** | **Context and/or Function** |
|---|---|
| Abrupt Roar | avian predator mobbing call; also emitted in some other contexts of high-level aggression (Fig. 4a) |
| Roar/Shriek Chorus | group advertisement call (Fig. 4b) |
| Growl-Snort | location marker in high arousal disturbances (Fig. 4c) |
| Growl | location marker in low/moderate arousal disturbances (Fig. 4d) |
| Pulsed Squawk | mammalian predator mobbing call; may signal high-urgency desire for group reaggregation (Fig. 4e) |
| Wail | 'all clear' call in antipredator contexts; may signal low-urgency desire for group reaggregation (Fig. 4g) |

## RESULTS

### Responses of Ringtailed Lemurs to Avian Predators

**Natural Encounters.** The initial response to the sight of large hawks or to their calls was to emit one or more kinds of vocalizations (Table 1). 'Gulps' (Fig. 2a) typically were given first, and were elicited in other contexts by any kind of startling stimulus (e.g., loud or unusual sounds, sudden movements). In the case of airborne birds, if an antiraptor call (see below) was not forthcoming, gulps often continued to be issued until the bird passed from view. Individuals responded to gulps by looking quickly toward the caller while becoming poised for locomotion. This response presumably allows individuals to track the direction of the caller's gaze and to observe the caller's reaction to the eliciting stimulus. The low emission amplitude and very broad range of eliciting stimuli suggest that the gulp functions as a generalized alerting/warning call.

'Rasps' (Fig. 2b-d), were given by one or several individuals when a large aerially- approaching bird reached a proximity to the group where its identity (apparently) could be ascertained as a potential threat. Although gulps usually preceded rasps, gulps sometimes were foregone when raptors appeared suddenly. For example, on one occasion a pair of red-tailed hawks appeared just above the tree tops in NHE-2 grappling with their talons in courtship. The first Lc1 Group member to vocalize emitted a brief rasp (Fig. 2b). The courting hawks, now below the canopy, then turned in the direction of the lemur group. The same caller responded again, this time with three consecutive brief rasps (Fig. 2c), followed approximately two seconds later with a long rasp (Fig. 2d).

When raptors were detected at some distance from the group (including high-altitudes directly overhead), usually only one or two rasps were uttered. Very distant raptors elicited, at most, gulps. In contrast, raptors suddenly appearing close to the group elicited a flurry of rasps and rasp-shriek intermediates (Fig. 2e; see below) from numerous individuals as they ran for cover. The rasp's broadband acoustic structure, its low-to-moderate amplitude, and its restriction as a vocal response to large aerial stimuli (Macedonia and Polack, 1989), suggests that this call functions as an alerting/warning call specific for aerial predators.

A third vocalization elicited by large avian stimuli was the 'shriek' (Andrew, 1963; Fig. 2f). Unlike gulps and rasps, shrieks were emitted at very high amplitudes (Sauther, 1989; Macedonia, 1990). Shrieks sometimes were emitted in synchrony (Fig. 2g) in cases where group members had been aware of a raptor's presence for some time (e.g., > 5s). For example, on one occasion the members of Lc1 Group were feeding on items in the leaf litter, or were grooming or sitting quietly. One individual began to emit gulps as it stared skyward. In the distance was a large soaring bird, about five meters above the treetops, that was moving on a trajectory toward the group. About ten seconds later, when the bird (a red-tailed hawk) was almost directly overhead, single brief rasps were emitted by two unidentified individuals. Before the second rasp had terminated, the group erupted in a synchronous shriek, during which time group members stared skyward toward the soaring hawk. All individuals remained still and appeared calm throughout the entire event, and continued to track the flight of the hawk until it disappeared from view before resuming prior activities. The restriction of this call as a response to large avian stimuli, its high amplitude level, and the demeanor of individuals emitting it suggests that the shriek functions as a perception advertisement call specific to aerial predators.

If Lc1 Group was on the ground when detecting a raptor whose proximity seemed potentially threatening, the group often proceeded quickly to a new location (usually one with increased overhead cover). Hurried group relocation always was accompanied by the emission of 'chirps' (Table 1; Fig. 2h) from most group members. If in trees when a raptor was seen or heard, the ringtailed lemurs moved from the peripheral branches toward the trunk. Often, this was followed by the group climbing down to the ground and moving to a new location.

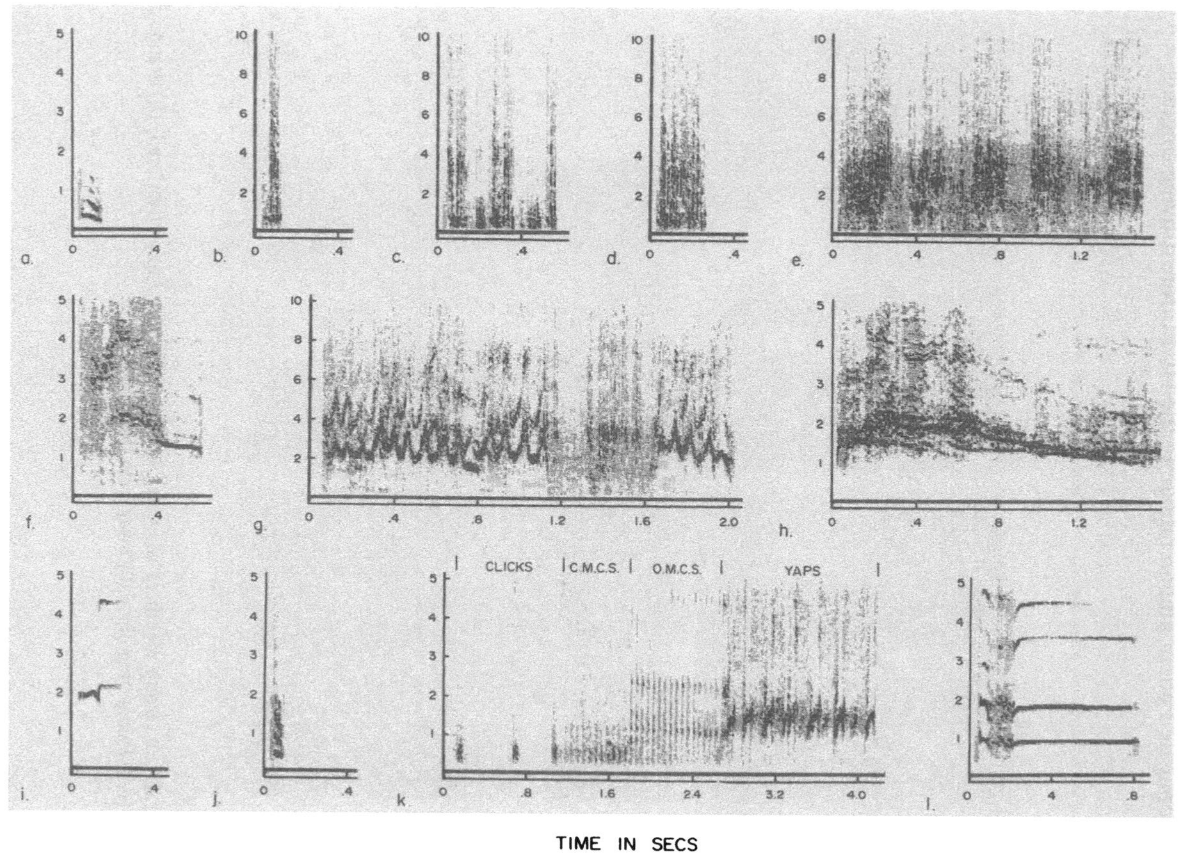


**Figure 2.** Antipredator calls and related vocalizations of the ringtailed lemur (*Lemur catta*): (a) gulp; (b) chort rasp; (c) three consecutive rasps; (d) long rasp; (e) five rasp/shriek intermediates; (f) typical shriek (Type I); (g) multiply-frequency-modulated shriek (Type II); (h) group synchronous shriek; (i) chirp; (j) plosive bark; (k) anticarnivore call series: click, closed-mouth click series, open-mouth click series; yaps; (l) howl.

A peculiar behavior sometimes exhibited by Lc1 Group in the context of avian predators was 'bipedal locomotion' (Fig. 3a-d). When a large hawk was sighted or its call heard, a number of individuals would assume a bipedal stance before walking or trotting bipedally away from the area (Macedonia and Yount, 1991; Pereira and Macedonia, 1991). This behavior has not been reported for *L. catta* in Madagascar (Jolly, 1966; Sauther, 1989) nor did it occur in other lemur species at the DUPC. Bipedal locomotion in response to raptors seems to be a culturally-propagated phenomenon unique to the DUPC forest-living ringtails.

**Figure 3.** Several Lc1 Group members when (a) sighting a soaring raptor overhead, (b) standing bipedally, then (c-d) locomoting bipedally away from the area.

The responses of infant ringtailed lemurs to the antipredator calls of conspecific adults also deserve mention. The emission of gulps by any group member would cause mothers and infants (that were not in close proximity) to begin searching for one another. Young infants gave 'infant contact calls' (Macedonia, in press) at this time, followed with running to and leaping onto their mothers' backs, once they located them. If the gulps had been elicited by a visible carnivore, young infants did not climb into trees along with adults and older immatures, but waited for their mothers to retrieve them. By the age of three months, however, infants first would leap into trees and then begin trying to locate their mothers. On several occasions a stereotyped response to antiraptor calls was observed in young infants (about 4-8 wks of age) that were clambering about in saplings or on sturdy ground vegetation at a height of 1-2m. These infants responded to rasps or shrieks by immediately letting go of the branches and dropping to the ground (pers. obs.; M. Pereira, pers. comm.). Once on the ground, the infants remained motionless while calling for retrieval.

**Experimental Manipulations.** Responses to the two presentations of the 'stooping hawk' model were similar in all important respects. In the first presentation, numerous rasps and Type II shrieks (Fig. 2g) were emitted upon its release. One adult female stood bipedally and shrieked directly toward the descending model, and an unidentified immature was observed pushing itself underneath an adult. Other individuals crouched close to the ground and rasped loudly before running for cover. As Lc1 Group scattered, chirping began. Several individuals trotted and hopped bipedally away from the test site. Though many group members ran rapidly from the immediate area, some individuals looked back toward the apparatus, after having taken positions behind tree trunks, and emitted plosive barks (Fig. 2i). A few subjects leapt a short distance up tree trunks while clinging to the trees' far sides, but did not move up to the level of the lowest branches. Chirping reached choral proportions as the lemurs regrouped some distance away, continuing to move rapidly on the ground away from the test site.

During presentations of the perched raptor specimens to Lc1 Group, subjects emitted plosive barks toward the birds as a threat. Typically, several group members (often adolescent and adult males) would approach cautiously to within 1-2m of the raptors and lunge toward them, without making contact, while emitting plosive barks. Each time a plosive bark was uttered in this context, many of the more timid individuals, some 5-15m distant, would begin to emit chirps and move further away, sometimes bipedally. Numerous males engaging in this threat behavior, as well as those that remained at a distance, were observed rubbing their tails with scent secretions (see also Sauther, 1989). In nonpredator contexts, male tail-rubbing preceeds tail-waving in an assertive display (Jolly, 1966; Evans and Goy, 1968; Schilling, 1979). Elsewhere, plosive barks were emitted most often by young infants (frequently from their mother's backs) to threaten closely-approaching adult conspecifics and humans.

**Responses of Ruffed Lemurs to Avian Predators**

**Natural Encounters.** Whenever large hawks were seen or heard nearby, ruffed lemurs responded vocally with 'abrupt roars' (Fig. 4a). These are high-amplitude, noisy calls (Macedonia, 1990) that appear to serve as threat signals. The equally powerful 'roar/shriek chorus' (Fig. 4b), an extended series of roars and shrieking sounds that appears to function as a group advertisement call (Pereira et al., 1988), often was mixed with abrupt roars at the outset of calling. Abrupt roars were used as hawk mobbing calls, and continued to be produced at irregular intervals long after the eliciting stimulus had disappeared from view. Once high-arousal levels began to subside, bouts of abrupt roars were punctuated with 'wails'(Fig. 4g). In this context, wails seemed to serve an 'all clear' function (Pereira et al., 1988). The duration of wail emissions (from less than a minute to more than 45min) appeared to reflect the level of arousal reached in response to a given predator.

If on the ground when an airborne raptor was detected, ruffed lemurs typically emitted explosive rounds of abrupt roars while assuming a posture in which the head was held low, the back arched, and the shoulders hunched. Nearby group members, unaware of the location of the stimulus, exhibited a 'scan-and-roar' behavior (Macedonia, 1990). This consists of turning the body sharply about the horizontal plane in short jumps of roughly 45 to 90 degree arcs (often in alternating directions), while producing one or more abrupt roars during each shift in position. When engaged in this behavior, individuals appeared as if they were attempting

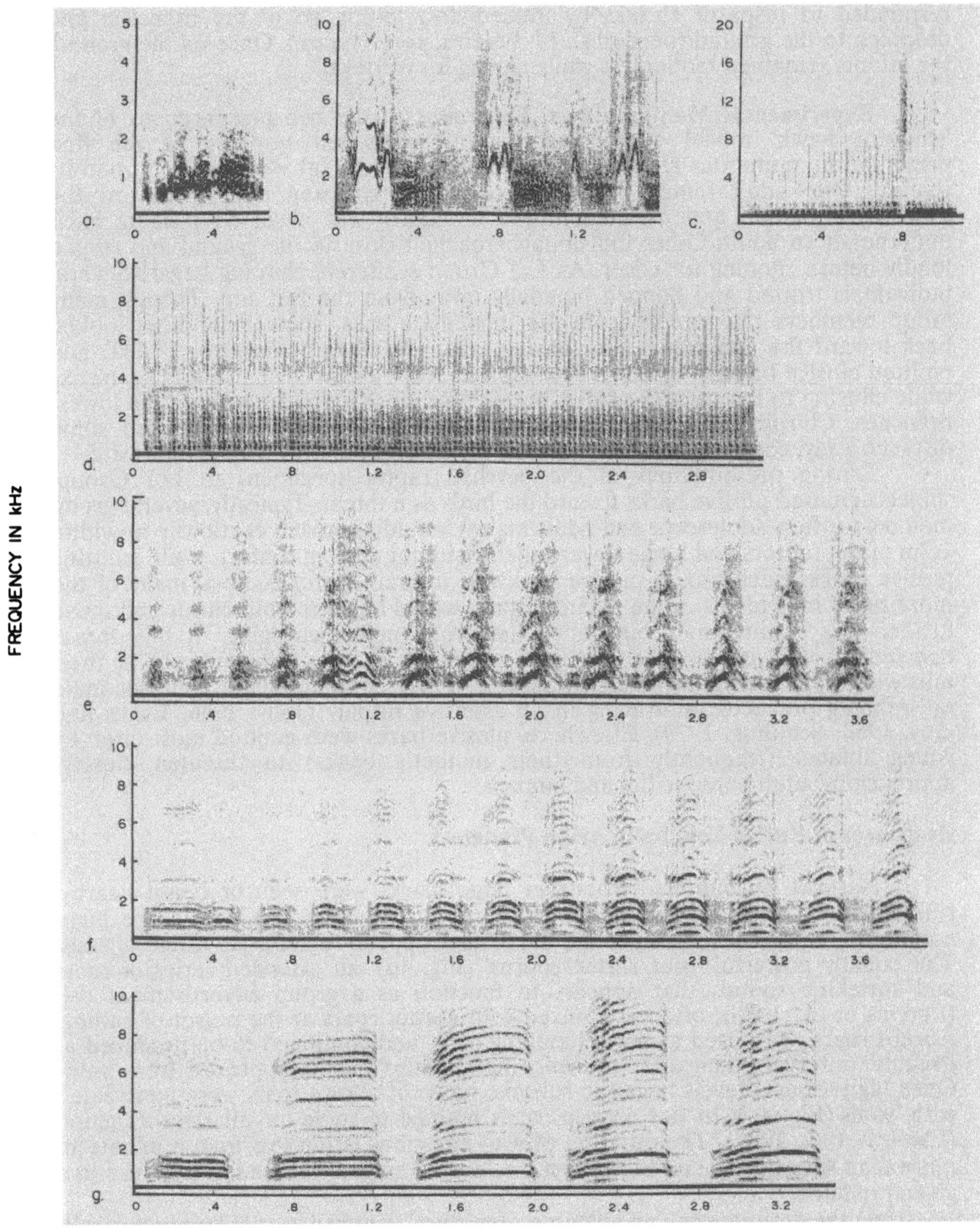


**Figure 4**. Antipredator calls and related vocalizations of the black-and-white ruffed lemur (*Varecia variegata variegata* ): (a) abrupt roar; (b) roar/shriek chorus; (c) growl-snort; (d) growl; (e) pulsed squawk; (f) pulsed squawk-wail intermediate; (g) wail.

to confront an unseen threatening stimulus from whichever direction it might be approaching. Foot movement was more restricted when exhibiting this response in trees. Occasionally, airborne raptors also were 'pursued' by ruffed lemurs as they climbed toward the treetops while emitting abrupt roars in the direction of the bird.

**Experimental Manipulations.** Two Vv1 Group members were present at the test site for the first stooping hawk silhouette presentation. Both individuals emitted abrupt roars in the direction of the model immediately upon its release. Other group members, between 50 and 150m away, responded to the abrupt roars with the same call as they ran in the direction of the first callers. The relocation of these more distant individuals was interrupted periodically as each would stop to emit one or more abrupt roars while scanning the environment, run several meters further, stop and roar again, and so on, until they arrived at the test site. One of the two adults present at the start of the trial had a 20 day-old infant. Upon release of the hawk silhouette, the mother rushed to the infant, put her mouth around its midsection, and transported it orally approximately 4m up into a tree where she 'parked' it in the fork of a large branch (see Pereira et al., 1987). The adult female then returned to the ground to join other group members (all seven of which now were at the test site) who were emitting abrupt roars toward the model. When the silhouette was lowered for removal and reached the ground, one ruffed lemur leapt out from behind a tussock of grass directly onto the model's 'back'. Two other group members then lunged toward the model while they emitted abrupt roars. These calls continued for five to ten minutes before wails (and 'pulsed squawk-wail' intermediates: Fig. 4f) began to punctuate the bursts of abrupt roars.

Adult ruffed lemurs usually approached the perched raptor specimens closely, whereas if immatures approached at all they did so briefly and maintained a distance of several meters. An adult male and two juveniles were present for the presentation of the perched red-shouldered hawk specimen. The adult first saw the bird from approximately 10m away, and began moving slowly toward it, stopping occasionally to emit 'growl-snorts', a high-arousal location indicating call (Fig. 4c; Pereira et al., 1988). When just over a meter from the stimulus, the adult paused while averting his gaze, and then lunged suddenly toward the hawk giving it a powerful, sharp shove with his hands followed with an abrupt roar. The immatures joined in with abrupt roars, but did not approach the bird. Frequent brief bouts of scratching with the 'grooming claw' of the foot also occurrend in all individuals throughout the presentation, apparently a reflection of anxiety (Diezinger and Anderson, 1986; Easley et al., 1987).

### Responses of Ringtailed Lemurs to Mammalian Predators

**Natural Encounters**. Mammalian stimuli elicited antipredator vocalizations from ringtailed lemurs consisting of four call types. 'Clicks' (Andrew, 1963; Fig. 2k) are uttered in circumstances of minor mammalian disturbances, including disturbances caused by arboreal movement of the lemurs themselves. Emission of these calls appears to reflect a mixed motivational state of 'curious-but-wary' on the part of the caller. Clicks preceded the 'closed-mouth click series' (CMCS: Fig. 2k), which also occurs during rapid (individual) locomotion, particularly when in trees. Unlike clicks and the closed-mouth click series, the 'open-mouth click series' (OMCS: Fig. 2k) and 'yaps' (Fig. 2k; Jolly, 1966) were observed, with rare exception, only as a response to mammals that seemed to be perceived as threatening. Occurrences of OMCS' outside the context of mammalian stimuli were limited to a few instances of equivalently-high arousal.

The transition from closed-mouth to open-mouth clicks may inform nearby group members that yapping is about to begin, thus allowing participants to synchronize their yaps. The accuracy of call synchronization appeared to covary with the proximity of callers. Only carnivores elicited yaps in this study, but one researcher (D. Wheeler, pers. comm.) observed Lc1 Group emitting yaps upon their first sighting of Lc2 Group in the trees of an adjacent enclosure. Unfamiliar humans also elicited yaps from free-ranging *L. catta* in Madagascar (Jolly, 1966;

Sauther, 1989). The group-coordinated use of the yap, its high emission amplitude, and its restriction to mammalian stimuli perceived as threatening indicates the function of this vocalization as a carnivore mobbing call.

A characteristic of the ringtailed lemur anticarnivore call sequence (clicks, CMCS, OMCS, yaps) is its repetitive nature. Emissions of these calls frequently continued until either the predator or the lemurs left the vicinity. As pointed out by Jolly (1966), individuals out of visible range of the stimulus often yapped 'sympathetically' with the rest of the group. In contrast to ruffed lemurs, however, repetitive calling dissipated quickly in ringtailed lemurs once the carnivore was no longer in view. Jolly (1966) also noted that extended bouts of yapping sometimes stimulated male ringtailed lemurs to howl (Fig. 2l). This occurred in Lc1 and Lc2 Groups as well, as did the rare, reverse case where several yaps were evoked by enthusiastic howling.

When on the ground, the primary non-vocal response of ringtailed lemurs to the appearance of a carnivore was to leap immediately into the trees. Once above the ground, clicks and CMCS' accompanied movement through the branches and between trees. On many occasions ringtailed lemurs were observed to mob raccoons, and young raccoons were mobbed with particular fervor. Raccoons never were observed to lunge toward or attack the ringtails; either they fled high into the crowns of trees or they remained motionless, appearing to ignore the vocal threats. Although the DUPC ringtailed lemurs became increasingly habituated to the presence of raccoons from spring through fall of each year, large adult raccoons generally continued to be treated as potential threats whenever detected.

**Experimental Manipulations.** When a large dog (Fig. 1d) was walked slowly toward Lc1 Group from a distance, group members stayed on the ground for 5-6s, stared at the approaching carnivore, and emitted gulps. All group members then leapt into the trees and began emitting clicks and CMCS' while continuing to monitor the dog's approach. Emission of yaps began just over a minute from the time the first gulps had been issued. The first clearly-audible OMCS was heard 2min after the first gulps. Several howls also occurred at this time and again 9min after the dog first had been seen approaching. Individuals nearest the dog stared at it while vocalizing, whereas more distant group members that were calling were not necessarily facing the carnivore. Calling continued for about a minute after the dog was removed from the enclosure (approx. 20min after it had been introduced). Less than 10min later, most Lc1 Group members were back on the ground engaged in normal daily activities (e.g., foraging, grooming, resting).

Upon seeing the dog in the 'ambush' presentation, all Lc1 Group members dashed toward the trees without vocalizing, and were between 10 and 15m up into the trees 5s or so after the dog's appearance. The ringtailed lemurs began to emit clicks and CMCS's as soon as they were arboreal, but did not start to emit yaps for about 40s after the start of the presentation. The dog was in view of the lemurs for approximately two minutes, and the ringtails ceased emitting anticarnivore calls several minutes after the dog had been led out of the enclosure.

**Responses of Ruffed Lemurs to Mammalian Predators**

**Natural Encounters.** When detecting a potentially-threatening mammal, ruffed lemurs typically first responded with growls and growl-snorts, followed by their mammalian predator mobbing call, the pulsed squawk (Fig. 4e). Like ringtailed lemur yaps, group emission of pulsed squawks was more or less synchronized, and accuracy of call synchrony seemed to depend on interindividual proximity (ruffed lemur groups housed in smaller quarters at the DUPC frequently exhibited tight synchrony in the emission of these calls).

Ruffed lemurs typically leapt into the trees before emitting any pulsed squawks. Once in the trees, an arched-back posture was assumed while calling. All group members did not always leap into the trees, however; ruffed lemur group matriarchs sometimes remained on the ground and confronted carnivores. An unusually large influx of raccoons into the DUPC enclosures in the summer of

1989 evoked a remarkable level of aggression from the matriarch of Vv2 Group. This female regularly chased the raccoons from the provisioning areas and, on several occasions, leapt onto the backs of raccoons and slapped them with her hands (L. Balko, pers. comm.).

It is intriguing that after pouncing on raccoons, this female often would run 10m or more away, stop, and lie still on the ground (L. Balko, pers. comm.). This behavior gave the impression that she was attempting to incite these carnivores to chase her, in the sense of predator distraction displays of nesting birds (Armstrong, 1949). The matriarch of Vv1 group exhibited similar 'luring' behavior, minus the attack, in response to the approach of museum specimens of mammals mounted on a radio-controlled toy vehicle (Macedonia and Shedd, unpubl. data). In contrast to *L. catta*, avian-like distraction displays would seem particularly adaptive for *Varecia*, given their means of rearing young (see Discussion).

**Experimental Manipulations.** In response to the introduction of the slowly-approaching dog, many Vv1 Group members responded initially with growls. Most (or all) individuals then leapt into the trees, emitting growl-snorts. Approximately 30s later, the ruffed lemurs began to respond with pulsed squawks. These three call types persisted throughout the 20-minute stimulus period and, unlike the ringtailed lemurs which had ceased calling less than a minute after the dog had been removed, the ruffed lemurs continued to call for at least 20min post-presentation (at which time observations were terminated). Interestingly, the group matriarch either had remained on the ground or had returned to it shortly after the dog had reached the lemurs' location. At one point the dog was allowed to approach to within 10m of the adult female. This lemur did not flee, but kept her head lowered and her shoulders hunched forward as she emitted growls toward the dog (which was more than seven times the lemur's weight).

In response to the dog's ambush attack, Vv1 Group members ran rapidly and silently up into the nearest trees. Emission of growl-snorts began about 10s later and pulsed squawks at about 15s. Although the dog was in view of the lemurs for only two minutes, the ruffed lemurs continued calling for more than 45min post-presentation.

### Responses of Ringtailed and Ruffed Lemurs to Snakes

Encounters by ringtailed and ruffed lemurs in the NHE's with snakes corroborate similar observations of ringtailed lemurs in Madagascar (Jolly, 1966; Budnitz and Dainis, 1975; Sauther, 1989). The lemurs responded to snakes as minor disturbances (emission of clicks by ringtailed lemurs, and growls or growl-snorts by ruffed lemurs). As of 1992, only two of the DUPC semi-captive lemurs had been killed by snakes (see Methods). Were this number to increase dramatically, however, the lemurs might be expected to begin responding to snakes as a predatory threat, as has been seen for some other primates (Gouzoules et al., 1975; Masataka, 1983; see also Mineka et al., 1984).

## DISCUSSION

One way to understand the adaptive differences in ringtailed and ruffed lemur antipredator behavior is to compare them with other primates for which relevant data are available. The nearby African continent harbors some monkey species that are analogous in several respects to ringtailed and ruffed lemurs. Ecologically, the vervet monkey (*Cercopithecus aethiops*) is a good anthropoid analog to the ringtailed lemur: both species are relatively small in comparison to their primary predators, both species live in large groups and typically live in open habitats, and although both spend much time in trees, they are the most terrestrial members of their respective genera (Kingdon, 1988; Ward and Sussman, 1979). The ruffed lemur does not have such a direct ecological analog in Africa, though it is in some ways comparable to the arboreal and frugivorous mangabeys (*Cercocebus* sp.).

Although these monkey species are larger than ringtailed and ruffed lemurs in absolute size, one variable of concern here is primate body size relative to predator body size. Also, whereas cercopithecids are sexually dimorphic in body size, with males often being much larger than females, there is no significant difference in body size between the sexes in lemurs (see Kappeler, 1993). This aspect of phylogenetic heritage could explain the lack of a male bias in predator defense among lemurs.

Because exposure to predation from raptors is severe for most mammals when foraging in the terminal branch milieu, moving toward tree trunks and/or out of trees in response to raptors or antiraptor calls is widespread among relatively small diurnal primates (capuchins: van Schaik and van Noordwijk, 1989; crowned lemurs: Wilson et al., 1989). Accordingly, when arboreal, the typical response of ringtails and vervets to raptors is to move away from peripheral branches and out of trees; likewise, these primates move away from open areas in response to raptors when on the ground (Struhsaker, 1967; Jolly, 1966; Seyfarth et al., 1980; Sauther, 1989; Macedonia, 1990; Macedonia and Yount, 1991). Notably, adult male vervets are the only age/sex class that is far above the average prey size taken by one of the common large raptors that are encountered by this species: the martial eagle, *Polemaetus bellicosus* (Cheney and Seyfarth, 1981). Yet, adult male vervets are not large enough to be too bold toward these raptors, and only once was one seen by Struhsaker (1969) to lunge aggressively toward a martial eagle. Experiments with raptor models showed that ringtailed lemurs also respond with 'contained aggression' toward avian predators. Like vervets, this aggression was limited primarily to perched raptors which, once detected, may pose less of a threat than raptors already on the wing.

The fact that ringtailed lemur and vervet monkey infants cling to their mothers from birth confers an advantage against predation over species whose young do not cling. Once infants of clinging species begin to explore and play at a distance from the mother, however, they become more vulnerable to predation. In Madagascar, for example, some species of large hawks (e.g. *Polyboroides radiatus* and *Buteo brachypterus*) perch motionless on the low branches of dead trees, apparently in wait to ambush passing prey (Sauther, 1989; pers. obs.). Young infants who are engaged in play or are exploring on peripheral branches of saplings and bushes must be prime targets for raptors hunting in this manner. Observations of infant ringtailed lemurs dropping from exposed low branches to the ground in response to antiraptor calls suggests that this may be an adaptive response against raptorial predation, once infants begin to distance themselves from direct maternal protection. Intriguingly, a similar response to antiraptor calls has been reported for vervet monkeys (Struhsaker, 1967).

Vervet monkeys and ringtailed lemurs both respond to potentially threatening mammalian predators by running up into the branch network of trees, where their grasping hands and feet give them an advantage in maneuverability over that of clawed carnivores (Struhsaker, 1967; Seyfarth et al., 1980). When on the ground, however, this advantage is lost and the probability of being killed by a carnivore is increased (Seyfarth et al., 1980). The fact that carnivores can place themselves between these primates and access to trees poses an additional level of threat that is less commonly experienced by forest-living primates. The increased predator pressure imposed on relatively small-bodied, highly-terrestrial primates like vervet monkeys (Cheney and Wrangham, 1987) and ringtailed lemurs (Sauther, 1989), in conjunction with the fact that avoiding predation from raptors and carnivores often involves responses that are diametrically opposed, may explain the high level of stimulus specifity witnessed in some of the antipredator calls of these primates (Macedonia and Evans, in press). For ringtailed lemurs at least, larger group size also can be seen as a response to increased predator pressure, given that this species is descended from bamboo-lemurs that live in smaller groups (Macedonia and Stanger, unpubl. data).

Because adult ruffed lemurs weigh about 75% more than adult ringtailed lemurs, the former undoubtedly are less easily dispatched by raptors. Like large monkeys, adult ruffed lemurs not only threaten but probably attack raptors. Yet, at 3.5kg, the ruffed lemur is not a large primate and its aggressiveness towards predators seems out of proportion for its size. This behavior may be a consequence of phylogenetic constraints in ruffed lemur reproduction and infant care.

Unlike other diurnal lemurs, ruffed lemurs retain ancestral features of prosimian reproduction, including a comparatively short gestation period, possession of three functional pairs of mammary glands, and giving birth to litters of young (Foerg, 1982; Tattersall, 1982). Infant ruffed lemurs are kept in nests, and later are 'parked' in tree forks by the mother when she leaves them (Pereira et al., 1987; Morland, 1990). During this time, these infants must be exceedingly vulnerable to predation. Moreover, simultaneous escape from predators via transport by the mother is not an option for ruffed lemur infants, as they must be carried orally to safety one at a time. Thus, if ruffed lemurs are not to loose their reproductive investments they must defend their sessile progeny by confronting predators rather than fleeing from them. Consequently, aggressive defense against raptors year-round might maintain a level of deterrence toward these predators that would enhance the survivorship of infants when present (Klump and Shalter, 1984). In this study, ruffed lemur parents and their adult offspring chased and threatened raptors year-round. Clearly, both parents and siblings would stand to gain fitness by deterring predation on infant kin.

Although adult ruffed lemurs may be largely immune to raptorial predation, their demeanor toward carnivores (excluding that of one group matriarch) was rather similar to that of ringtailed lemurs in being defensive rather than aggressive. This may stem in part from the fact that carnivores not only take larger prey per their own body weight than do raptors, but also because the size range of prey taken is much broader (Cheney and Seyfarth, 1981). The lack of sexual size dimorphism in lemurs (Kappeler, 1993), however, still raises the question of why there should be a matriarchal bias in defense against carnivores in ruffed lemurs. It is, perhaps, that the potential cost associated with confronting carnivores may be too high for individuals other than multiparous adult females. Ruffed lemur fathers were, in fact, among the first to flee during naturally-occurring and staged encounters with carnivores, which seems to highlight the vast differential in reproductive investment between the sexes that is characteristic of mammals.

Finally, with regard to vocal antipredator behavior, ruffed lemurs called repetitively after encounters with predators for a far longer period than did ringtailed lemurs. Though many predators give up the hunt after being discovered by their intended prey (see Klump and Shalter, 1984, for a review), such resignation is not guaranteed, and extended bouts of antipredator calling may provide a margin of safety against renewed hunting efforts from discovered predators (Owings and Hennessy, 1984; Loughry and McDonough, 1988). Considering the helplessness of ruffed lemur neonates, extending the period of predator awareness through repetitive calling could be crucial to their survival.

## SUMMARY

The responses of semi-captive, forest-living ringtailed lemurs and ruffed lemurs to avian predators differed considerably. Whereas ringtailed lemurs responded to the sight or sound of airborne raptors by fleeing peripheral branches and by moving downward and/or out of trees, ruffed lemurs entered the trees and pursued raptors and raptor models with overt displays of aggression. Comparative data on responses of some African monkeys to raptors suggest that the differences in antiraptor behavior between these two lemur species stem partly from differences in their body sizes. The overall responses of ringtailed and ruffed lemurs to carnivores were more similar than their responses to raptors, and this may be due to the larger prey size per predator body weight that carnivores can accommodate as compared to raptors. Because the hallmark traits of grasping hands and feet confer to primates an advantage over carnivores when in the branch network of trees, most primates habitually seek or maintain an arboreal location when dealing with dangerous carnivores. Indeed, both lemur species typically entered the trees before beginning to mob carnivores vocally. An exception to this rule would be expected when, as in ruffed lemurs, progeny do not cling to their mothers and when prior to locomotor independence infants are kept in nests or parked on branches. Adult ruffed lemurs therefore may not always flee into trees in response to carnivores because deterring predators that are still on the ground (and thus who still are at a distance from arboreally-located immatures) may be more effective than waiting until the carnivore has entered the trees before attempting to fend it off. Unlike ringtailed lemurs, ruffed lemur group matriarchs confronted, chased, and sometimes attacked carnivores on the ground, and performed what seemed to be predator distraction displays. This suggests that, like many avian species, the ruffed lemur has evolved such antipredator behavior as an adaptation to the phylogenetic constraint of possessing sessile young.


## ACKNOWLEDGEMENTS

I thank DUPC Director, K.E. Glander, and DUPC Scientific Director, E.L. Simons for the opportunity to conduct this research. I am particularly grateful to M.E. Pereira for the help and insights that he provided throughout the course of this work. D. Lee at the North Carolina Museum of Natural Sciences loaned the perched raptor specimens, and R.H. Wiley gave unlimited access to his Uniscan II audiospectrograph. I am grateful to M.E. Pereira, two anonymous reviewers, and the editors of this volume for constructive comments on earlier versions of this manuscript. This research was supported by an NSF Dissertation Improvement Grant BNS 8912589. This is DUPC publication no. 552.


## REFERENCES


Alexander, R.D., 1974, The evolution of social behavior, *Ann. Rev. Ecol. Syst.* 5:324.

Andrew, R.J., 1963, The origins and evolution of calls and facial expressions of the primates, *Behaviour* 20:1.

Armstrong, E.A., 1949, Diversionary display: the nature and origin of distraction displays. *Ibis* 91:88, 179.

Bartecki, U., and Heyman, E.W., 1987, Field observations of snake-mobbing in a group of saddle-back tamarins, *Saguinus fuscicollis nigrifrons*, *Folia Primatol.* 48:199.

Budnitz, N., and Dainis, K., 1975, *Lemur catta*: ecology and behaviour, *in*: "Lemur Biology," I. Tattersall and R.W. Sussman, eds., Plenum Press, New York.

Boinski, S., 1988, Use of a club by a wild white-faced capuchin (*Cebus capucinus*) to attack a venomous snake (*Bothrops asper*), *Am. J. Primatol.* 14:177.

Chapman, C.A., 1986, *Boa constrictor* predation and group response in white-faced cebus monkeys, *Biotropica* 18:171.

Charles-Dominique, P., 1977, "Ecology and Behaviour of Nocturnal Prosimians," Duckworth, London.

Cheney, D.L., and Seyfarth, R.M., 1981, Selective forces affecting the predator alarm calls of vervet monkeys, *Behaviour* 76:25.

Cheney, D.L. and Wrangham, R.W, 1987, Predation, *in*: "Primate Societies", B.L. Smuts, D.L. Cheney, R.M. Seyfarth, R.W. Wrangham, and T.T. Struhsaker, eds., University of Chicago Press, Chicago.

Diezinger, F., and Anderson, J.R., 1986, Starting from scratch: A first look at "displacement activity" in group-living rhesus monkeys, *Am. J. Primatol.* 11:117.

Easley, S.P., Coelho, A.M., and Taylor, L.L., 1987, Scratching, dominance, tension, and displacement in male baboons, *Am. J. Primatol.* 13:397.

Eason, P., 1989, Harpy eagle attempts predation on adult howler monkey, *Condor* 91:469.

Evans, C.S., and Goy, R.W., 1968, Social behaviour and reproductive cycles in captive ring-tailed lemurs (*Lemur catta*), *J. Zool. Lond.*, 156:181.

Foerg, R., 1982, Reproductive behaviour in *Varecia variegata*, *Folia Primatol.* 38:103.

Gautier-Hion, A., and Tutin, C.E.G., 1987, Simultaneous attack by adult males of a polyspecific troop of monkeys against a crowned hawk eagle. *Folia Primatol.* 51:149.

Goodman, S.M., in press, The enigma of anti-predator behavior in lemurs: evidence of a large extinct eagle on Madagascar, *Int. J. Primatol.*

Goodman, S.M., O'Connor, S., Langrand, O., 1993, A review of predation on lemurs: implications for the evolution of social behavior in small, nocturnal primates, *in*: "Lemur Social Systems and Their Ecological Basis," P.M. Kappeler, and J.U. Ganzhorn, eds., Plenum Press, New York.

Gouzoules, H., Fedigan, L.M., and Fedigan, L., 1975, Responses of a transplanted troop of Japanese macaques (*Macaca fuscata*) to bobcat (*Lynx rufus*) predation, *Primates* 16:335.

Hall, K.R.L., 1965, Behaviour and ecology of the wild patas monkey, *Erythrocebus patas*, in Uganda, *J. Zool.* 148:15.

Hasson, O., 1991, Pursuit-deterrent signals: communication between prey and predator, *Trends Ecol. Evol.* 6:325.

Heyman, E.W., 1987, A field observation of predation on a moustached tamarin (*Saguinus mystax*) by an anaconda, *Int. J. Primatol.* 8:193.

Isbell, L.A., 1990, Sudden short-term increase in mortality of vervet monkeys (*Cercopithecus aethiops*) due to leopard predation in Amboseli National Park Kenya, *Am. J. Primatol.* 21:41.

Jolly, A., 1966, "Lemur Behavior," University of Chicago Press, Chicago.

Kappeler, P.M., 1991, Patterns of sexual dimorphism in body weight among prosimian primates, *Folia Primatol.* 57:132.

Kappeler, P.M., 1993, Sexual selection and lemur social systems, *in*: "Lemur Social Systems and Their Ecological Basis," Kappeler, P.M. and J.U. Ganzhorn, eds., Plenum Press, New York.

Kingdon, J., 1988, Comparative morphology of hands and feet in the genus Cercopithecus, *in*: "A Primate Radiation: Evolutionary Biology of the African Guenons," A. Gautier-Hion, F. Bourliere, J.-P. Gautier, and J. Kingdon, eds., Cambridge University Press, Cambridge.

Klump, G.M., and Shalter, M.D., 1984, Acoustic behaviour of birds and mammals in the predator context, *Z. Tierpsychol.* 66:189.

Loughry, W.J., and McDonough, C.M., 1988, Calling and vigilance in California ground squirrels: a test of the tonic communication hypothesis, *Anim. Behav.* 36:1533.

Macedonia, J.M., 1990, What is communicated in the antipredator calls of lemurs: Evidence from antipredator call playbacks to ringtailed and ruffed lemurs, *Ethology* 86:177.

Macedonia, J.M., in press, The vocal repertoire of the ringtailed lemur (*Lemur catta*), *Folia Primatol.*

Macedonia, J.M., and Polak, J.F., 1989, Visual assessment of avian threat in semi-captive ringtailed lemurs (*Lemur catta*), *Behaviour* 111:291.

Macedonia, J.M., and Yount, P.L., 1991, Auditory assessment of avian threat in semi-captive ringtailed lemurs (*Lemur catta*), *Primates* 32:169.

Macedonia, J.M., and Evans, E.S., in press, Variation among mammalian alarm call systems and the problem of meaning in animal signals, *Ethology*.

Masataka, N., 1983, Psycholinguistic analyses of alarm calls of Japanese monkeys (*Macaca fuscata fuscata*), *Am. J. Primatol.* 5:111.

Mineka, S., Davidson, M., Cook, M., and Keir, R., 1984, Observational conditioning of snake fear in rhesus monkeys, *J. Ab. Psychol.* 93:355.

Morland, H.S., 1990, Paternal behavior and infant development in ruffed lemurs (*Varecia variegata*) in a Northeast Madagascar rain forest, *Am. J. Primatol.* 20:253.

Owings, D.H., and Hennessy, D.F., 1984, The importance of variation in sciurid visual and vocal communication, *in*: "The Biology of Ground Dwelling Squirrels," J. Murie, and G. Michener, eds., University of Nebraska Press, Lincoln.

Pereira, M.E., and Macedonia, J.M., 1991, Response urgency does not determine antipredator call selection by ringtailed lemurs, *Anim. Behav.* 41:543.

Pereira, M.E., Klepper, A., and Simons, E.L., 1987, Tactics of care for young infants by forest-living ruffed lemurs (*Varecia variegata variegata*): Ground nests, parking, and biparental guarding, *Am. J. Primatol.* 13:129.

Pereira, M.E., Seeligson, M.L., and Macedonia, J.M., 1988, The behavioral repertoire of the black-and-white ruffed lemur (*Varecia variegata variegata*), *Folia Primatol.* 51:1.

Sauther, M., 1989, Antipredator behavior in troops of free-ranging *Lemur catta* at Beza Mahafaly Special Reserve, Madagascar, *Int. J. Primatol.* 10:595.

Schilling, A., 1979, Olfactory communication in prosimians, *in:* "The Study of Prosimian Behavior," G.A. Doyle and R.D. Martin, eds., Academic Press, New York.

Seyfarth, R.M., Cheney, D.L., and Marler, P., 1980, Vervet monkey alarm calls: Semantic communication in a free-ranging primate, *Anim. Behav.* 28:1070.

Sherman, P.T., 1991, Harpy eagle predation on a red howler monkey, *Folia Primatol.* 56:53.

Struhsaker, T.T., 1967, Auditory communication among vervet monkeys (*Cercopithecus aethiops*), *in:* "Social Communication Among Primates," S. Altmann, ed., University of Chicago Press, Chicago.

Struhsaker, T.T., 1969, Correlates of ecology and social organization among African cercopithecines, *Folia Primatol.* 11:80.

Struhsaker, T.T., and Leakey, M., 1990, Prey selectivity by crowned hawk-eagles on monkeys in the Kibale Forest, Uganda, *Behav. Ecol. Sociobiol.* 26:435.

Tattersall, I., 1982, "The Primates of Madagascar," Columbia University Press, New York.

Terborgh, J., 1983, "Five New World Primates: A Study in Comparative Ecology," Princeton University Press, Princeton.

Terborgh, J., and Janson, C.H., 1986, The socioecology of primate groups, *Ann. Rev. Ecol. Syst.* 17:111.

van Schaik, C.P. 1983, Why are diurnal primates living in groups? *Behaviour* 87:120.

van Schaik, C.P., and Mitrasetia, T., 1990, Changes in the behavior of wild long-tailed macaques (*Macaca fascicularis*) after encounters with a model python, *Folia Primatol.* 55:104.

van Schaik, C.P., and van Hooff, J.A.R.A.M., 1983, On the ultimate causes of primate social systems, *Behavior* 85:91.

van Schaik, C.P., and van Noordwijk, M.A., 1989, The special role of male *Cebus* monkeys in predator avoidance and its effect on group composition, *Behav. Ecol. Sociobiol.* 24:265.

van Schaik C.P., van Noordwijk, M.A., Warsono, B., and Sutriono, E., 1983, Party size and early detection of predators in Sumatran forest primates, *Primates* 24:211.

Ward, S.C., and Sussman, R.W., 1979, Correlates between locomotor anatomy and behavior in two sympatric species of *Lemur*, *Am. J. Phys. Anthrop.* 50:575.

Wilson, J.M., Stewart, P.D., Ramangason, G.-S., Denning, A.M., and Hutchings, M.S., 1989, Ecology and conservation of the crowned lemur, *Lemur coronatus*, at Ankarana, N. Madagascar, *Folia Primatol.* 52:1.

# TERRITORIALITY IN *LEMUR CATTA* GROUPS DURING THE BIRTH SEASON AT BERENTY, MADAGASCAR

Alison Jolly[1], Hantanirina R. Rasamimanana[2], Margaret F. Kinnaird[3], Timothy G. O'Brien[3], Helen M. Crowley[4], Caroline S. Harcourt[5], Shea Gardner[6], and Jennifer M. Davidson[6]

[1]Department of Ecology and Evolutionary Biology
Princeton University, Princeton, NJ, 08544-1003, U.S.A.
[2]École Normale III. Univ. d'Antananarivo, B.P.883, Antananarivo, Madagascar
[3]Wildlife Conservation International, New York Zoological Society, Bronx, NY 10460, U.S.A.
[4]Berenty Reserve,B.P. 54, Fort Dauphin, Madagascar
[5]World Conservation Monitoring Centre, 219c Huntingdon Road, Cambridge, UK
[6]University of California, Davis, CA 95616, U.S.A.

## ABSTRACT

Seven adjacent groups of *Lemur catta* were studied for three five-day blocks: two in 1990 and one in 1991, during the Sept-Oct birth season. Females confront females of other troops with staring, lunges, and occasional physical aggression. Confrontations apparently defended against territorial intrusion: troops retreated from confrontations toward their territorial centers. Mean day range (1377m, N = 101) relative to home range diameter (319m for concave polygons) is one of the highest in primates, indicating high resource defensibility. Fifteen-day home ranges were as large as yearly ranges. Intergroup relations fell into two categories, High Confrontation (HC), (4 of the possible neighbor pairs, ≥ 1.0 confrontations/day) and Low Confrontation (LC), (eight pairs, ≤ 0.3 confrontations/day). HC pairs have nearer ranging patterns, more close approaches, and a higher proportion of close approaches which escalate into confrontations. LC pairs have little range overlap, fewer close approaches, and a low proportion of close approaches which lead to confrontations. Overall, both HC and LC pairs avoid each other. Confrontations were preceded and followed by high troop velocity and large turning angles, indicating approach and retreat. Male howls had no immediate effect on troop distance or turning angle. Tourists provisioned lemurs with bananas in 1990; resources dropped sharply in 1991 as provisioning ceased in a drought year. The same neighbor pairs remained as HC and LC in 1991: provisioning did not change the relatively high or low antagonism between particular neighbors. We suggest that territoriality is fundamental to *Lemur catta* female resource access, (at least at high population density), that large troops do not necessarily gain ground in intertroop conflict, that female targeted aggression is related to the constraints of the territorial system, and that males play little role in troop spacing during the birth season.

## INTRODUCTION

*Lemur catta* at Berenty, Madagascar, live in home ranges which average only six to eight hectares in the richest area of closed canopy gallery forest (Jolly, 1966; Budnitz and Dainis, 1975; Budnitz, 1978, Mertl-Millhollen et al., 1979.) It is possible to define a frontier zone defended by territorial confrontations, and also by scent marking (Mertl-Millhollen, 1988). Most troops have a small core area of exclusive use, although this is not an area of particularly concentrated use. Aggressive encounters between troops are called "confrontations" because opponents face each other, each side staring outward from its own position toward its opponents. Confrontations may be limited to tense stares and glares, which may last more than an hour, or be interrupted and resumed. They can escalate to lunging, cuffing, grappling and occasional slashes with the canines. At high intensity, females can loose infants or be seriously wounded themselves (Koyama, 1991; Wood, Pereira, Hood, pers. comm; pers. obs.).

Adult females take the most active role in such confrontations: males may or may not challenge the opposite troop males, by swaggering, chases and by ritualized scent-marking, but this male behavior is often apart from and apparently incidental to the females' overt hostilities. On the other hand, males have a long-distance howl, or song, which is commonly answered by males of other troops. This howl may sometimes be accompanied by a chorus of moans from the females, in which case it could be considered a duet, but the female calls do not carry so far (to human ears). The males' howling is the sort of call commonly assumed to be a spacing mechanism.

This data set on *Lemur catta* covers seven troops over 15 days' observation during the birth seasons of 1990 and 1991. For those days, it allows one to answer the following questions: Do troops confront at territorial boundaries or throughout the range? Do they retreat toward the center of the range after confrontations? Do all pairs of troops interact in similar fashion? What is the typical day range and home range size? Is the range defensible? Do troops interact or avoid each other more than expected? Does the pattern of intergroup aggression correlate with defense of resources? Does the loud, long distance male howling affect troop movement or intergroup spacing?

This in turn lets one ask, Does lemur territorial defense at this season stem from resource defence? (Clearly, yes). Does it benefit large troops over small troops? (Apparently not much). Does it reflect the immediate influences of either tourist provisioning or a first year of serious drought? (Apparently not so much as the antagonism between particular pairs of troops rather than other pairs of troops). From qualitative observations we then speculate that resource constraint is reflected first by intra-troop targeting, rather than by changes in inter-troop behavior.

Female *Lemur catta* are invariably dominant over males, both for food priority and in aggressive encounters (Kappeler, 1990; Pereira, et al., 1990). One to four males travel and feed with the body of the females, and in at least some troops there is a clear alpha or control male (Sauther, 1991a). Subordinate males, the "Drone's Club" (Jolly, 1966) tag along behind, feeding after the main troop. Troops range in size from 4 to about 25 adults and juveniles, not counting infants of the year. A stable troop apparently has a minimum of two adult females and one male, but there may be up to 10-12 adults of either sex. Sex ratio varies widely (Budnitz and Dainis, 1975; Sussman, 1991; Koyama, 1991).

Lemur behavior varies dramatically by season. Most mating is compressed into two weeks in April; most births into about three weeks in September. Late matings, however, produce a few offspring born as late as November. About a quarter of the males migrate between troops during September-November. Day ranges also vary seasonally: they are shortest during the dearth of the dry season, when females are pregnant, and longest during the early wet season, when females are lactating but there is abundant food. Day ranges lengthen during the mating and birth seasons (Jolly, 1966; Koyama, 1988; Budnitz and Dainis, 1975; Jones, 1983; Sauther, 1991b, Sussman, 1974; Rasamimanana and Rafidinaririvo, 1993).

Females become very aggressive during the birth season both within and between troops (Vick and Pereira, 1989). Sauther (1991a) argues that the birth season and early lactation is the season of greatest female need: low food supplies at the end of the dry season combine with the energy demand of lactation and escalating inter- and intra-troop competition. Pereira and Weiss (1991) argue that there is simultaneously a risk of infanticide by immigrant males, which the females and resident males counter by aggressive chasing. Koyama (1991, pers. comm.) feels this is the likely time for most troop fission. Thus, studies during the birth season sample a peak of female aggression.

## METHODS

### Data Collection

This paper results from a collaboration extending far beyond the many authors. Sixty-six dedicated Earthwatch volunteers collected the data analyzed here. They were joined by five Malagasy from the École Normale III, the branch of the University of Antananarivo which teaches future secondary school teachers. Sheila O'Connor, Mark Pidgeon, Gioia Theler, and John Walker acted as staff memebers. Gwendolyn Wood, Shea Gardner, and Jennifer Davidson found spare energy in the course of their own studies to help Earthwatchers. The 1990-91 field work was supervised by Jolly, Rasamimanana, Crowley, and Harcourt.

This research was originally planned as four blocks, two in each year. Political unrest in Madagascar in 1991 condensed the study to one block in that year, carried out by one Earthwatch group who monitored the effect of near-total dearth of banana-toting tourists. Troops A, C, and G (see Mertl-Millhollen et al., 1979 and O'Connor, 1987 for troop letters) had been provisioned for the previous five years. Simultaneously, lemurs were undergoing the first year of the 1990-1991 El Nino, a drought which brought famine to the surrounding human population. This drought may have exacerbated the effects of the drop in tourism--for the lemurs as for the humans.

Earthwatch volunteers trained in following lemurs and filling out maps and checksheets for one week, and then embarked on a five-day dawn to dusk follow of seven adjacent troops. Each troop had its own team of three to four watchers, who picked it up as soon as possible after 05:00 (the troops were sometimes not found until 06:00 or 07:00), followed in shifts of two watchers per shift through the day, and left the troop between 18:30 and 19:00. Every 10 minutes the watchers recorded the position of the core of troop females both on a map of the reserve, and on a checksheet, noting nearest trail tag number (trails numbered every 25m), and direction and approximate distance from last observation point. During the subsequent 10-min block, watchers recorded whether any of the troop fed, whether they ate fruit, new leaves, old leaves, buds, flowers, or other, whether they made loud calls (especially the male howl), whether they met another troop, and whether intergroup interactions occurred. In 1991, the watchers also flagged trees where their troop ate in three consecutive 10-min blocks, as a pilot attempt to identify preferred feeding trees. In this forest a two-dimensional grid is adequate, as the canopy is only 10-20m, and troops almost never cross one above the other.

The troops studied all live in high quality closed canopy gallery forest (Budnitz and Dainis, 1975; O'Connor, 1987). Troops A, C, G, and at times G2 received a large quantity of bananas from tourists in 1990, though not in 1991. Troops D, E, F, and H were rarely provisioned. All the troops ranged and fed principally from the forest. F troop, a "low-banana troop", was watched only in 1990. In 1991 we dropped F and included H, which seemed to be a troop attracted to the high conflict area of G and G2 by the restaurant, but which turned out to have a "normal" home range of its own. Troops were observed for 177-193h , except for F with 130h and H with 77h.

The disadvantages of using Earthwatch volunteers are: 1) novices can become confused about where they are, especially in marking on a very small scale map which covers the ranges of all troops watched. They marked the location of the center of gravity of troop females, which can be vague at times of high troop spread, necessitating judgement calls by differing observers. One hopes that the quantity of data will compensate for some inevitable blurriness. 2) The timing of feeding and other activities has to be done in rough categories to be sure of comparable data.

On the other hand, volunteers who have given their vacation (and a great deal of money) to be allowed to take data are very dedicated, and many of the group have high professional qualifications. The level of output compares to an excellent field class. The advantage is that a team can actually monitor an entire neighborhood of troops, a task beyond any one or two researchers, for a total of 1337 contact hours of observation over 15 days.

**Statistical Analysis**

The data were analyzed by Kinnaird, O'Brien, Gardner and Davidson. Day range, activity, and random gas model analyses used d-Base programs developed by O'Brien and Kinnaird. Home range area and mapping used Surfer and McPaal packages as well as hand calculations. Further details of the analyses are presented in the separate sections of results.

## RESULTS

**Intergroup Conflict: High and Low Confrontation Rates Between Neighbors**

Table 1 shows the total of 163 confrontations seen when we were observing at least one participant, over all three five-day blocks. If confrontations continued through consecutive 10-min time blocks, only one confrontation was counted. If the confrontation ceased for one or more 10-min block and hostilities then resumed, two confrontations were counted. Thirty of these conflicts were associated with tourists who either offered bananas or who arrived noisily so that two or more troops converged upon them, even though there were actually no bananas.

**Table 1.** Total Confrontations Between Groups[1].

| | A | C | D | E | F | G | G2 | H | Other |
|---|---|---|---|---|---|---|---|---|---|
| A | - | 31(14) | 0 | 0 | 0 | 0 | 16(2) | 2 | 1(1) |
| C | | - | 3 | 0 | 0 | 0 | 0 | 0 | 15 |
| D | | | - | 20(3) | 0 | 0 | 1 | 0 | 3(1) |
| E | | | | - | 3(2) | 0 | 0 | 0 | 7(3) |
| F | | | | | - | 0 | 0 | 0 | 14 |
| G | | | | | | - | 22(3) | 19(1) | 1 |
| G2 | | | | | | | - | 3 | 2 |
| H | | | | | | | | - | 1 |

[1]Totals inlude confrontations around tourists (in parentheses)

Table 2 gives the confrontation rates by troop per 12-hour day. The mean, even without conflicts directly provoked by tourists, is 2.1/day. This is high for primates two orders of magnitude higher than some of the southeast Asian langurs cited by van Schaik et al. (1992).

Five pairs of neighbors stand out as having high numbers of confrontations (HC), A-C, G-G2, D-E, G-H and A-G2. If our study had included troop C's northern neighbor or F's eastern one, these pairs would also appear as highly confrontational. In contrast, seven other pairs of neighbors had much lower confrontation rates (LC) even though their ranges overlapped: A-H, A-G, C-D, C-G2, D-G2, E-F and G2-H. The remaining 16 troop combinations were not neighbors and so had no opportunity to confront (see Figs. 1-3).

**Table 2.** Confrontation rates.

| Troop | #Days | Confrontation/Day | Non-Tourist Confrontation/Day |
|---|---|---|---|
| A | 15 | 3.3 | 2.1 |
| C | 15 | 3.3 | 2.3 |
| D | 15 | 1.8 | 1.5 |
| E | 15 | 2.0 | 1.5 |
| F | 10 | 1.7 | 1.5 |
| G | 15 | 2.8 | 2.5 |
| G2 | 15 | 2.9 | 2.5 |
| H | 8 | 2.8 | 2.8 |
| Mean | | 2.5 | 2.1 |

The same troop pairs continued as HC in all three five-day blocks, except for G-H. This means that the same neighbors were antagonists in 1991 as in 1990, in spite of the drop in banana provisioning as well as the first year of drought, which may have lowered resources throughout the forest. Either for reasons of resource distribution, such as long-term geographical barriers or trails, or for historical reasons, the same neighbors stuck to high agonistic levels in spite of changed environmental circumstances (see later section on resources).

**Table 3.** Rates of confrontation and near approaches between groups.

| Pairs | Confrontations per Day | Approach Rate to 80m | Approach Rate to 120m |
|---|---|---|---|
| High Confrontation Pairs | | | |
| A-C | 1.93 | 6.10 | 11.47 |
| C-G2 | 1.47 | 6.00 | 9.67 |
| D-E | 1.07 | 4.80 | 8.40 |
| A-G2 | 1.00 | 4.93 | 9.00 |
| Low Confrontation Pairs | | | |
| C-D | 0.27 | 1.53 | 2.47 |
| G-H | 0.25 | 1.75 | 4.75 |
| G2-H | 0.25 | 1.25 | 3.88 |
| A-H | 0.25 | 0.00 | 0.25 |
| D-G2 | 0.07 | 0.67 | 1.00 |
| E-F | 0.02 | 1.20 | 2.70 |
| A-G | 0.00 | 0.73 | 1.53 |
| C-G2 | 0.00 | 0.00 | 0.25 |

Table 3 gives the confrontation rates compared to the near approach rates observed for neighbor pairs. This table and subsequent calculations are based on 95 confrontations where both sides were observed before and after conflict, and on 108 full and half day follows. The troop relations are the same as for total conflicts, with one exception. Troops G and H separated their ranges somewhat between 1990 and 1991 when we incorporated Troop H in the five-day study. This means that for the period when we can calculate ranging patterns, G-H had become an LC pair.

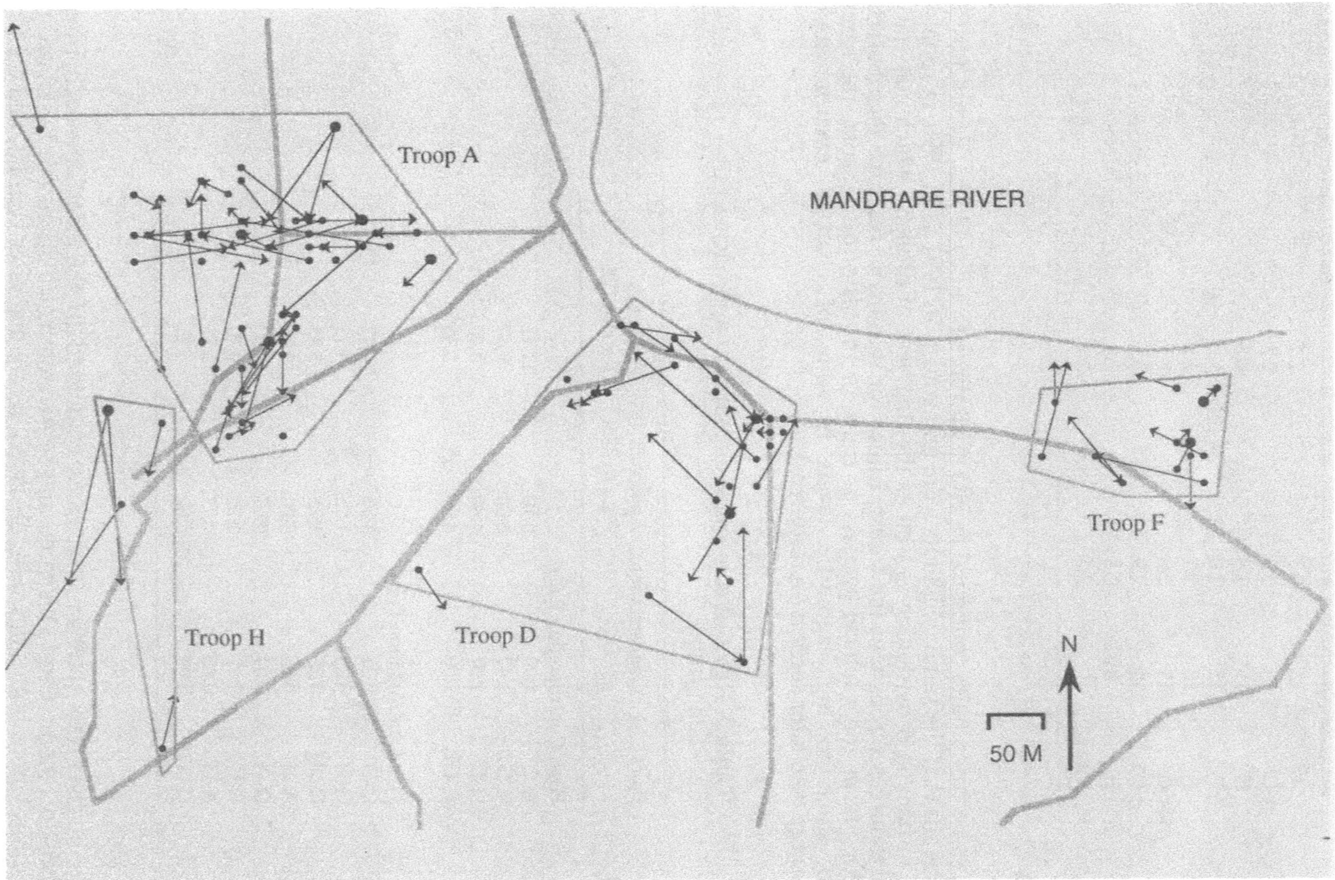


**Figure 1.** Position of troops at scan following start of confrontation, with direction and distance moved during next half hour. Troops A,D,F,H.

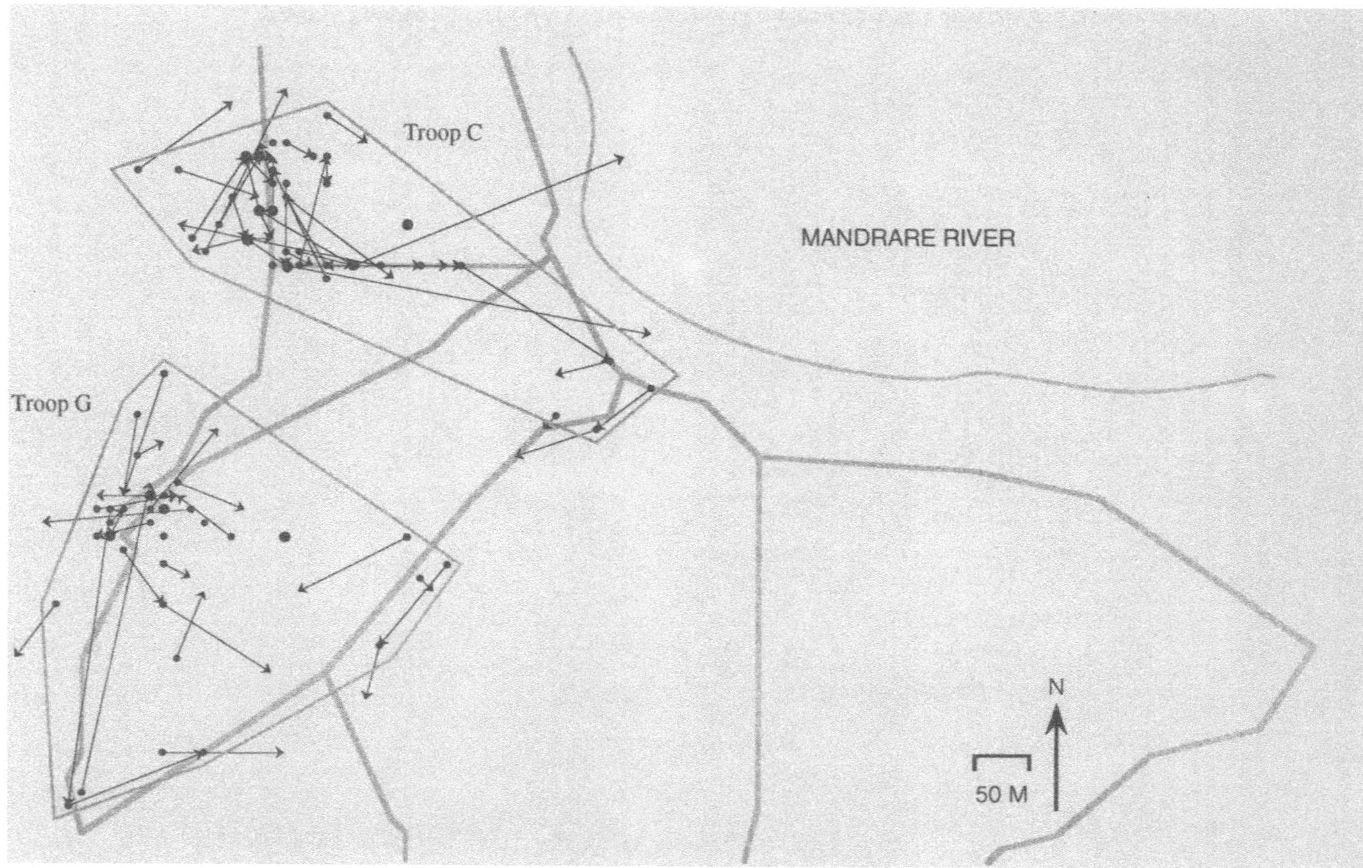


**Figure 2.** Position of troops at scan following start of confrontation, with direction and distance moved during next half hour. Troops C,G.

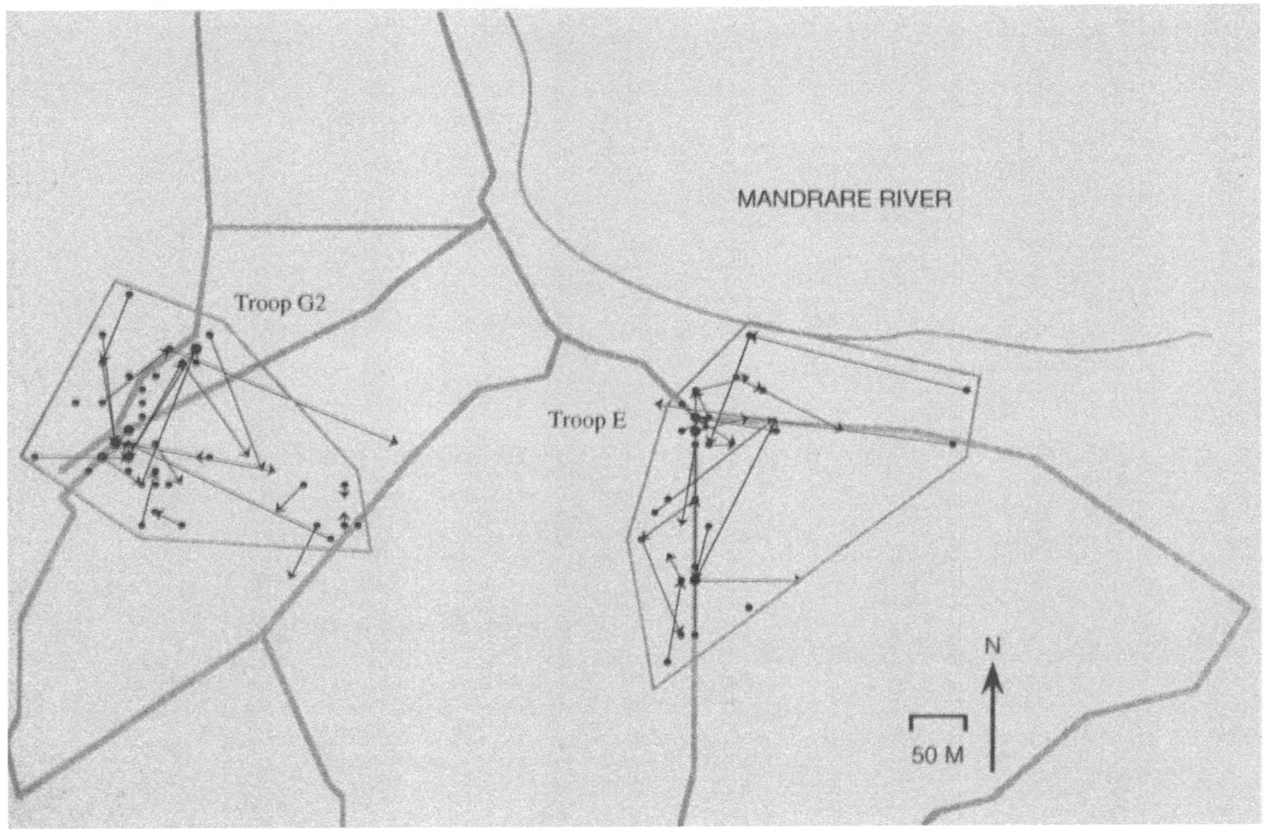


Figure 3. Position of troops at scan following start of confrontation, with direction and distance moved during next half hour. Troops E,G2.

The four HC pairs fought ≥ 1.0 times per day (mean = 1.37, range 1.00-1.93) while the eight LC combinations fought 0 to 0.27 times per day (mean = 0.14). Further, the four HC turned a higher proportion of their approaches into confrontations: a mean of 25%, range 20.3-31.6% of 80m approaches, compared to a mean of 10.5%, range 0-20% of 80m approaches among the LC. The only exception to the ranking of HC above LC for confrontations per approach is troops A and H, which approached each other only twice to within 120m, but had confrontations both times.

Approaches are one example where multiple observers correct the subjective impression of single observers, who are more likely to see a nearby troop in the forest if it is fighting or facing off--that is, giving some level of confrontation. This can give the impression that *Lemur catta* always confront when they are near. In fact the number of confrontations is significantly less than the number of near approaches at both 80m, (two troop diameters) and 120m (three troop diameters).

Confrontations as a behavior pattern did not differ between HC and LC pairs, either in duration and or intergroup distance. Mean duration for all pairs was 1.8 10-min time blocks (range 1-10), with no significant difference in mean or range.

The distances of confronting troops range from about 20 to 120m, with one aberrantly high outlier that we prefer to consider a map confusion. (When intergroup distance varied during a single confrontation, mean of high and low distance was used for calculations). Mean low distance over all confrontations was 60m, mean high distance 97.7m. The distances were the positions of troops recorded on the 10-min scan point before the confrontation, as well as scans taken during the course of more prolonged confrontations (Fig. 4). As the later section on velocity shows, the troops often speed up just before a confrontation, so these distances were not indicative of actual "battle stations", but of lemurs coming into each other's sphere of influence. Further, confrontations could involve only a few of the troop females, while the body of the troop was further away. It is not surprising that there is no significant difference between HCs and LCs in this measure.

**Day Range, Home Range, Defensibility, and Overlap**

Table 4 shows day range length for each troop, as well as distance moved per half hour. The mean was 1377m (N = 101), which is typical for primates (Martin, 1981). The one markedly smaller troop, G2, had the next-to highest day range. The one troop with markedly smaller day range and speed of movement was troop F, which had more transitional, open-canopy forest in its home range, and little closed canopy forest. This shorter range concurs with the study by Rasamimanana and Rafidinarivo (1993) who find shorter ranges in *L. catta* in poorer habitat or seasons. These lemurs appear to conserve energy by moving less (cf. Budnitz, 1978; Richard, 1985; Jolly, 1985).

Mean home range size for a five day block was 3.95ha based on minimum concave polygons, and 9.67ha based on minimum convex polygons (Table 5). Home range size for all days observed (8-15 days per troop) was 7.97ha for minimum concave polygons, 18.92ha for minimum convex polygons. The smallest home range belonged to G2, the troop with only five animals, even though their day ranges were second to longest (Fig. 5).

To compare this with home ranges taken in other studies, over five months to two years, one can turn to earlier work in the same forest. Jolly (1966) reported 5.7ha home range for a group in the ranges of present troops D and E, Budnitz (1978) reported 8.1ha for present group C and 6.0ha for a group in the ranges of present troops D and E, Budnitz (1978) reported 8.1ha for present group C and 6.0ha for group D, Mertl-Millhollen (1988) reported 9.1ha for Troop C and 7.9ha for D, O'Connor (1987) 5.7ha for Troop A and 6.4ha for Troop C, and Sussman (1974) 6.0ha for a troop adjacent to the present study. Given that each observer used his or her own methods to calculate range, this is remarkable concurrence. The estimates are specific to the very rich, closed canopy tamarind forest; Budnitz, for instance, found 23ha and 12.9ha for troops only 500m away from the river in open canopy scrub.

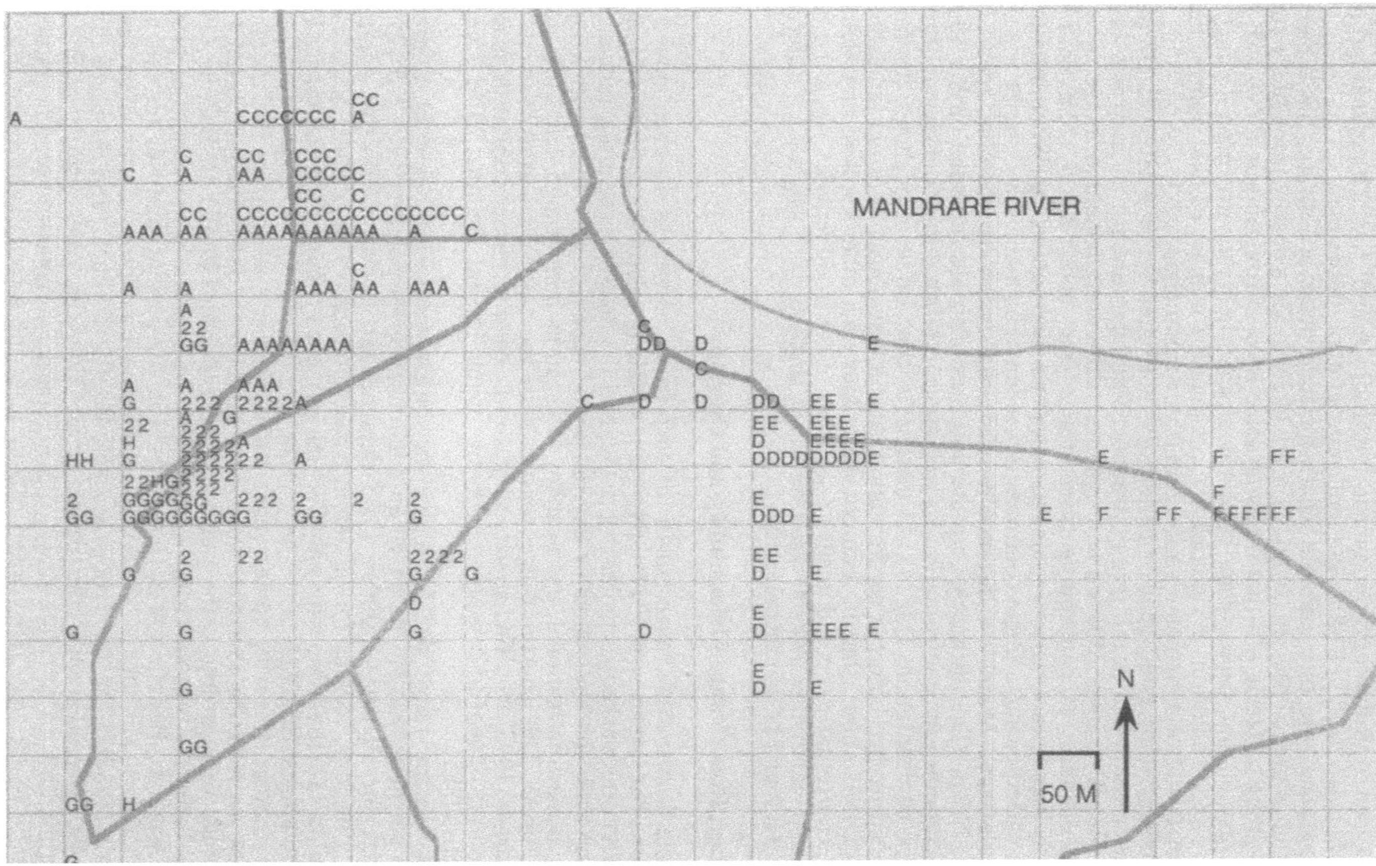


**Figure 4.** Positions of troops at scan following start of confrontations for all confrontations. 1/4 Hectare grid.

**Table 4.** Group size, day range, and velocity[1]

| Group | No. Adult Females in 1991 | Mean Day Range ± SD (m) | Mean Distance Moved per ½h ± SD (m) | N (Days) |
|---|---|---|---|---|
| A | 9 | 1503 ± 1124 | 56 ± 43 | 15 |
| C | 8 | 1712 ± 1732 | 67 ± 73 | 15 |
| D | 8 | 1165 ± 1060 | 45 ± 42 | 15 |
| E | 10 | 1404 ± 1201 | 49 ± 49 | 15 |
| F | 5 | 878 ± 890 | 33 ± 35 | 10 |
| G | 8 | 1372 ± 1289 | 54 ± 55 | 15 |
| G2 | 2 | 1543 ± 1478 | 58 ± 54 | 15 |
| H | 8 | 1441 ± 1186 | 53 ± 40 | 8 |

[1]Data from days when follows began before 07:00 and ended after 17:30. If observers lost groups for more than four hours, days were omitted from calculation.

**Table 5.** Estimates of home range size of *Lemur catta* troops.

| Group | Day Sample | Concave Polygon (ha) | Min. Convex Polygon (ha) | # Location Points |
|---|---|---|---|---|
| A | 1-5 | 3.59 | 8.80 | 385 |
| | 6-10 | 3.53 | 7.14 | 393 |
| | 11-15 | 3.12 | 9.27 | 390 |
| | Total: 15 | 6.59 | 12.37 | 1158 |
| C | 1-5 | 5.99 | 18.31 | 374 |
| | 6-10 | 4.29 | 11.44 | 372 |
| | 11-15 | 4.20 | 9.92 | 402 |
| | Total: 15 | 10.74 | 29.20 | 1141 |
| D | 1-5 | 3.98 | 8.80 | 384 |
| | 6-10 | 1.80 | 4.02 | 387 |
| | 11-15 | 4.73 | 13.14 | 387 |
| | Total: 15 | 6.84 | 20.14 | 1150 |
| E | 1-5 | 4.69 | 8.54 | 390 |
| | 6-10 | 2.67 | 6.35 | 396 |
| | 11-15 | 2.56 | 6.22 | 284 |
| | Total: 15 | 6.77 | 11.99 | 1062 |
| F | 1-5 | 2.95 | 5.59 | 389 |
| | 6-10 | 3.56 | 5.45 | 392 |
| | Total: 10 | 5.08 | 13.04 | 785 |
| G | 1-5 | 7.54 | 26.55 | 369 |
| | 6-10 | 8.58 | 20.73 | 388 |
| | 11-15 | 7.19 | 16.08 | 372 |
| | Total: 15 | 16.77 | 44.40 | 1121 |
| G2 | 1-5 | 2.48 | 6.63 | 387 |
| | 6-10 | 1.97 | 4.24 | 379 |
| | 11-15 | 1.94 | 3.08 | 384 |
| | Total: 15 | 6.53 | 12.21 | 1141 |
| H | 6-10 | 1.88 | 4.13 | 172 |
| | 11-15 | 3.59 | 8.34 | 309 |
| | Total: 8 | 4.40 | 7.96 | 464 |

The lemurs of this study apparently were visiting their whole ranges in 15 days, often in as little as one day. This means that the observed ranges can be taken as a basis for calculating defensibility. Mitani and Rodman (1979) and Martin (1981) argued that primates can effectively exclude others only if their day range allows effective monitoring of the home range. Mitani and Rodman's Defensibility Index (D) is d/d', where d is mean day range and d' is the diameter of an ideal circle equal in area to the observed home range. For this study with d = 1377m, d' equals 318.6m for concave polygons or 491.7m for convex polygons. The defensibility index calculated from these 15 day ranges is 4.32 or 2.80 for concave and convex areas respectively. This is extraordinarily high for primates. Anything over 1 allows territoriality, according to Mitani and Rodman.

**Table 6.** Home range overlap and activity in overlap zones.

| Pairs | % Range Overlap | % Time in Overlap Area |
|---|---|---|
| High Confrontation Pairs | | |
| A - C, A | 43 | 63 |
| A - C, C | 18 | 61 |
| G - G2, G | 14 | 48 |
| G - G2, G2 | 50 | 95 |
| D - E, D | 19 | 42 |
| D - E, E | 32 | 50 |
| A - G2, A | 28 | 59 |
| A - G2, G2 | 28 | 68 |
| Means ± SD | 29 ± 11 | 61 ±15 |
| Low Confrontation Pairs | | |
| C - D, C | 12 | 13 |
| C - D, D | 17 | 32 |
| G - H, G | 17 | 57 |
| G - H, H | 94 | 99 |
| E - F, E | 32 | 20 |
| E - F, F | 29 | 39 |
| G2 - H, G2 | 13 | 60 |
| G2 - H, H | 21 | 10 |
| A - H, A | 8 | 1 |
| A - H, H, | 5 | 1 |
| D - G2, D | 1 | 3 |
| D - G2, G2 | 13 | 18 |
| A - G, A | 27 | 29 |
| A - G, G | 8 | 4 |
| C - G2, C | 9 | 3 |
| C - G2, G2 | 21 | 15 |
| Mean ± SD | 20 ± 21 | 25 ± 26 |
| without G-H | 19 ± 9 | 18 ± 17 |

The defensibility index supports the hypothesis that the range is defensible, which concurs with the field observation that the lemurs typically cross their home range two or three times a day. Does this mean that they successfully exclude neighbors from a well-defended core area? On the contrary: there is a high degree of overlap between troops, even within five-day blocks.

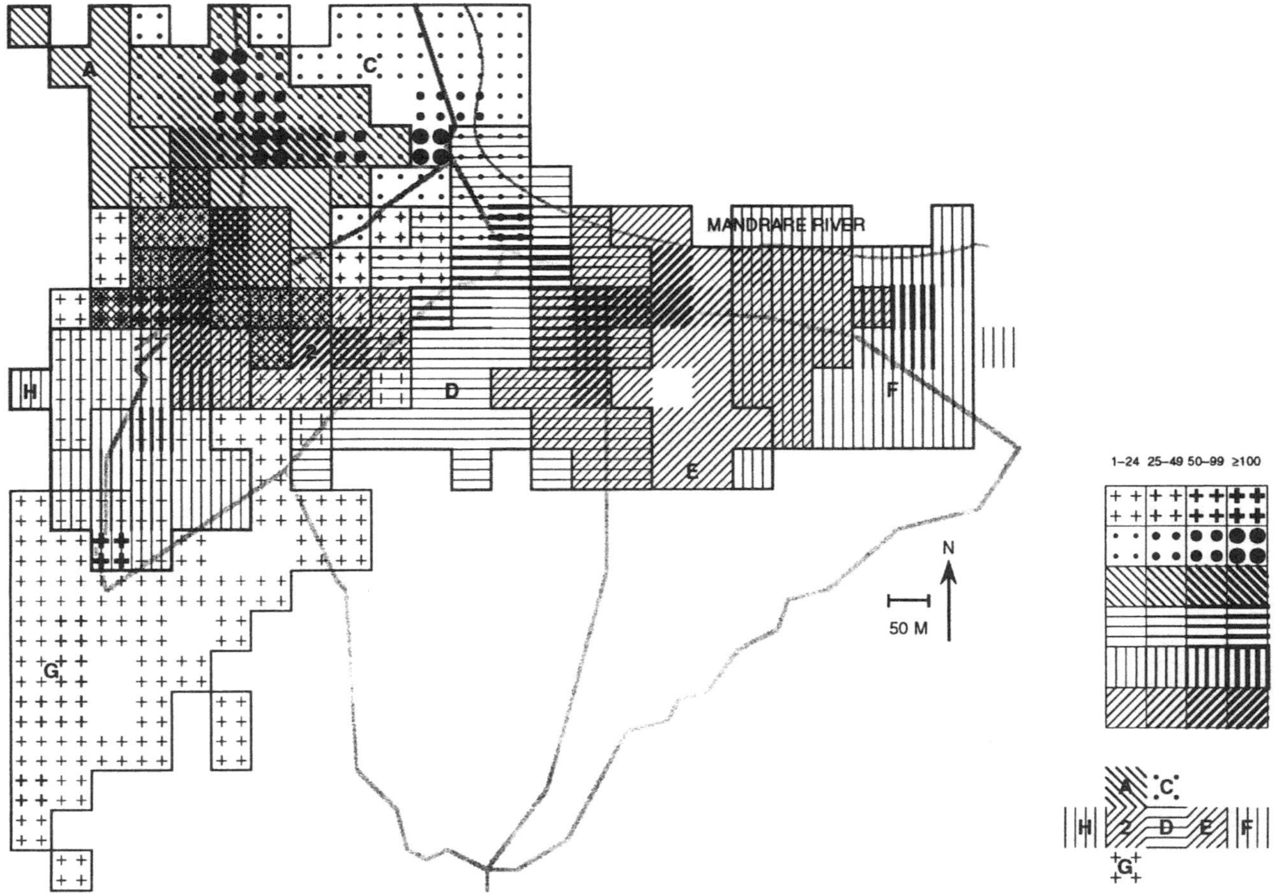


**Figure 5.** Total range use and intensity of use by the seven troops; 1/4 hectare grid.

The degree of overlap between neighbors is calculated from minimum convex polygons (Table 6). This table also shows percent of activity in the zones of overlap compared to the exclusive area. The mean percent of home range overlap in five day blocks is 42% ± 19%, and 55% ± 31% of the troops' time is spent in these overlap zones. HC pairs had 29% ±1% area overlap, and spent 61% ± 15% of their time in overlap zones. Thus the overlap zones were attractive to the HC lemur pairs, either for food and water supply, for vigilance in territorial defense, or for other reasons (Table 7). LC pairs varied widely. G and H by a fluke of geometry, and by G extending its range to drink and feed by the restaurant had 95% and 99% overlap of area and activity, respectivly. LC pairs other than H-G averaged 15% ± 9% overlap of area, and 18 ± 17% activity in the overlap zones.

**The Random Gas Model: Did Troops seek Each Other Out Or Avoid Each Other?**

Were lemur troops near each other more or less often than one would expect by chance? A variety of tests, including actually looking at animations of the data, indicate no reaction to mere proximity of another troop without confrontation, beyond what would follow from approaching the edge of the normal home range. The standard measure to test general approach or avoidance is the random gas model, which Waser (1976, 1977) adapted for use with primate ranging. The null hypothesis is that troops ricochet around inside their joint home range at random, like gas molecules in a two dimensional container.

What does the random gas model indicate when the two troops are not moving freely in a single overlapping space, but in two partially overlapping home ranges? The model is most appropriate for polyspecific encounters of troops with superimposed ranges, or such animals as orang-utans who seem to traverse each other's ranges at will (Whitesides, 1989; Mitani et al., 1991). Applied to the lemurs, it shows that troops are at a larger distance than expected, but it does not distinguish minute-to-minute avoidance of each other, from longer-term ranging in distinct areas. It does, however, show at what distance the avoidance is or is not significant. The distance turns out to differ sharply between HC and LC pairs.

The random gas model is calculated as:

$$Z = [8(p)(v)(d+s)]/3.14$$

Z is the expected frequency of troops being at or less than a given intergroup distance, d, where:

v, velocity = mean distance travelled per day (km) for each neighboring group pair
p, range size = two groups/cumulative home ranges (calculated from the concave polygons) of each neighboring pair
s, troop diameter = 40m (the model was relatively insensitive to changes in group spread so 40m was used throughout.)

Expected values were calculated for a range of intergroup distances (40-480m) in 40m intervals. Observed values were tallied from the observed intergroup distance data, and divided by the number of days both groups were simultaneously followed. Continuous records with the same measure of intergroup distance resulted from two groups sitting for long periods. Only the first record was used for such situations: the random gas model tests for "collisions" and does not allow for "sticking." Observed and expected values are cumulative over distance: therefore if A and H approach twice to 40 and once to 80m, the observed value to within 40m will be two and the observed value within 80m will be three.

Observed frequencies of intergroup approaches were tested against expected frequencies for 12 measures of intergroup distance (40-480m) for each neighboring pair, using a chi-squared test. Attraction between the troops, either for social reasons, protection, sharing or disputing the same resources at the same time, would results in figures which were greater than expected at near distances. The lemur figures all indicate troops which are not attracted to each other at very close range.

The different lemur troops space out to different degrees. Four of the LC troops are significantly infrequently near each other at close range, from 400m and under for A-H down to only 160m and under for G2-H. These troops, A-H, D-G2, C-G2, and G2-H, all have very little overlap of range. One would not expect them to often have the opportunity to meet or pass close to each other, the "avoidance" is merely an expression of divergent ranges. In contrast, the four HC pairs are near each other not significantly different from random expectation, even though their home ranges overlap only in part. Further, the two most confrontational pairs are significantly likely to be found as little as 160m, or four troop diameters distant from each other, as well as at greater distances. The HC again emerge as having a very different pattern from the LC, and to have far more overlap in both space and time (Table 8, Figs. 5, 6). It cannot be said, though, that they seek each other out in general, even though when they are in sight they may actually run together to confront.

In short, these are not animals who are mutually attracted in the sense that they associate, actively patrol the same boundaries at the same time, or that they are more likely than random to be found feeding in adjacent trees, even though the HC pairs have some 60% of their activity in overlap zones. In summary, lemurs interact and confront each other less than we thought. Given their small home ranges and the distance they move per day, they should bump into each other at a fairly high rate. However, this rate is not higher than expected by chance, even for HC troops. More importantly, when they do come together, groups avoid confrontations 70-80% of the time. They may indeed run toward each other at the start of a confrontation (see below) or linger afterward, but the safest conclusion is that they are found in proximity mainly as they travel to resources in their joint home range.

**Movements Associated With Intergroup Encounters**

How are intergroup distances and territories maintained? The most obvious means is by the confrontations themselves. On average, groups spent 76.1% of their time at distances greater than 160m from each other (the equivalent of 4 group diameters). It again appears that lemur groups generally avoid one another, in agreement with the random gas model. In the half-hour before and after a group encounter, distributions of intergroup distances differ from the profile of the whole population of intergroup distances (Figs. 8 and 9). Groups are closer together at 160m or less at 30min before an encounter, but also apparently linger in the vicinity in the 30min after an encounter. It is interesting that groups remain in very close proximity (≤40m) after an intergroup encounter in 20% of the encounters while, in general, groups spend only 3% of their time in close proximity.

When intergroup distances are separated into HC and LC pairs, the distances before confrontations are clearly different (Figs. 7 and 8). Twenty minutes beforehand LC pairs were more than 160m apart over 60% of the time, and less than 10% of the time at 80m, never at 40m. HC pairs were at 80m over 45% of the time, and at 40m over 5%. After a confrontation, both HC and LC were likely to remain at 40 or 80m in the first 10min, but by 30min after, 40% of the LC had retreated to over 160m, while the HC were still evenly distributed between 40, 80, 120, 160 and over 160m.

The population of turning angles illustrates the high rate of sitting or straight line movement overall (68%). For turning angles greater than 0, animals tend to move in a forward direction 1.5 times more often than they turn around overall. This distribution is the same in the time period 30min before and after an encounter (Fig. 9). At 20min and 10min before and after, groups are as likely to retreat as they are to move forward. At 10min before and after an encounter, there is a much lower proportion of zero turning (still or straight ahead), and a higher proportion of complete reversals (180° turns) (Fig. 9). Turning angles, however, do not seem to differ significantly when one separates confrontations of HC and of LC. Once committed to a confrontation the interaction has its own structure, whether the participants are HC or LC pairs.

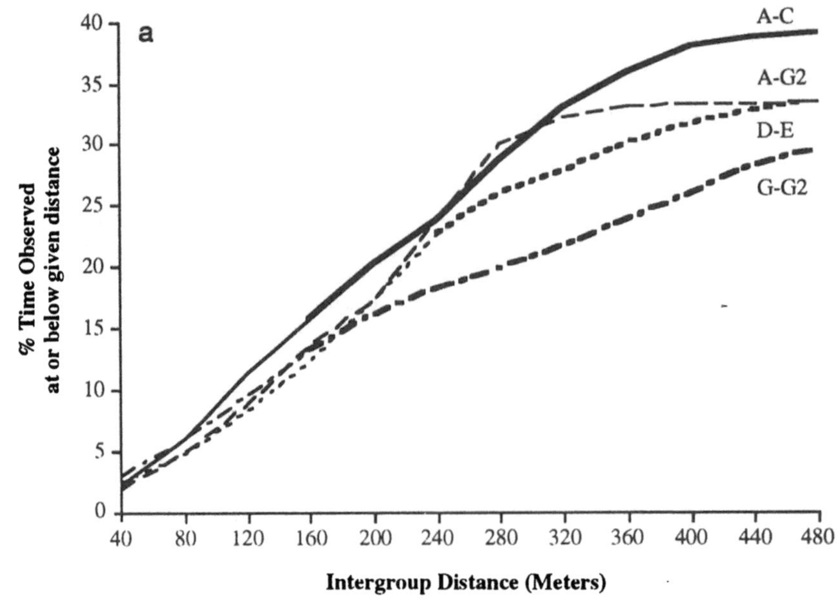


Bold line: Frequency of intergoup distance is significantly higher than expected.

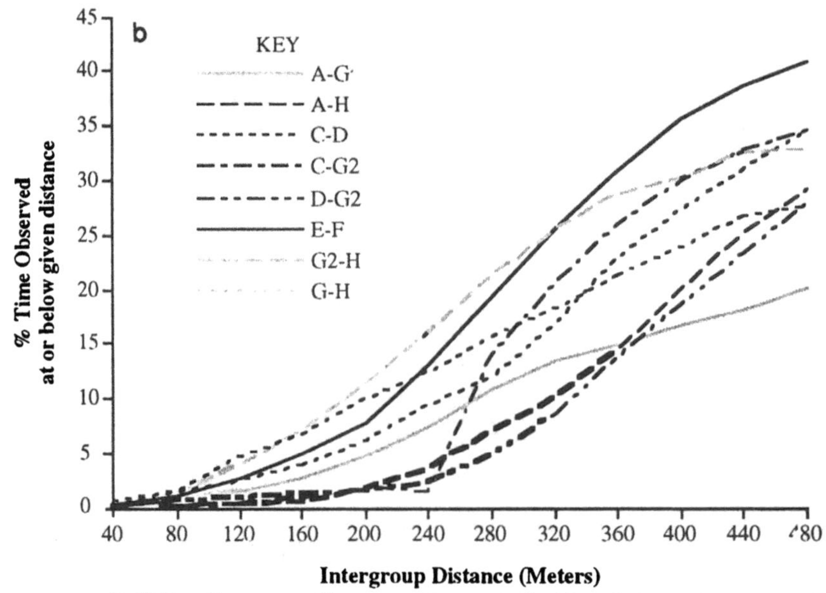


Bold line: Frequency of intergoup distance is significantly lower than expected.

**Figure 6.** **(a) High Confrontation Pairs: Frequency of Intergroup Distances. Bold line: Frequency of intergroup distance is significantly higher than expected.**
**(b) Low Confrontation Pairs: Frequency of Intergroup Distances. Bold line: Frequency of intergroup distance is significantly lower than expected.**

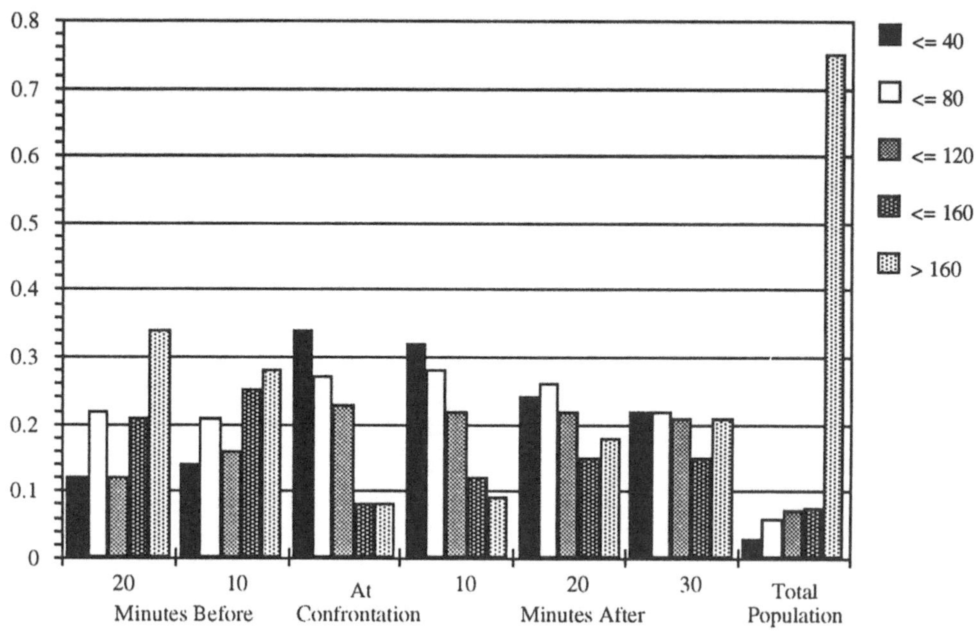


**Figure 7.** Intergroup Distance Before, During and After Confrontation: HC Pairs

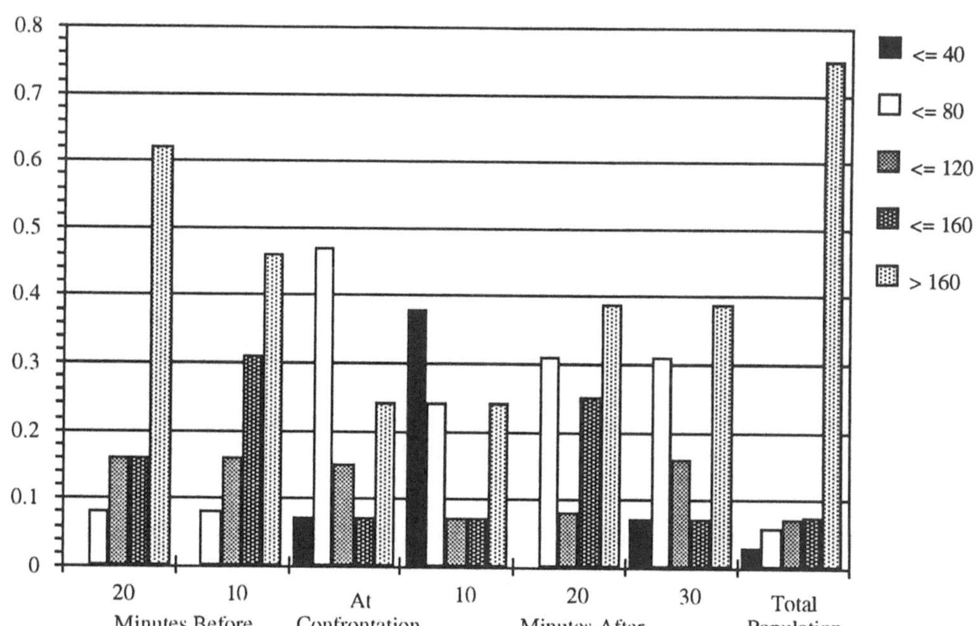


**Figure 8.** Intergroup Distance Before, During and After Confrontation: LC Pairs

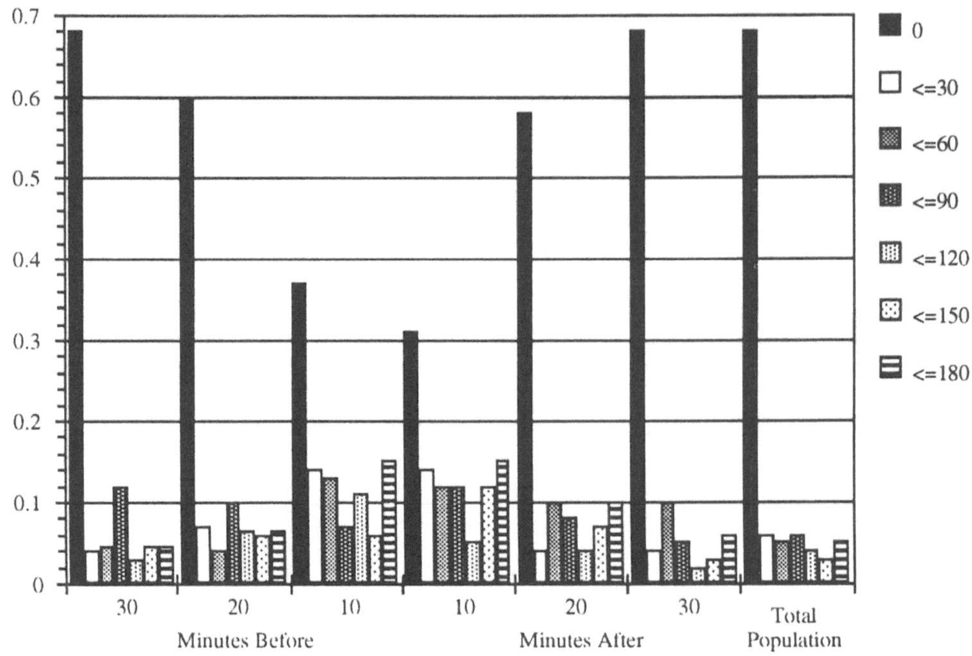


**Figure 9.** Turning Angles Before and After Confrontations

## Lack Of Movements Associated With Howls

Male ringtails give frequency modulated howls. The call is almost exclusively male, though females separated from their troops have twice been seen giving this call (Jolly, 1966; Sauther, 1991a,). The call is given most often at dusk, just before the troops' final or next-to-final move to sleeping trees. It may, however, be given at any time of day. The frequency varies by season, being commonest in the mating and immigration (birth) seasons (Jones, 1983). These are seasons both of troop conflict and of individual inter-male conflict between troops.

The howl is a long-distance communication or song of the sort that ethologists tend to assume to be a spacing call. It is usually followed by female calls from within the troop, and is commonly "answered" by males in other troops. However intergroup distances in the time periods before and after howls are essentially identical to the total population of distances. When one separates HC and LC groups, the HC groups tend to be at nearer distances, which we have already seen with the random gas model, but there is no movement either nearer or farther away after the howl. Turning angles similarly seem to be wholly unaffected by howling, LC or HC pairs do not differ in this respect. This large data set (for a primate study) gives the luxury of arriving at a negative conclusion, namely that howls do not seem to influence troop movement at short or intermediate time-frames.

## Resource Distribution in the Home Range: Water, Trees, Bananas and Drought

This study did not attempt to analyze resource distribution separately from ranging patterns of the lemurs themselves. However, animal territorial boundaries may occur at natural barriers or, in an opposite case, they may occur in rich resource areas which are worth both sides contesting. Klopfer and Jolly (1970) stated that *L. catta* range overlaps tend to be in rich areas. The map of activity within the range (Fig. 4) confirms that highly used areas lie in overlap zones.

First, drinking troughs (artificially provided for these troops) occur in the overlap between G, G2, and H, between A and C, C and D, and between D and E. Only one water trough was unequivocally "owned", by Troop C. The troughs do not occur only between HC rather than LC troop pairs, but they can clearly be the focus of troop battles, particularly between D and E. Although these artificial water points are highly patchy, they are not much more so than the natural water access. The reserve borders a river, but the points where lemurs can descend the steep bank to drink under cover of bushes or vines are also limited and contested between troops (Jolly, 1972).

Second, in a pilot attempt to identify highly favored trees, during the third five-day block, 119 trees were flagged where observed lemur troops fed in the course of three or more 10-min observation blocks. Only eight of these trees, or 7% were flagged for two different troops. Perhaps a more relevant measure is the proportion of flagged trees where two or more such long feeding bouts were recorded, since the trees had to be rich enough to be worth a repeat or continued visit by one or more troops: eight (20%) of multiple-flagged trees were shared and 31 others (80%) flagged more than once for a single troop. In terms of time spent feeding, 20% of the troops' time feeding in flag trees (52 of 258 flags) was spent on shared flag trees. This measure of resource use overlap in highly preferred food sources compares with the 42% home range overlap area or 55% of time spent in this overlap area. By any measures, the resources in shared zones were a substantial factor for the lemurs.

Results of the 1991 season are particularly interesting, because there were two major changes in circumstance from the previous year. There was no rain in the normal growing season. This was a serious El Nino drought year for the whole of southern Africa, including the semi-arid south of Madagascar. However, the fruiting and leafing of the gallery forest trees did not change dramatically in this first year of drought. There was no change in the pattern or distance of lemur ranging, although animals were visibly thin, with increased intra-troop aggression (see Discussion).

In 1991, there was also a seven month strike of most civil servants in Madagascar, leading to fears of violence and travel advisories by most foreign embassies. This meant that tourism dropped to near zero. Instead of 9000 tourists per year visiting Berenty, most armed with two to five bananas to feed the lemurs, the most favored troops might pick up five bananas per week.

What this meant in terms of percent of 10min intervals when troops fed on bananas was that in 1990, days one to five, A troop had bananas in 21% of observed intervals, C Troop 11%, D troop 5%, and the others 0-1%. In the second block of 1990, days 6-10, A troop had bananas in 6% of feeding intervals, C Troop 11%, D Troop 8%, G2 Troop 5%, and the rest 0-2%. In 1990, therefore three of the four highly agonistic pairs involved banana-provisioned troops. A proportion of the fights between A and C were directly provoked by tourists and their bananas, though A-G2 and G-G2 fought without the bananas, and D-E fought round their drinking trough.

However, the same troops (other than G-H) continued to be highly agonistic in 1991, particularly A-C and A-G2, even though the observed banana provisioning was 0-1% for every troop in the reserve. It thus appeared that the level of aggression was not related to immediate banana provisioning, or indeed to much provisioning over the previous six months. The total level of intertroop conflict dropped, though non-significantly, as the fights over bananas disappeared, but the sites of high inter-troop conflict remained the same.

## DISCUSSION AND CONCLUSIONS

### The Territorial System of *Lemur catta*

Territoriality may be defined as defense of a certain area in space. More precisely, the defender wins when the battle is on its own side of a recognized frontier, and loses when outside the frontier (Burt, 1943). Alternatively, territory may be defined as exclusive access to a particular area (Pitelka, 1949). Both versions are used by primatologists, as well as measures that approximate to "aggression fields" and "activity fields" (Waser and Wiley, 1979; Jolly, 1985).

Jolly (1985) remarks that "*Lemur catta*, the ringtailed lemur is a dubious case. Ringtails have exclusive core areas and intertroop aggression, but the range overlap varies widely from season to season, so they are best considered intermediates [between territoriality and non-spatial dominance]. They fit Wranghams (1980) argument that troops evolved as a means of dominating neighbors and forcing them off food patches in the short term." Mertl-Millhollen et al. (1979) also remark that *Lemur catta* can be called territorial but "one wonders at the ecological advantage afforded by a defended but not exclusively used territory."

Why, then, do we now consider the *Lemur catta* at Berenty territorial, when even with five day blocks the troops maintained only a mean of 58% exclusive area, and spent only 45% of their time in that exclusive area? The reason is that the zones of overlap were highly contested, and after battles troops retreated predictably toward the center of their home ranges. The ringtails thus attempted to prevent incursions toward their own territorial center and repulsed opponents toward the opponents own center rather than simply supplanting each other randomly at food resources.

The underlying purpose of the territorial system at Berenty does seem to be securing access to resources. Dunbar (1988) contrasts resource defense territoriality with mate defense territoriality in primates. As elaborated by van Schaik et al. (1992, p. 234):

> "If resource defense is the primary function of territoriality, several testable predictions follow. First, both sexes, but especially females, are expected to be involved in advertisement of occupancy and in range defense. Between-group encounters will usually be hostile, but approach-retreat interactions not involving overt aggression are also compatible with resource defence function. Second, range overlap should be small where ranges are defensible, resources are limiting, and costs of defense are moderate (cf. Mitani and Rodman, 1979)...Finally, as population density (and hence the importance of resource defense) increases, ranges should tend to become more exclusive (Dunbar, 1988). However, since high intruder pressure makes it impossible to maintain exclusive territories at the highest densities, a test of this prediction should ideally control for a measurement of intruder pressure."

In several respects, the Berenty *Lemur catta* conform to this description: overt female defense, inter-group hostility, high defensibility index. The exclusivity of range cannot be answered from these five day samples, but when put together with earlier studies (Mertl-Millhollen et al., 1979), it seems that the overlap at Berenty is over 50% for highly confrontational neighbors, much less for low conflict neighbors. Population density is about 250 per $km^2$ in the section of very rich gallery forest where this study was done (O'Connor, 1987). It seems that the lemurs are on the high side of both intrusion and defensibility. This is reasonable: Berenty is a naturally rich habitat island, a gallery forest on alluvial soil, which was originally surrounded by Madagascar's spiny desert. It has been modified by being both isolated and protected for the past 55 years, which means that the lemur populations have been able to grow until reaching their food limits without the possibility of either immigration or emigration. There is some predation by endemic raptors and by feral dogs and cats, but not by people--instead, people have if anything increased population by banana provisioning. Berenty quite plausibly represents an extreme upper limit to *Lemur catta* population density, intruder pressure, and resource territoriality.

This high-conflict territoriality may not occur in lower density sites, such as Beza-Mahafaly (Sauther and Sussmann, 1993). It might even differ in the lower-density scrub habitat at Berenty (Budnitz, 1978). Within the gallery forest areas, though, there were no differences in conflict rates between banana-provisioned troops and low-banana troops, or between provisioned troops in 1990 and the same troops when provisioning ceased in 1991. The system is very robust, continuing in quantitatively similar fashion from 1990 through 1991, and in at least qualitatively similar fashion from 1963 to 1992.

Instead of conflict obviously provoked only by spatially and temporally limited resources on particular frontiers but not on others, there may also be a historical or individual component to high-conflict zones. Four out of twelve of the neighbor pairs had significantly higher conflict rates, closer ranging, and more overlap in both 1990 and 1991 than the remaining groups. Perhaps the antagonism of particular neighbors relates to the phases of growth and expansion of troops as much as to disputes over particular resources, although there may be something in the long-term geography of these frontiers that also provokes conflict.

It is not possible to decide from these data whether any of these troops was winning ground over the long run. It is clear that most confrontations were standoffs in the short run. Even though one troop might retreat behind a fairly clear territorial line, the same troops would likely confront in the same area the next day or later on the same day. What we recorded was maintenance of a system, not expansion.

The behaviors that maintained the system were confrontations, including active approach prior to confrontations, and lingering within sight after active hostilities stopped. However, it does not seem that lemurs actively seek out other troops when out of sight: rather they meet frequently because at Berenty they cover so much of their home range each day. The defensibility index of day range divided by home range diameter was 4.32 or 2.80 for home ranges calculated as concave or convex polygons, respectively, either way one of the highest in primates (Mitani and Rodman, 1979).

It may be that lemurs as a group tend to defensibility. In Mitani and Rodman's original table, *L. catta's* defensibility index calculated from Sussman's 1974 study is 2.88, within the range of this study. Of the five species with higher indices than *L. catta* cited in Mitani and Rodman, three are lemuriforms. Of 27 species with lower indices, only two are lemuriforms.

**Between-Group and Within-Group Female Competition**

Large troops did not necessarily win ground in intertroop confrontations, *pace* Wrangham (1980). Within *L. catta* troops, usually only two or three of the females actively confront opposing troops (Gardner, pers. comm;, Wood, pers. comm.). This means that a small troop with two pugnacious females, such as G2, may repeatedly drive off larger troops such as G or A, with four to nine females. G2 must be considered a "successful" group, since it lasted at least since 1989, and raised infants. However, G2 spent much energy in confrontation, and had the least exclusive core area. The costs of territoriality for smaller or newer troops may be high. This may also be reflected in the fact that G2's day range was as high as the larger troops, although theoretically one might expect small troops to have smaller day ranges in this sort of patch-feeding species (Isbell, 1991).

The closed canopy forest at Berenty has much larger troops than those in the drier forest, or at the University of Madagascar study site at Beza Mahafaly (O'Connor, 1987; Sussman, 1991). Why have these larger troops formed, or stayed together, rather than splitting into "normal" lemur troops in even more minute territories, on the lines of G2 troop? The obvious reason is that the forest is rich, particularly where supplemented by tourist bananas, and that the ancient tamarind trees offer large patches where a troop of twenty lemurs can feed simultaneously.

During the drought and banana-dearth year of 1991, there was little difference in between-group conflict levels between troops from the preceding year. However, there was a marked increase in within-troop conflict. This within-troop conflict took the form of active harassment of subordinates, particularly over the point water sources, when the two lowest females would be actively chased back from the trough, often by a second-or third-ranking female. Jolly and Rasamimanana observed a sharp increase in this sort of behavior in five troops: Wood (pers. comm.) chronicled it to the point of possible incipient troop split in one troop. Harassed subordinate females were mostly allowed to follow the rest of the troop at a distance, or to drink following the last of the subordinate males, but in one incident observed by Jolly, the two exiled subordinates simply sat on a log and cried for half an hour (high-intensity miaows) while the troop retreated. This behavior is known as targeted aggression (Taylor and Sussman, 1985; Vick and Pereira, 1989). It has repeatedly been observed during the birth and premating seasons in the free-ranging groups at Duke Primate Center. Among well-fed troops at Duke, targeting occurred spontaneously as *Lemur catta* troops grew larger. Its simultaneous onset in so many troops at Berenty during the drought year suggests that it was a way of reducing troop size in response to shortage.

Assuming subordinates are usually related to dominants in the troop, the dominants have an interest in the subordinates' reproductive success as long as it does not threaten their own. Crudely (cruelly?) the subordinates are an optional extra for the dominants. For the dominants it may be better to get rid of a subordinate or two by targeting aggression than to have a real troop split, perhaps losing territory to a new and challenging neighbor (Koyama, 1991). Note all this is how it would work in a full, bounded habitat, with highly aggressive and effective between-troop resource defense. In areas where there was somewhere to migrate, or which were predator or disease-limited, the dynamics could be quite different, because territorial competition would be less important.

This picture of *Lemur catta* society does not quite fit either the Wrangham (1980) or van Schaik (1989) arguments for the effects of inter-troop versus intra-troop competition. The lemurs form female-bonded troops with high competition for patchy resources both within and between troops, so they ought to fit the theories developed for just this situation. However, they do not fit Wrangham's model because larger troops do not ordinarily win over smaller troops: territorial defense seems to depend on the energy of two or three females in each troop. Small fugitive troops with females losing ground just don't exist, unless one happens to see one on its way to oblivion; or one could call this Wrangham's size-advantage model with the losers' category usually empty.

Beyond this minimum, defense is not actually related to troop size. Above the threshold of troops existing and successfully defending a territory, we see something more like van Schaik's model, where subordinates remain in larger groups even though they have fewer resources, because something worse will happen to them if they leave. van Schaik, however, bases his argument on predation risk, while the chief risk for Berenty lemurs may be that if they leave, they will have no territory at all.

**Territoriality and Female-Male Relations**

*Lemur catta* are the extreme example of female dominance among primates. Lemuriforms as a group tend to have some degree of female dominance over males, but *Lemur catta* females win 100% of disputes with males, in whatever context (Kappeler, 1990; Pereira et al., 1990). Although female dominance was noticed long ago (Jolly, 1966), it is only recently being clarified as an evolutionary phenomenon. "Clarified" is perhaps too strong: there is continuing controversy over the "female need" hypothesis (Jolly, 1984; Richard, 1987; Young et al., 1990; Sauther, 1991a), the "female choice" hypothesis (Pereira and Weiss, 1991), and various "male deference" hypotheses (Hrdy, 1981; Kappeler, 1992). This symposium volume should further the discussion; in the meantime we all agree that female dominance happens.

Since we do not yet understand the relative importance of these various aspects of female dominance, we cannot assess the respective benefits and costs of male-female relations (O'Brien, 1991). It is clear that males do not decide territorial success, as for instance in wedge-capped capuchins (Robinson, 1988); *Lemur catta* territoriality seems to be largely female driven.

If the *Lemur catta* females have here a relatively "pure" system, the intensity of territorial conflict can reflect their own priorities. Territorial intrusions and conflict increase sharply in the birth season, when this study was done. Sauther (1991a) and Rasamimanana and Rafidinarivo (1993) conclude from their meticulous studies of energy and feeding that this season is the most demanding for female *Lemur catta*: they are in early lactation at the close of the dry season. The direct relation to new infants' survival is also suggested by an observation of Koyama's (1991). Koyama followed the origin of a new troop, C2, which fissioned from Troop C. C2 subsequently took over B Troop's territory, with the death of all B Troop infants.

**Do Males Have Any Role In Troop Spacing?**

Male *Lemur catta* are highly aggressive animals. They attack and wound each other during the mating season. They howl, swagger, scent-mark, and stink-fight at all times of the year. During the birth season a quarter of them may change troops, resisted by the resident males. There is some controversy and much yet to learn about how this translates into reproductive success, but there is no question that *Lemur catta* males are active and interesting animals, not simply led around by the dominant females.

The male howl is the sort of call which we automatically assume functions in spacing. It is a falling, frequency-modulated call, individually recognizable to other lemurs. It is structurally much like indri and gibbon wails, wolf howls, and apparently even the song of the grasshopper mouse (Hafner and Hafner, 1979). In a few cases, notably wolf howls and gibbon song, mangabey whoop-gobbles and mantled howlers' howls, it has been shown that such loud calls do function in spacing (Waser, 1977; Harrington and Mech, 1979, 1983; Mitani, 1985; Whitehead, 1987, 1989). On the other hand, red howler howling may primarily convey male advertisement, as do the male giggles of female-dominant spotted hyaenas (Sekulic, 1982; East and Hofer, 1991).

It is likely that many species or even populations have a mixed response strategy. One well-worked out case is the Tana River Mangabey (Kinnaird, 1992), where response to the male long call or whoop-gobble depended on the season. In months of patchy resources when territorial defense was both possible and profitable, troops retreated from whoop-gobbles, or at close range, converged and challenged each other. In seasons of abundant and shareable resources, troops merged to feed, and troop movement did not change on hearing a whoop-gobble. Even when troops merged, however, males at times herded receptive females away from the other group, and on first hearing a whoop-gobble would rapidly rejoin the females if they were feeding at a distance. Thus, underlying troop (female) relationship to food both used the male long call as a cue to spacing, and at times conflicted with male interest in herding the females.

The results of this study show that *Lemur catta* troops do not change direction or distance on hearing a male howl, even though this is a season of high territorial conflict. Neither do the males herd the females, who would not pay much attention. It seems that the "spacing" call plays no minute-to-minute role in spacing.

What does happen after a howl is that, first, females of the howler's troop may add on a chorus of high intensity meows. This would fit van Schaik's (1989) prediction that females would advertise that the troop is in its territory, for resource territoriality. Second, males of other troops may answer with howls of their own. At high intensity, several males in one troop may chorus together.

The simplest conclusion is that howling is almost wholly a male to male communication. Although it would not be ethical to experiment with this species, our prediction is that if the males of a troop suddenly lost their voices, then other males would invade within 24 hours. The call's function, then, would be to maintain the status quo among males; basically an indication that this troop is occupied. This would then be mate defense. If the females and males run separate but parallel inter-group systems, you might even argue for (or do research on) a whole parallel system of male mate-defense territoriality.

## ACKNOWLEDGEMENTS

First and always we thank the de Heaulme family for preserving Berenty and its lemurs for the past 55 years. We thank them also for welcoming researchers since 1963, and for their efforts to integrate nature preservation with human interests in Madagascar. Earthwatch and the 66 Earthwatch volunteers supported the field work financially, as well as doing most of the data collection. The Ecole Normale III of the University of Madagascar released the time of Dr. Rasamimanana and the Malagasy students. Wildlife Preservation Trust International helped support Dr. Crowley and the Malagasy students. Princeton University provided facilities for data analysis. Ben

Bolker analysed data from pilot studies in 1983. Some of the other people who contributed their time are listed in the "Methods" section; those whose names do not appear may recognize their troops' data points. We thank them all.

## REFERENCES


Budnitz, N., 1978, Feeding behavior of *Lemur catta* in different habitats, *in* "Perspectives in Ethology, Vol. 3," P.P.G. Bateson and P.H. Klopfer, eds., Plenum, New York.

Budnitz, N. and Dainis, K., 1975, *Lemur catta*: ecology and behavior, *in* "Lemur Biology," I. Tattersall and R.W. Sussman, eds., Plenum Press, New York.

Burt, W.H., 1943, Territoriality and home range concepts as applied to mammals, *J. Mammal.*, 24:346.

Dunbar, R.I.M., 1988, "Primate Social Systems", Croom Helm, London.

East, M.L., and Hofer, H., 1991, Loud calling in a female-dominated mammalian society: II. Behavioural contexts and functions of whooping of spotted hyaenas, *Crocuta crocuta*, *Anim. Behav.* 42:651.

Hafner, M.S., and Hafner, D.J., 1979, Vocalisations of grasshopper mice (Genus *Onychomus*), *J. Mammal.*, 60:85-94.

Harrington, F.H. and Mech, L.D., 1979, Wolf howling and its role in territory maintenance, *Behaviour*, 68:207.

Harrington, F.H. and Mech, L.D., 1983, Wolf pack spacing: howling as a territory-independent spacing mechanism in a territorial population, *Behav. Ecol. Sociobiol.*, 12:161.

Hrdy, S.B., 1981, "The Woman that Never Evolved," Harvard Univ. Press, Cambridge.

Isbell, L.A., 1991, Contest and scramble competition: patterns of female aggression and ranging behavior among primates, *Behav. Ecol.*, 2:143.

Jolly, A., 1966, "Lemur Behavior: A Madagascar Field Study," Chicago Univ. Press, Chicago.

Jolly, A, 1972, Troop continuity and troop spacing in *Propithecus verreauxi* and *Lemur catta* at Berenty, Madagascar, *Folia Primatol.*, 17:335.

Jolly, A., 1984, The puzzle of female feeding priority, *in* "Female Primates: Studies by Women Primatologists," M. Small, ed., Liss, New York.

Jolly, A., 1985, "The Evolution of Primate Behavior", MacMillan, New York.

Jones, K.D., 1983, Inter-troop transfer of *Lemur catta* males at Berenty, Madagascar. *Folia Primatol.*, 39:115.

Kappeler, P.M., 1990, Female dominance in *Lemur catta*: more than just female feeding priority? *Folia Primatol.*, 55:92.

Kappeler, P.M., 1992, "Female Dominance in Malagasy Primates," PhD Thesis, Department of Zoology, Duke Univ., Durham, N.C.

Kinnaird, M.F., 1992 Variable resource defence by the Tana River crested mangabey, *Behav. Ecol. Sociobiol.*, 31:115.

Klopfer, P.H. and Jolly, A., 1970, The stability of territorial boundaries in a lemur troop, *Folia Primatol.*, 12:199.

Koyama, N., 1988, Mating behavior of ring-tailed lemurs (*Lemur catta*) at Berenty, Madagascar, *Primates* 29: 163.

Koyama, N, 1991, Troop division and inter-troop relationships of ring-tailed lemurs (*Lemur catta*) at Berenty, Madagascar, *in* "Primatology Today", A. Ehara, T. Kimura, O. Takenaka, and M. Iwamoto, eds. Elsevier, Amsterdam.

Martin, R.D., 1981, Field studies of primate behaviour, *Symp. Zool. Soc. Lond.*, 46:287.

Mertl-Millhollen, A.S., 1988, Olfactory demarcation of territorial but not home range boundaries by *Lemur catta*, *Folia Primatol.*, 50:175.

Mertl-Millhollen, A.S., Gustafson, H.L., Budnitz, N., Dainis, K., and Jolly, A., 1979, Population and territory stability of the *Lemur catta* at Berenty, Madagascar, *Folia Primatol.*, 31:106.

Mitani, J.C., Grether, G.F., Rodman, P.S., and Priatna, D., 1991, Associations among wild orang-utans: sociality, passive aggregations or chance? *Anim. Behav.*, 42:33.

Mitani, J.C. and Rodman, P.S., 1979, Territoriality: the relation of ranging pattern and home range size to defendability, with an analysis of territoriality among primate species, *Behav. Ecol. Sociobiol.*, 5:241.

O'Brien, T.G., 1991, Female-male social interactions in wedge-capped capuchin monkeys: benefits and costs of group living, *Anim. Behav.*, 41:555.

O'Connor, S.M., 1987, The Effect of Human Impact on the Vegetation and the Consequences to Primates in Two Riverine Forests, Southern Madagascar. PhD Thesis, Department of Applied Biology, Cambridge University, Cambridge.

Pereira, M.E., Kaufman, R., Kappeler, P.M., and Overdorff, D.J., 1990, Female dominance does not characterize all of the Lemuridae, *Folia Primatol.*, 55:96.

Pereira, M.E. and Weiss, M.L. 1991., Female mate choice, male migration, and the threat of infanticide in ringtailed lemurs, *Behav. Ecol. Sociobiol.*, 28:141.

Pitelka, F.A., 1949, Numbers, breeding schedule and territoriality in pectoral sandpipers in northern Alaska, *Condor*, 61:233.

Rasamimanana, H.R. and Rafidinarivo, E., (1993), Feeding behavior of *Lemur catta* females in relation to their physiological state, *in*: "Lemur Social Systems and Their Ecological Basis," P.M. Kappeler, and J.U. Ganzhorn, eds., Plenum Press, New York.

Richard, A.F., 1985, "Primates in Nature," Freeman, New York.

Richard, A.F., 1987, Malagasy prosimians: female dominance, in "Primate Societies," Smuts, B.B., Cheney, D.L., Seyfarth, R.M., Wrangham, R.W., Struhsaker, T.T., eds, Chicago Univ. Press, Chicago.

Robinson, J.G., 1988, Group size in wedge-capped capuchin monkeys *Cebus olivaceus* and the reproductive success of males and females, *Behav. Ecol. Sociobiol.* 23:187.

Sauther, M.L., 1991a, Reproductive behavior of free-ranging *Lemur catta* at Beza Mahafaly Special Reserve, Madagascar, *Amer. J. Phys. Anthropol.*, 84:463.

Sauther, M.L., 1991b, Reproductive costs in free-ranging ring-tailed lemurs, *Am. J. Primatol.* 24:133 (Abstract).

Sauther, M.L, and Sussman, (1993), A new interpretation of the social organization and mating system of the ringtailed lemur (*Lemur catta*), *in*: "Lemur Social Systems and Their Ecological Basis," P.M. Kappeler, and J.U. Ganzhorn, eds., Plenum Press, New York.

Sekulic, R., 1982, The function of howling in red howler monkeys (*Alouatta seniculus*), *Behaviour*, 81:38.

Sussman , R.W., 1974, Ecological distinctions of sympatric species of Lemur, *in* "Prosimian Biology," R.D. Martin, G.A. Doyle and A.C. Walker, eds., Duckworth, London.

Sussman, R.W., 1991, Demography and social organization of free-ranging *Lemur catta* in the Beza Mahafaly Reserve, Madagascar. *Am. J. Phys. Anthropol.*, 84:43.

Taylor, L.L. and Sussman, R.W., 1985, A preliminary study of kinship and social organization in a semi-free ranging group of *Lemur catta*, *Int. J. Primatol.*,6:601.

van Schaik, C.P., 1989, The ecology of social relationships amongst female primates. *in* "Comparative Socioecology," V. Standen and R. Foley, eds., Blackwell, Oxford.

van Schaik, C.P., Assink, P.R., and Salafsky, N., 1992, Territorial behavior in southeast Asian langurs: resource defense or mate defense? *Amer. J. Primatol.*, 26:233.

Vick, L.G., and Pereira, M.E., 1989, Episodic targeting aggression and the histories of *Lemur* social groups. *Behav. Ecol. Sociobiol.*, 25:3.

Waser, P.M., 1976, *Cercocebus albigena*: site attachment, avoidance, and intergroup spacing. *Amer. Nat.*, 110:911.

Waser, P.M., 1977, Individual recognition, intragroup cohesion, and intergroup spacing: evidence from sound playback to forest monkeys, *Behaviour*, 60:28.

Waser, P.M. and Wiley, R.H., 1979, Mechanism and evolution of spacing in animals, *in*: "Handbook of Behavioral Neurobiology", vol. 3, P. Marler and J.G. Vandenberg, eds., Plenum, New York.

Whitehead, J.M., 1987, Vocally mediated reciprocity between neighboring groups of mantled howling monkeys (*Alouatta palliata palliata*), *Anim. Behav.* 35:1615.

Whitehead, J.M., 1989, The effect of the location of a simulated intruder on responses to long-distance vocalizations of mantled howling monkeys (*Alouatta palliata palliata*), *Behaviour*, 108:73.

Whitesides, G., 1989, Interspecific associations of Diana monkeys, *Cercopithecus diana* in Sierra Leone, West Africa: biological significance or chance? *Anim. Behav.*, 37:760.

Wrangham, R.W., 1980, An ecological model of female-bonded primate groups, *Behaviour*, 75:262.

Young, A.L., Richard, A.F., and Aiello, L.C., 1990, Female dominance and maternal investment in strepsirhine primates. *Amer. Nat.*, 135:473.

# A NEW INTERPRETATION OF THE SOCIAL ORGANIZATION AND MATING SYSTEM OF THE RINGTALED LEMUR (*LEMUR CATTA*)

Michelle L. Sauther and Robert W. Sussman

Department of Anthropology
Washington University
St. Louis, MO 63130
U.S.A.

## ABSTRACT

Long-term and intensive studies of identified individuals have enabled us to reexamine the social structure and organization of *Lemur catta*. Ringtailed lemurs have highly overlapping home ranges which vary in size depending upon season and habitat, and should not be considered strictly territorial. This species lives in female-resident, multimale groups centered around one dominant female who appears to be the focal point of other group members. Groups average around 13 individuals (range: 5-27) with generally equal adult sex ratios. Upon reaching adulthood, all males emigrate from their natal group and older males transfer between groups at an average of every 3-5 years, though some males have remained in the same group for at least 6 years. Males within a group can be differentiated based on natality, social status, and relationships with females. In groups intensively studied, a single, non-natal "central" male has been identified. Such males interact with females at greater rates, and are the first to mate. Females exhibit mate choice, rejecting closely related males, and actively mating with non-troop males, which results in a number of male mating strategies.

## INTRODUCTION

Recent longterm studies of groups of ringtailed lemurs at Beza Mahafaly Special Reserve, Madagascar have allowed us to update and clarify certain aspects of the social structure and organization of *Lemur catta*. Beza Mahafaly reserve has been maintained as a guarded reserve since 1978. It consists of two parcels, one which contains a continuous deciduous and semideciduous riverine forest. This parcel is inhabited by 9 groups of ringtailed lemurs which have been collared and tagged so that all adults are individually identifiable. The habitat is very seasonal, with a specific hot/wet season and a cool/dry season. Social groups of ringtailed lemurs were initially characterized as multi-male, multi-female, with a central core of females, juveniles, and high-ranking males moving together, and a close-knit subgroup of subordinate adult males ("the Drones' Club") lagging behind the core of the group (Jolly, 1966). Now, studies of marked and identified groups at Beza Mahafaly have allowed us to refine this description. *Lemur catta* has also been characterized as territorial (e.g., Klopfer and Jolly, 1970; Mertl-Millhollen et al., 1979; Mertl-Millhollen, 1988). We believe this is not true, and involves a problem of

*Lemur Social Systems and Their Ecological Basis*, Edited by
P.M. Kappeler and J.U. Ganzhorn, Plenum Press, New York, 1993

definition. Recently, it has been theorized that the social organization of ringtailed lemurs is the result of adaptations to prevent infanticide (Pereira and Weiss, 1991). We discuss an alternative theory which we believe to be more representative of what is currently known of ringtailed lemur mating systems (Sauther, 1991).

## Social Structure

**Spatial Relationships of *Lemur catta* Groups.** By social structure, we are referring to the size and composition of groups (group structure) and the population demography and dispersal of groups within the population (population structure) (Rowell 1972, 1976). *Lemur catta* groups average 13 to 15 individuals and range in size from 5 to 27. Adult sex ratios are close to 1.0 (Jolly, 1966, 1972; Sussman, 1974, 1991; Budnitz and Dainis, 1975; Mertl-Milhollen et al., 1979; Jolly et al., 1982). At Beza Mahafaly, where we have followed 9 groups for 6 years, the number of both adult males and females within groups have ranged from 2 to 8 individuals. In 53 groups censused in the wild, the average number of adult males per group is 4.0 and that of females is 4.75 (Jolly, 1966; Sussman, 1974, 1991; Budnitz and Dainis, 1975). Eighty to eighty-five percent of females give birth each year, and mortality rate for the first year is between 30 and 50%, depending upon the year and locality (Mertl-Millhollen et al., 1979, Jones, 1983; Sussman, 1991, 1992).

Home ranges of ringtailed lemurs overlap, with little or no areas of exclusive use (Sussman, 1991). The size of ranges varies with habitat and location, with averages at different localities ranging from 10 to 32ha (range: 6-35ha). At different sites, population densities vary from 17.4 to 350 animals per km$^2$, and biomass has been estimated between 50 and 700kg per km$^2$.

Larger home ranges and lower densities are found in drier or more disturbed habitats (Budnitz and Dainis, 1975; O'Connor, 1987; Sussman, 1991). For example, at Beza Mahafaly groups living in drier habitats have home ranges averaging 32ha, whereas those in wetter habitats average 17ha. Group sizes in the two habitats are not significantly different. However, groups with small home ranges expand their range during certain months to feed on particular trees that come into fruit and which are absent from their normal home range (Fig. 1).

There has been some confusion as to how to define the spatial relationships of ringtailed lemurs. For example, using earlier reports from Berenty, Mitani and Rodman (1979) listed *Lemur catta* as the only primate species with multimale groups that is territorial. Because groups at Berenty have highly overlapping ranges, Jolly (1985:151) considered ringtailed lemurs to be a "dubious case" for territoriality and that "they are best considered intermediates like the Nairobi baboons (Mertl-Millhollen, et al., 1979, Jolly, 1966)". Further she stated (Jolly 1985:145): "We keep returning to the degree of overlap of range as the final criterion of territoriality". Currently, however, Jolly et al. (1993) consider ringtails to be territorial. We believe that this confusion is related to the lack of a reliable operational definition of territoriality, which often makes the meaning of this concept quite different in different contexts. In this light, we propose that territoriality should refer to "the active defense of individual or group home range boundaries by actual or ritualized agonistic encounters, thereby maintaining essentially *exclusive use* of the home range." (Waser and Wiley, 1980). Using this definition, we would not consider ringtailed lemurs to be territorial. There is considerable and in some cases almost total home range overlap among both Berenty and Beza Mahafaly groups, and, as Jolly (1972) has described, a number of troops may time-share sites for feeding, sleeping or resting. Furthermore, at Beza Mahafaly, groups will expand their home ranges into other group's ranges when certain rare resources are unavailable in their own home range. However, as Jolly et al. (1993) have pointed out there are core areas which are more intensely used by *L. catta* groups at Berenty, and at Beza Mahafaly these can change seasonally (Sauther, unpubl. data), but even these are used by other groups.

**Intergroup Encounters.** Intergroup encounters do occur but are most likely related to the defense of important but seasonally available resources (Sauther, 1992), and the maintenance of group integrity (Sussman, 1974). Table 1 depicts the

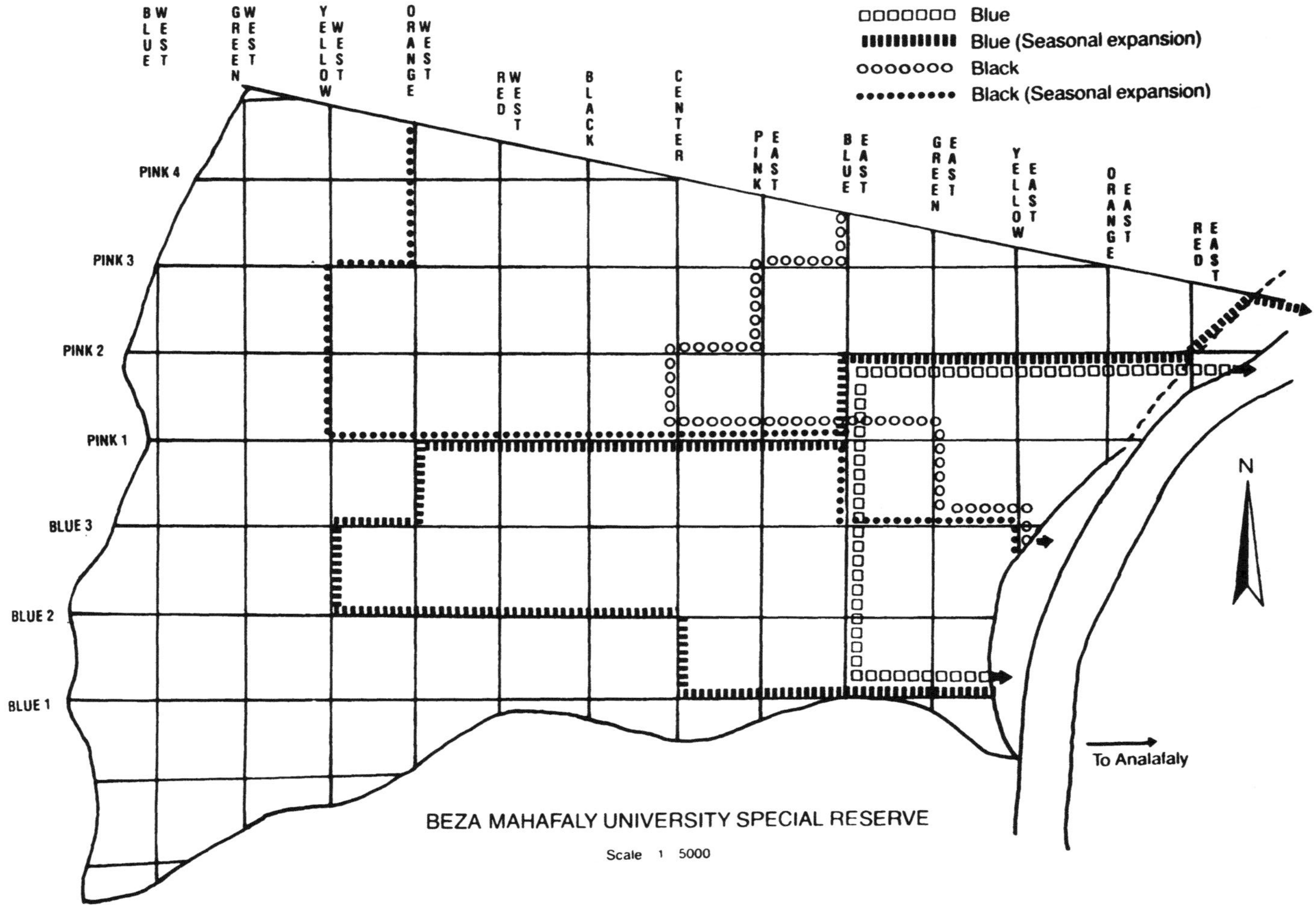


**Figure 1.** Map Showing Seasonal Home Range Expansion During October and November for Two Groups at Beza Mahafaly Reserve.

results of intergroup encounters between ringtailed lemurs at Beza Mahafaly during a 13 month period. The percentage of wins are based on where the encounter occurred. Of the 58 intergroup encounters the majority of wins (57%), occurred within the winner's most intensively (but not exclusively) used portion of their own home range. "Draws" (28%), where neither troop wins (i.e., feeds at the contested site), most often occurred in less commonly used areas. Group size had little effect on the outcome of such encounters. Larger troops only won when the groups fought in equally used portions of the ranges.

**Table 1.** Results of intergroup encounters between ringtailed lemurs at Beza Mahafaly.

| Location | Result of Intergroup Encounters (N=58) | |
|---|---|---|
| | Win | Draw |
| Own Home Range | 57% | 0 |
| Another Troop's Home Range | 5% | 0 |
| Area Where Home Ranges Overlap: | | 28% |
| Bigger Group | 9% | 0 |
| Smaller Group | 1% | 0 |

## Social Organization

**Spatial Relationships of Individuals Within Groups.** By social organization, we refer to the pattern of social interactions which occur between individuals and between groups. Like many cercopithecine monkeys, *Lemur catta* lives in multi-male, female resident groups in which females remain in their natal group and males migrate (Sussman, 1992). Groups are centered around a core of adult females and their offspring. Furthermore, females are organized along matrilineal kinship lines with more friendly social interactions occurring between close relatives and more serious aggressive encounters taking place between nonrelatives (Taylor and Sussman, 1985; Taylor, 1986).

Males regularly migrate into and out of the marked study population. We assume many missing males have moved into unmarked groups (see Sussman, 1992). Males migrate from their natal group between 3 to 5 years of age. On average, mature males migrate every 3.5 years, with approximately 30% migrating each year (Sussman, 1992). At Beza Mahafaly, after 6 years (between 1987 and 1992) only 7 adult males out of 43 (16%) are still in their original group, and only 14 originally tagged males (33%) are still in the census population (as opposed to 62% of the originally tagged females, all of which are in their original group).

For the Beza Mahafaly groups, there are a number of males which can be differentiated by their closeness to the central core of females and by a higher percentage of friendly interactions with group females. These include lower-ranking, adult-sized natal males (3-year olds) who are often found in close association with their mothers. One or more non-natal "central males" are also found in this core group. Among these males, there is one agonistically dominant male that has feeding and mating priority over other males. In all groups studied at Beza Mahafaly, the dominant male has been in the prime adult age category (see Sussman, 1991). Peripheral males are not a cohesive subgroup but include lower-ranking males and males attempting to migrate into the group. The latter are the most peripheralized animals and are challenged by other group members of both sexes for many months.

**Dominance Relationships.** There are dominance hierarchies within both captive and free-ranging groups of *L. catta* (Taylor and Sussman, 1985; Taylor, 1986; Sauther, 1992) and ringtailed lemurs have been characterized as female dominant (Jolly, 1966; Budnitz and Dainis, 1975; Taylor, 1986). Dominance hierarchies were based on winners/losers in agonistic encounters and approach-retreat behavior, which both gave the same hierarchies. A single, top-ranking female appears to be the focal point for the rest of the troop. She often initiates the direction of group progressions, although as Jolly (1966) has pointed out, other individuals (including low-ranking adult males) may lead the actual travel.

Top-ranking males and females may leave the troop on occasions, and travel to where other troops are located. At Beza Mahafaly the importance of the top-ranking female was highlighted when on two occasions she left the troop and travelled several hundred meters alone. She was separated from her troop for 45 minutes. During this time the rest of the troop became visibly agitated, and began to frantically contact call when she disappeared. Departure of the dominant male did not elicit such behavior, nor did the temporary absence of several lower ranking females who on one occasion became separated from the rest of the troop.

Ringtailed lemur groups characteristically contain more than one matriline. For example, in the two intensively studied groups at Beza Mahafaly, two or more adult females were accompanied by one adult offspring, a juvenile and an infant. In captivity some matrilines of a single *L. catta* group are dominant to others, and if a group becomes too large one matriline may be evicted by another (Taylor and Sussman, 1985; Taylor, 1986). In several macaque species groups can fission, with low-ranking matrilines forming new groups (Chepko-Sade and Sade, 1979; Koyama, 1970). It is possible that group fission, which has been observed at Beza Mahafaly (Sussman, 1991, 1992), also may be along matrilines. Alternatively, if a matriline is too small to establish itself as a new group, it may transfer into another group. This may explain a single case of female transfer at Beza Mahafaly involving a mother-daughter pair who entered a new group (which contained only four adult females) and immediately became the two top-ranking females in the group (Kaufmann, pers. comm.). After this transfer two of the original females disappeared. In all cases of male transfer, the males hold low-ranking and peripheral positions upon entering a new group.

Among males, although there is one male who dominates all other males in priority of access to food and who has mating priority, there does not seem to be a consistent linear hierarchy (Budnitz and Dainis, 1975; Taylor, 1986; Kappeler, 1990; Sauther, 1992; L. Gould, pers. comm.). This may be because the adult male component of a group is constantly changing. For example, over a six year period, only three males have maintained their dominant position in the nine study groups at Beza Mahafaly. Furthermore, there are a number of different "types" of males in a group at any one time. A group might contain young adult natal males, young adults after their first transfer, prime adult multi-transfer males, old subordinate males, and males in the process of transferring into the group. From our preliminary data at Beza Mahafaly we suspect that the "central males" include prime adult and some natal young adult males from high status matrilines, whereas the peripheral males are mainly young first transfers, recent multi-transfer male immigrants, and older males. The fact that three males were able to maintain their dominance position for 6 years shows a pattern more similar to male rhesus macaques than to male baboons, with some macaque males maintaining alpha position within a group for 5 to 10 years (Bercovitch, 1991). Males currently attempting to immigrate into the groups are the most peripheralized animals and are challenged by both group females and males for many months. Thus, a cohesive subgroup of subordinate males, the "Drones Club" described by Jolly (1966) for the Berenty groups have not been seen at Beza Mahafaly. Although males form partnerships while transferring (Jones, 1983; Sussman, 1992), we do not know if any male-male alliances continue once group membership is attained.

**Aggressive Interactions.** Although Budnitz and Dainis (1975) report that males at Berenty were involved in more agonistic encounters than females, the actual percentage based on their Table 3 shows that males and females were involved in similar percentages of agonism overall. At Beza Mahafaly, females were involved in significantly higher mean percentages of agonism than males. This held true for both feeding agonism as well as agonism over resting, sunning and drinking sites (Table 2). This suggests that not only are females dominant in this species, but that they are actively involved in a greater percentage of agonistic encounters than are males. Male aggression, on the other hand, was greater than that of females only for the month of April, just prior to the mating season in May.

**Table 2.** Percentage of agonism involving female and male ringtailed lemurs at Beza Mahafaly.

| Green Troop | Feeding | Non-Feeding |
|---|---|---|
| Females | 59 | 56 |
| Males | 41 | 44 |
| Females *vs* Males | $t=7.46$; $p<0.05$* <br> $N=1375$ | $t=4.76$; $p<0.05$ <br> $N=785$ |
| **Black Troop** | **Feeding** | **Non-Feeding** |
| Females | 71 | 64 |
| Males | 29 | 22 |
| Females *vs* Males | $t=14.21$; $p<0.05$ <br> $N=555$ | $t=3.77$; $p<0.05$ <br> $N=98$ |

*Randomization procedure (see Manly, 1991; Sauther, 1993). Sample sizes were Green Troop: 5 females and 5 males; Black Troop: 4 females and 2 males.

In the two groups studied intensively at Beza Mahafaly the top-ranking female was responsible for the greatest number of aggressive encounters overall. In both troops the top-ranking female also won all such encounters. The majority of these involved access to food.

Although adult females are dominant to males, at Beza Mahafaly, in depth observations of identified individuals revealed that female dominance is gradually established, and appears to become stable after a female's first mating season. Prior to their first estrus, young females of adult body size (aged 2 years) were sometimes displaced from feeding and drinking sites by adult males (Sauther, 1992). Approximately 3% of all male-female agonistic dyads involved males displacing females. These were non-aggressive displacements. Males would simply move females out of the way with their bodies. During the mating season these females were repeatedly approached by adult males, and young, non-receptive females quickly learned that the only way to curtail such unwanted advances was to aggressively cuff or chase the male away. After the mating season these same females dramatically increased the percentage of feeding displacements directed toward males (pre- *vs* post-mating: $x=0.08\%$ *vs* $3\%$, $N=3$ females, 7 males, $t=2.74$, $P<0.0005$, randomization procedure). Such consistent and large increases in feeding agonism directed toward males did not occur among older adult females (pre- *vs* post-mating: $x=3.2\%$ *vs* $2.7\%$, $N=6$ females, 7 males, $t=0.40$, $P=0.38$, randomization procedure).

**Mating Season.** Jolly (1966) and Budnitz and Dainis (1975) both report that subordinate males successfully mated with troop females at Berenty during the brief mating season. While this has also been observed at Beza Mahafaly, the actual order of mating reflected the dominance hierarchy of non-natal males. In all observed cases the most dominant, central male was the first to mate and ejaculate when a female initially became receptive. The second ranked male mated next. The next individuals to mate were transferring males and/or non-group males. Thus, the dominance hierarchy did not break down with respect to the *order* of mating for the two top ranking males (Sauther, 1991). Koyama (1988) observed a similar pattern at Berenty, where a single high-ranking male was the first to mate with receptive females. Male rank can change, however, and in one recent case during the mating season the third ranked male moved up to the top ranking position and was then the first to mate (L. Gould., pers. comm.).

Observations during the mating season at Beza Mahafaly support other reports from captive groups (Taylor and Sussman, 1985; Taylor, 1976, Pereira and Weiss, 1991) that female mate choice is an important reproductive strategy in ringtailed lemurs. Female mate choice is defined here as a female rejecting, (i.e., cuffing, chasing away, sitting down) copulation attempts of certain males, but a female approaching and presenting to other males after the onset of her receptivity. Of the 17 matings observed during the 1988 mating season, 47% of these involved an initial approach by the female, and in spite of vigorous and persistent harassment by group males, 35% of these matings involved mating with males of other groups.

**Mating Strategies.** It has been recently suggested that infanticide may be an adaptive aspect of the mating system of ringtailed lemurs (Pereira and Weiss, 1991). In semi-free-ranging groups of ringtailed lemurs at the Duke University Primate Center, females have been reported to repeatedly attack and chase males attempting to immigrate during the lactation period. Such female aggression is presumed to be a response to preventing infanticide by males, which is suggested to be an established male strategy which has developed in tandem with patterns of female mate choice in this species. The rationale behind such a strategy is that it may influence female reproductive success. If a female loses her infant one year, it is suggested that her next infant has a better chance at survival. "By killing current infants, then, male lemurs may often advance by a full year the time of females' next successful reproduction...When infants die, for whatever reason, their fathers become unlikely to be chosen again as mates ("incompetent fathers")" (Pereira and Weiss, 1991, pp. 149-150).

This hypothesis is problematic on a number of levels. First, despite thousands of hours of observation on identified, habituated individuals, the phenomena of increased agonism directed by females toward males during the lactation period has not been observed in most studies of free-ranging, or semi-free-ranging populations (Taylor, 1986; Gould, 1991, 1992; Sauther, 1992; 1993, Sussman unpubl. data). Furthermore, no relationship between parity or rank and agonsitic behavior by females toward migrating males has been observed. For example, from February to May, 1988 a new male (number 60) attempted to transfer into Green troop. The female showing the highest level of agonism toward this new male was number 53, who was lactating at the time. However female 13, who was also nursing, exhibited similar percentages of agonism toward this male as did non-lactating female 33. Furthermore female 93, also a non-lactating female, had the second highest percentage of agonism toward this male (Table 3).

Second, our census data indicate that 80-85% of females give birth annually, and between 30-50% of these infants perish each year. If infanticide is a viable male strategy, presumably a high proportion of these deaths should be due to *male* infanticidal activities. Recently, two instances of infant-killing have been reported at Berenty (M. Pereira, pers. comm.). Intergroup encounters and agonism between females of different groups peak during the birth season (Sauther, 1992). Because infants can be wounded and/or killed during aggressive encounters between *females* of different groups (S. O'Connor; M. Gould, pers. comm.) it is essential that such observations be made on clearly identified animals. Infanticide in ringtailed lemurs has not been observed among individually identified ringtails at Beza Mahafaly (Sussman and Sauther, unpubl. data; M. Pereira, pers. comm.). In fact, during a year of intensive observations of two groups of ringtailed lemurs at Beza Mahafaly we failed to observe a single episode of male aggression directed toward any infant. Furthermore, and perhaps most importantly, in the birth seasons between 1987 and 1990, 16 of 19 females (84%) followed for the 3 years successfully produced surviving infants (those living to one year of age) two years in a row. Six of these females (32%)

were successful in reproducing surviving infants for 3 consecutive years (Sussman, unpubl. data). As in many mammals, infant survival of ringtailed lemurs at Beza Mahafaly is related to the age of the mothers, with young prime and prime-aged females having higher infant survival rates and older females lower rates (Sussman, 1991).

**Table 3.** Percentage of agonism between transferring male, 60, and female members of Green troop.

| Dyad | Agonism[1] | Context | |
|---|---|---|---|
| | | Feed | Space |
| 53 & 60[2] | 37 | 34 | 2 |
| 93 & 60 | 16 | 11 | 5 |
| 33 & 60 | 11 | 3 | 8 |
| 13 & 60 | 11 | 8 | 3 |

[1]Agonism involving females only.
[2]Significantly greater at $p<0.05$ than all other dyads except "Space" for 93 & 60, t-test significance determined by the randomization procedure.

Third, because females regularly mate with a number of males both from within and outside of the group, it is difficult to conceive of either the female or the male recognizing the father of any particular infant, a prerequisite for the incompetant father hypothesis of Pereira and Weiss (1991). This pattern of multimale mating has been observed both at Berenty and Beza Mahafaly and is clearly a part of the female mating strategy. It is therefore difficult to understand the advantages of infanticide as a male reproductive strategy in *Lemur catta*.

We believe that an interplay between stochastic events, (predation, parasites, illness, falls) and the precocial development pattern in this species are more likely responsible for infant mortality in ringtailed lemurs. Although we agree, and have presented elsewhere (Sauther, 1991) that male mating strategies are adapted to female mate choice, we do not believe that infanticide is a viable tactic. Instead, we hypothesize that the dramatic seasonal fluctuation of resource availability in this species makes it essential for females to time their reproduction so that infants can be weaned during the period of relative food abundance (see also Martin, 1972). All females enter estrus during the same, short mating season and thus mating with more than one male helps insure fertilization during this short receptive period. This does not mean that females will mate with all males present, but rather that they will *selectively* mate with more than one male, and at least one of these mates may be from outside of the group (see also Sauther, 1991).

Female choice of multiple group and non-group males results in a number of male mating strategies (Fig. 2). One of the adaptive advantages of becoming a central male appears to be the ability of this male to establish close relationships with females throughout the year (Sauther, unpubl. data), and thus to gain first access during mating (c.f. Smuts, 1985). Becoming the female's first mating partner may be especially inportant in *L. catta* as there is evidence that mating with ejaculation may lead to a loss of receptivity (Evans and Goy, 1968; van Horn and Resko, 1977). Furthermore, Evans and Goy found that within 2 hours of the first male's ejaculation females began rejecting males and that "Subsequent males rarely achieved more than incomplete sequences with probable intromissions" (p. 188). Close relationships with group females allows the top-ranking central male to monitor the reproductive state of group females during the short breeding season and to limit sexual monitoring by other males. He also forms a consortship with females nearing estrus, thereby

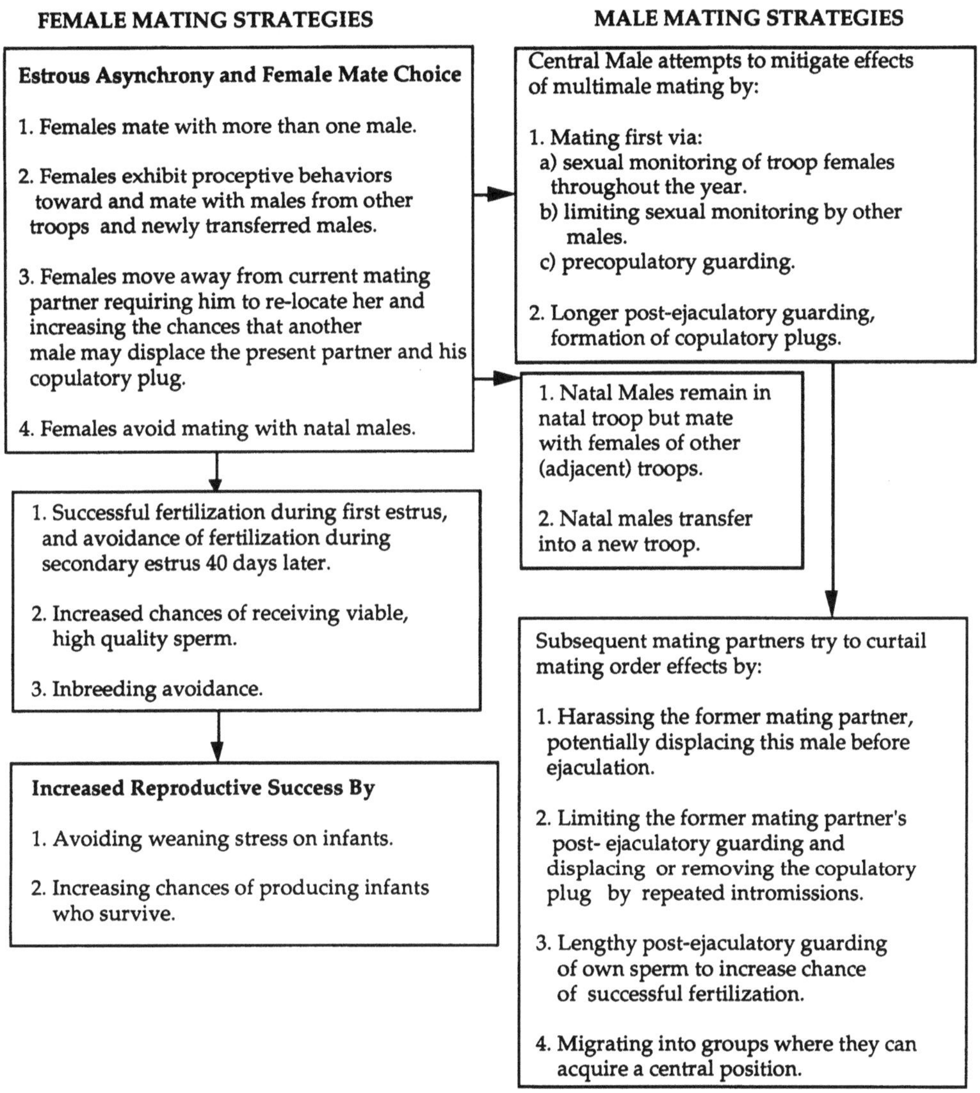


**Figure 2.** A model of mating strategies for ringtailed lemurs (modified from Sauther, 1991).

increasing his chance of being the first to mate if he is acceptable to the female when she becomes behaviorally receptive. The central male is also able to delay successive matings by other males, allowing the formation of a copulatory plug which might impede fertilization by other mating partners. Other males vigorously harass such males, potentially displacing them before ejaculation, or at least limiting the length of post-copulatory guarding. If successful, they attempt to replace the former mating partner as quickly as possible, which may inable them to displace previous copulatory plugs. They then guard the female for as long as possible after their ejaculation.

In any case, without evidence that these incidents do in fact lead to increased survival of the offspring of infanticidal males, the burden of proof for this hypothesis falls on its proponents. It should not be accepted as fact because it is an elegant hypothesis. Clearly further information is needed to prove or disprove either hypothesis.

## ACKNOWLEDGEMENTS

We would like to thank the following people for their assistance at Beza Mahafaly: Glen Green, Jeff Kaufmann, Alison Lake, Manjagasy, Armand Rakotozafy and Katie Reilly. The project would not be possible without the generous aide of the Ministry of Higher Education and the School of Agronomy of the University of Madagascar, especially Mme Berthe Rakotosamimanana, Benjamin Andriamihaja, Joseph Andriamampianina, and Pothin Rakotomanga. The hospitality and assistance of the people of Analafaly and the reserve guards in also gratefully acknowledged. The project was funded mainly by the World Wildlife Fund, a Fulbright Senior Research Grant, a Fulbright Collaborative Grant, the National Geographic Society, the Leakey Fund, and NSF Grant #BNS-8619240.

## REFERENCES

Bercovitch, F.B., 1991, Social stratification, social strategies and reproductive success in primates *Ethol. and Sociobiol.* 12:315.

Budnitz, N. and Dainis, K., 1975, *Lemur catta*: ecology and behavior *in*: "Lemur Biology," I. Tattersall and R.W. Sussman, eds., Plenum, New York.

Chepko-Sade, B.D. and Sade, D.S., 1979, Patterns of group splitting within matrilineal kinships groups: A study of social group structure in *Macaca mulatta* (Cercopithecidae: Primates), *Behav. Ecol. Sociobiol.* 5:67.

Evans, C.S. and Goy, R.W., 1968, Social behaviour and reproductive cycles in captive ringtailed-lemurs (*Lemur catta*), *J. Zool., Lond.* 156:181.

Gould, L., 1991, The social development of free-ranging infant *Lemur catta* at Berenty Reserve, Madagascar, *Int. J. Primatol.* 11:297.

Gould, L., 1992. Alloparental care in free-ranging *Lemur catta* at Berenty Reserve, Madagascar, *Folia Primatol.* 53:72.

Jones, K.C., 1983, Inter-troop transfer of *Lemur catta* at Berenty, Madagascar, *Folia Primatol.* 40:145.

Jolly, A., 1966, "Lemur Behavior," University of Chicago Press, Chicago.

Jolly, A., 1972, Troop continuity and troop spacing in *Propithecus verreauxi* and *Lemur catta* at Berenty (Madagascar), *Folia Primatol.* 17:335.

Jolly, A., 1985, "The Evolution of Primate Behavior," 2nd edition, Macmillan, New York.

Jolly, A., Oliver, W.L.R. and O'Connor, S.M. 1982, Population and troop ranges of *Lemur catta* and *Lemur fulvus* at Berenty, Madagascar: 1980 census, *Folia Primatol.* 39:115.

Jolly, A., Rasamimanana, H.R., Kinnaird, M.F. O'Brien, T.G., Crowley, H.M., Harcourt, C.S., Gardner, S. and Davidson, J.M., 1993, Territoriality in *Lemur catta* groups during the birth season at Berenty, Madagascar, *in*:"Lemur Social Systems and Their Ecological Basis,", P.M. Kappeler and J.U. Ganzhorn, eds., Plenum Press, New York.

Kappeler, P.M., 1990, Social status and scent-marking behavior in *Lemur catta, Anim. Behav.* 40:774.

Klopfer, P.H. and Jolly, A., 1970, The stability of territorial boundaries in a lemur troop, *Folia Primatol.* 12:199.

Koyama, N., 1970, Changes in dominance rank and division of a wild Japanese troop in Arashiyama, *Primates* 11:335.

Koyama, N., 1988, Mating behavior of ring-tailed lemurs (*Lemur catta*) at Berenty, Madagascar, *Primates* 29:163.
Manly, B.F.J., 1991, "Randomization and Monte Carlo Methods in Biology," Chapman and Hall, New York.
Martin, R.D., 1972, Adaptive radiation and behaviour of the Malagasy lemurs, *Phil. Trans. R. Soc. Lond. B* 264:295.
Mertl-Millhollen, A.S., Gustafson, H.L., Budnitz, N., Dainis, K. and Jolly, A. 1979, Population and territory stability of the *Lemur catta* at Berenty, Madagascar, *Folia Primatol.* 31:106.
Mitani, J.C. and Rodman, P.S., 1979, Territoriality: The relation of ranging pattern and home range size to defendability, with an analysis of territoriality among primates, *Behav. Ecol. Sociobiol.* 5:241.
Pereira, M.E. and Weiss, M.L., 1991, Female mate choice, male migration, and the threat of infanticide in ringtailed lemurs, *Behav. Ecol. Sociobiol.* 28:141.
O'Connor, S.M., 1987, The Effect of Human Impact on Vegetation and the Consequences to Primates in Two Riverine Forests, Southern Madagascar, Ph.D. thesis, University of Cambridge.
Rowell, T., 1972, "Social Behaviour of Monkeys," Middlesex: Penguin Press.
Rowell, T., 1976, How would we know if social organization were not adaptive, *in*: "Primate Ecology and Human Origins," I.S. Bernstein and E.O Smith, eds., Garland Press, New York.
Sauther, M.L., 1991, Reproductive behavior of free-ranging *Lemur catta* at Beza Mahafaly Special Reserve, Madagascar, *Am. J. Phys. Anthropol.* 84:463.
Sauther, M.L., 1992, The Effect of Reproductive State, Social Rank and Group Size on Resource Use Among Free-ranging Ringtailed Lemurs (*Lemur catta*) of Madagascar, Ph.D. thesis, Washington University, St. Louis.
Sauther, M.L., 1993, Resource competition in wild populations of ringtailed lemurs (*Lemur catta*): implications for female dominance, *in*:"Lemur Social Systems and Their Ecological Basis,", P.M. Kappeler and J.U. Ganzhorn, eds., Plenum Press, New York.
Smuts, B.B., 1985, "Sex and Friendship in Baboons," Aldine, New York.
Sussman, R.W., 1974, Ecological distinctions between two species of *Lemur*, *in* "Prosimian Biology," R.D. Martin, G.A. Doyle and A. C. Walker, Duckworth, London.
Sussman, R.W., 1991, Demography and social organization of free-ranging *Lemur catta* in the Beza Mahafaly Reserve, Madagascar, *Am. J. Phys. Anthropol.* 84:43.
Sussman, R.W., 1992, Male life history and intergroup mobility among ringtailed lemurs (*Lemur catta*), *Int. J. Primatol.* 13:395.
Taylor, L., 1986, Kinship, Dominance, and Social Organization in a Semi-free Ranging Group of Ringtailed Lemurs (*Lemur catta*), Ph.D. thesis, Washington University, St. Louis.
Taylor, L. and Sussman, R.W., 1985, A preliminary study of kinship and social organization in a semi free-ranging group of *Lemur catta*, *Int. J. Primatol.* 6:601.
Waser, P.M. and Wiley, R.H., 1980, Mechanisms and evolution of spacing in animals, *in*: "Handbook of Behavioral Neurobiology," P. Marler and J.G. Vanderbergh, eds., Vol. 3 Plenum, New York.

# FEEDING BEHAVIOR OF *LEMUR CATTA* FEMALES IN RELATION TO THEIR PHYSIOLOGICAL STATE

Hantanirina R. Rasamimanana and Elie Rafidinarivo

Ecole Normale Niveau III
University of Antananarivo
B.P. 881
Antananarivo 101
Madagascar

## ABSTRACT

For *Lemur catta* at Berenty Reserve in southern Madagascar, birth corresponds to the early wet season, lactation to the wet season, estrus to the early dry season, and gestation to the dry season, with marked differences in food available to females in each physiological state. Feeding behavior, food availability and palatability were compared in two *L. catta* troops living in different parts of the reserve over a 20-month period. Troop H's territory included many introduced tree species and access to tourists' bananas. Troop F lived in natural gallery forest far from the tourist circuit. Troop H ate more food items, from more species, was predominantly folivorous throughout the year and showed less change in diet and activity from season to season. Troop F ate less, from fewer plant species, reduced both activity and dietary diversity markedly in the dry season and was seasonally more frugivorous than folivorous. Activity, dietary selectivity, and "snacking" during the birth season may function to maximize energy gain. There are indications of need to balance dietary components, but quantity of foodstuffs may be more important to this adaptable species. Troop F's more limited natural diet did not translate into lower reproductive success because it increased by 50%, while Troop H increased by 17% in size.

## INTRODUCTION

Madagascar is world-famous for its exceptional riches of flora and fauna and its high endemicity of taxa. The lemurs form part of this natural heritage. They are among the species in danger of extinction through human predation and environmental degradation. The work we have undertaken must thus be situated in this context of the protection of nature, and specifically of lemur conservation. A better understanding of their ways of life can only contribute to their protection, and by extension to that of the environment.

*Lemur Social Systems and Their Ecological Basis*, Edited by
P.M. Kappeler and J.U. Ganzhorn, Plenum Press, New York, 1993

## METHODS

This study took place in Berenty Nature Reserve, where scientists of all nationalities have been welcome to carry out research during the last 30 years, and which has also welcomed increasing numbers of tourists over the past ten years. The reserve has an area of 240ha (Jolly et al., 1982). Its northern limit is the Mandrare river, which is in spate during part of the rainy season (December-April). In contrast, the river is low, or completely dry during the months of September and October. The types of vegetation include a gallery forest of about 100ha, with closed canopy composed of large trees such as *Tamarindus indica* or *Quivisianthe papinae*, and undergrowth composed largely of *Azima tetracantha*. To the south, there is 0.5ha of Didiereaceae forest (spiny desert), which is connected by open canopy forest to the gallery forest.

We chose the maki (*Lemur catta*) as study species. In spite of numerous research publications on its biology, much remains to be learned, especially about this species' feeding behavior. We have employed a method of group sampling. This consists of following one group of lemurs continuously from 6:00 hours to 18:00 hours. Within the group, we selected the first adult male or female that began feeding as our first focal animal. The beginning of a feeding session was recorded and we noted species and plant part of all food items. This individual remained the focal animal until it began a different activity. At that point, we chose a new focal animal of the opposite sex that began a feeding session. During sampling periods we also counted the individuals feeding in the same patch (trees or adjacent food sources of the same species) as the focal animal at the beginning and end of each session in order to compare individual preferences for a given food species. Immediately following each session, a sample of the food plant was collected for botanical identification and future chemical analysis. Identification was done on the spot within the limits of our competance or by specialists of the Service de Biologie Vegetale du Parc Tsimbazaza.

Diet was quantified by the number of mouthfuls of leaves, fruits, or seeds consumed during a session. The session began as soon as the focal animal brought a foodstuff of any sort to its mouth, and ended after a series of episodes of picking-swallowing-search-sniffing-picking-chewing-and-swallowing. Each session therefore was composed of feeding and search time.

We compared adults in two free-living troops of *Lemur catta* in the reserve: Group F in the northwest and Group H in the southeast. The troops were recognized by the territories they inhabit. Letter names follow those of previous researchers at Berenty (Mertl-Millhollen et al., 1979.) Group F's territory was located in natural primary gallery forest. Group H's range included natural forest and also planted areas. The troops could also be recognized by group size and composition, particularly the number of adult females. Finally each group had some marker animals, particularly an old, one-eyed male in Group H and an one-eyed infant in Group F.

Observations were made in four seasons, each corresponding to a physiological state of the females: birth, lactation, mating and gestation. There were 110 12-hour observation days, spread over two years. The birth season at Berenty coincides with the end of the dry season and the beginning of the rainy period (September to November). Most births occur in early September, just before the spring equinox, but birth in some females is delayed until the beginning of November. We consider this period separately, and do not include it in the period called "lactation", even though suckling itself begins a few minutes after birth. The lactation period coincides with the rainy season (November to March). The mating season falls into the end of the rainy season and the beginning of the dry season (April and May). In 1988, matings took place in mid-April, because most births occurred in early September. Within the mating season, receptivity of each female lasts only a few hours. Finally, the gestation period falls into the dry season (May-September). In this study, data for the gestation period were collected towards late gestation.

**Table 1.** Study periods and subjects.

| Observation Dates | | # 12h Samples | | Group Size* | | #Adult Females | |
|---|---|---|---|---|---|---|---|
| | | F | H | F | H | F | H |
| **Birth** (Late Dry, Early Wet Season) | | | | | | | |
| 29/10 - 10/11 | 1987 | 6 | 6 | 10 | 17 | 3 | 8 |
| 04/09 - 20/09 | 1988 | 8 | 8 | 13 | 19 | 4 | 8 |
| **Lactation** (Wet Season) | | | | | | | |
| 16/02 - 05/03 | 1988 | 7 | 8 | 13 | 24 | 4 | 8 |
| 28/03 - 31/03 | 1988 | 2 | 2 | 13 | 24 | 4 | 8 |
| 16/11 - 26/11 | 1989 | 5 | 5 | 15 | 20 | 6 | 8 |
| **Mating** (Late Wet Season) | | | | | | | |
| 01/04 - 14/04 | 1988 | 7 | 7 | 13 | 24 | 4 | 8 |
| **Gestation** (Dry Season) | | | | | | | |
| 25/06 - 06/07 | 1989 | 5 | 6 | 15 | 18 | 7 | 8 |
| 30/07 - 24/08 | 1989 | 13 | 12 | 15 | 18 | 7 | 8 |

* not including new infants. F and H are group names.

## RESULTS

### Group Size and Dynamics

At the beginning of our observations in October 1987, the two troops contained 10 and 17 individuals, respectively (Table 1). During this study, Troop F increased from 3 to 7 adult females, before loosing one female during the birth season of 1989. Total troop size increased steadily from 10 to 15. H remained stable at 8 adult females, and troop size fluctuated from an initial 17 up to 24, last seen at 20.

The fecundity rate is the number of adult females giving birth in a given year divided by the total number of adult females in the group. Infant survival rate here represents the proportion of infants that survived the first weeks of life. Juvenile survival rate is the proportion of weaned infants which survived to two years of age. The figures for 1987 show the proportion of 1-year-old juveniles counted at the beginning of the study which survived to the age of two (Table 2).

**Table 2.** Fecundity, infant and juvenile survival in two *Lemur catta* groups.

| | Group F | | | Group H | | |
|---|---|---|---|---|---|---|
| | 1987 | 1988 | 1989 | 1987 | 1988 | 1989 |
| % Fecundity | 100 | 50 | 75 | 63 | 50 | 75 |
| % Infant Survival | 100 | 100 | 100 | 100 | 100 | 83 |
| % Juvenile Survival | 33 | 33 | 33 | 50 | 25 | 40 |

**Table 3.** Female feeding behavior at different reproductive stages.

| | Birth | | Lactation | | Mating | | Gestation | |
|---|---|---|---|---|---|---|---|---|
| | F | H | F | H | F | H | F | H |
| **# Food Items Ingested per Day** | | | | | | | | |
| Fruit | 47 | 14 | 141 | 115 | 35 | 141 | 27 | 38 |
| Leaves | 49 | 54 | 46 | 189 | 105 | 253 | 141 | 137 |
| Flowers | 0 | 0 | 0 | 6 | 1 | 1 | 10 | 30 |
| Others | 0 | 0 | 29 | 16 | 16 | 0 | 24 | 20 |
| Total | 96 | 68 | 215 | 326 | 157 | 395 | 202 | 225 |
| **% Time Spent Feeding on (Including Pauses)** | | | | | | | | |
| Fruit | 60 | 30 | 62 | 40 | 49 | 45 | 36 | 23 |
| Leaves | 36 | 68 | 28 | 49 | 47 | 48 | 48 | 54 |
| Flowers | 1 | 0 | 0 | 1 | 0 | 0 | 6 | 8 |
| Others | 1 | 1 | 10 | 8 | 4 | 7 | 10 | 15 |
| **Mean Percent Pause Time During Each Feeding Session** | | | | | | | | |
| Fruit | 63 | 44 | 42 | 43 | 35 | 30 | 13 | 32 |
| Leaves | 47 | 36 | 33 | 34 | 27 | 26 | 27 | 17 |
| Flowers | 0 | - | - | 19 | 0 | 1 | 39 | 0 |
| Others | 36 | 24 | 26 | 45 | 1 | 1 | 16 | 29 |

**Table 4.** Food plant species diversity.

| | Birth | | Lactation | | Mating | | Gestation | |
|---|---|---|---|---|---|---|---|---|
| | F | H | F | H | F | H | F | H |
| **Total # Plant Species Eaten** | 5 | 7 | 12 | 22 | 10 | 21 | 6 | 20 |
| **# Plant Species Eaten** | | | | | | | | |
| Fruit | 4 | 6 | 9 | 8 | 4 | 8 | 3 | 6 |
| Leaves | 3 | 3 | 8 | 12 | 7 | 14 | 3 | 7 |
| Flowers | 1 | 1 | 0 | 5 | 1 | 2 | 1 | 3 |
| Other | 0 | 0 | 5 | 3 | 10 | 1 | 3 | 2 |
| **Plant Species Accounting for More Than 10% Feeding Time** | *Tamarindus indica*<br>*Acacia rovumae*<br>*Celtis phillipensis* | *T. indica*<br>*Cordia rothii* | *T. indica*<br>*Rinorea greveana*<br>*C. phillipensis*<br>*Crateva excelsa* | *T. indica*<br>*C. rothii*<br>*R. greveana* | *T. indica*<br>*Celtis bifida* | *T. indica*<br>*C. rothii*<br>*Caesalpina pulcherrima* | *T. indica*<br>*Capparis sepiaria* | *T. indica* |
| **Plant Species Common to Both Troops** | *T.indica*<br>*A.rovumae*<br>*C.phillipensis*<br>*R.greveana*<br>*Azima tetracantha* | | *T.indica*<br>*A.rovumae*<br>*C.phillipensis*<br>*R.greveana*<br>*A.tetracantha*<br>*C.bifida* | | *T.indica*<br>*A.rovumae*<br>*A. tetracantha*<br>*Boerhavia diffusa*<br>*Commicarpus commersonii* | | *T.indica*<br>*C. sepiaria*<br>*Cordia varo*<br>*Enterospermum pruinosum*<br>*Quivisianthe papinae* | |

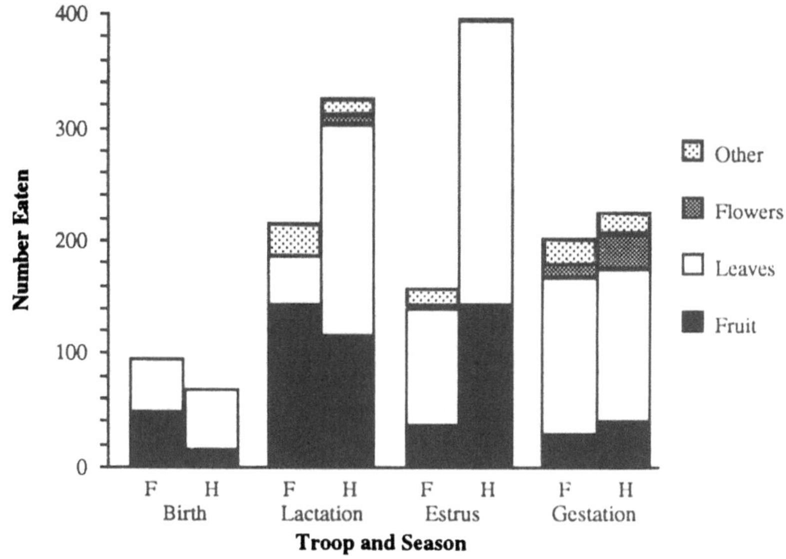


**Figure 1.** Number of food items ingested per day. Estrus refers to the mating season.

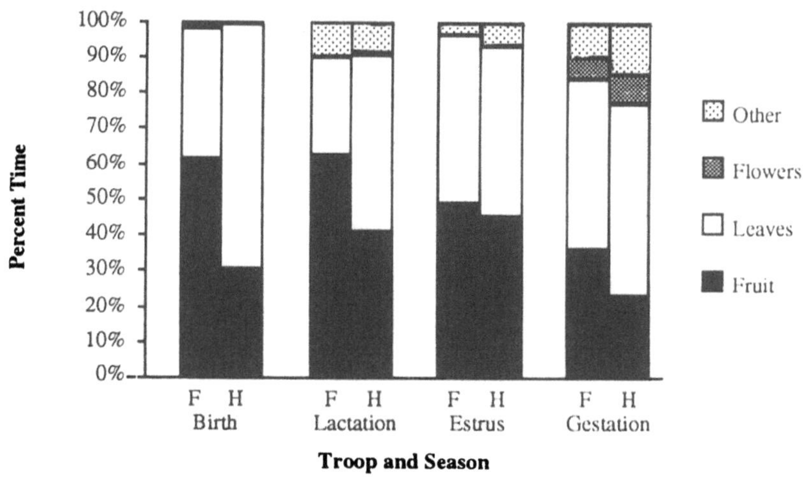


**Figure 2.** Percent session feeding on different food types, including pauses.

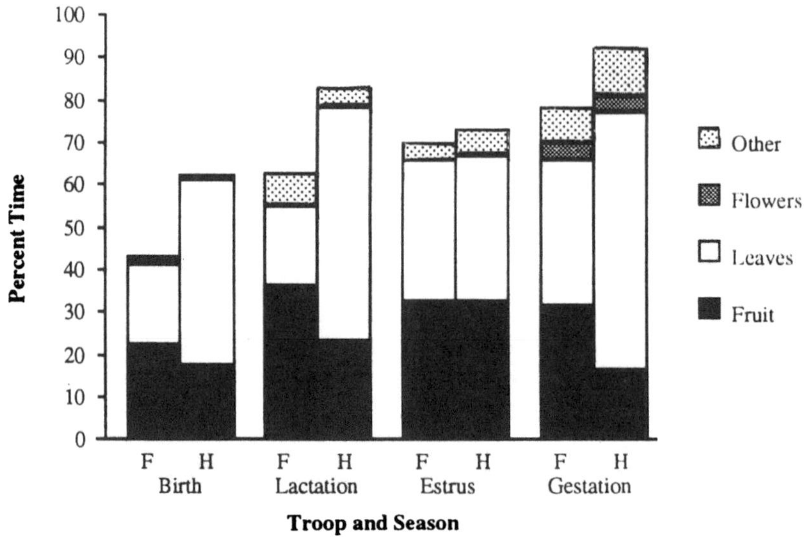


**Figure 3.** Percent time actually eating food types, without pauses.

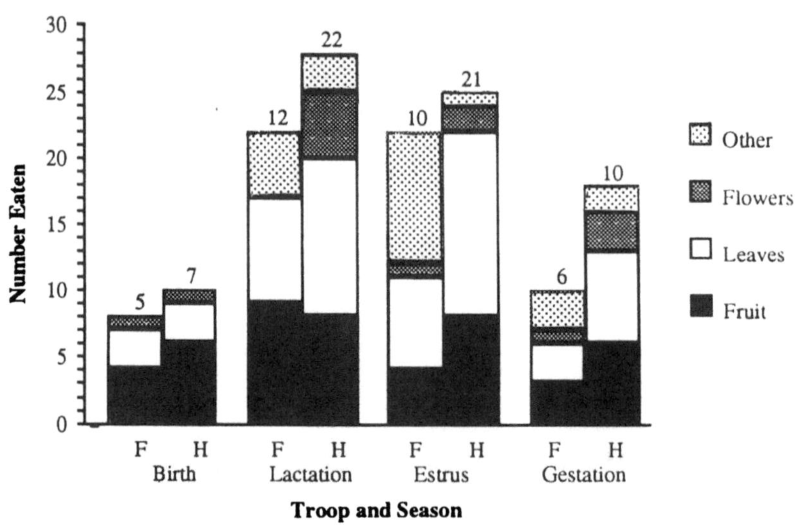


**Figure 4.** Number of species of food plant parts eaten per season. Total food plant species entered above colums.

## Female Behavior During Gestation

During the gestation period there were few species of fruit available. There was, however, a ready supply of mature leaves. The principal food source for Troop F were the mature leaves of *Tamarindus indica*. Tamarind leaves alone represented 90% of all leaves eaten. The females of troop F spent 35.5% of their feeding time on fruit, of which 50% was green tamarind fruit, and the rest was spread between only two other species (Table 1, Fig. 1). Thus, their diet was more folivorous than frugivorous (Table 2, Figs. 2-4). The percentage of the day spent travelling, in search for food or for other reasons, is at a minimum during this season. The troop fed in only six trees, staying virtually in the same place throughout the cold dry period. The troop was particularly difficult to find in the early morning, because it woke late and moved so little. In the other three seasons feeding activity was reduced between 11:00 and 15:00h. During the gestation period, however, feeding increased throughout the day, to reach its peak at the day's end. The normal mid-day siesta was reduced to one hour, instead of the usual 3-4 hours during other seasons (Fig. 5).

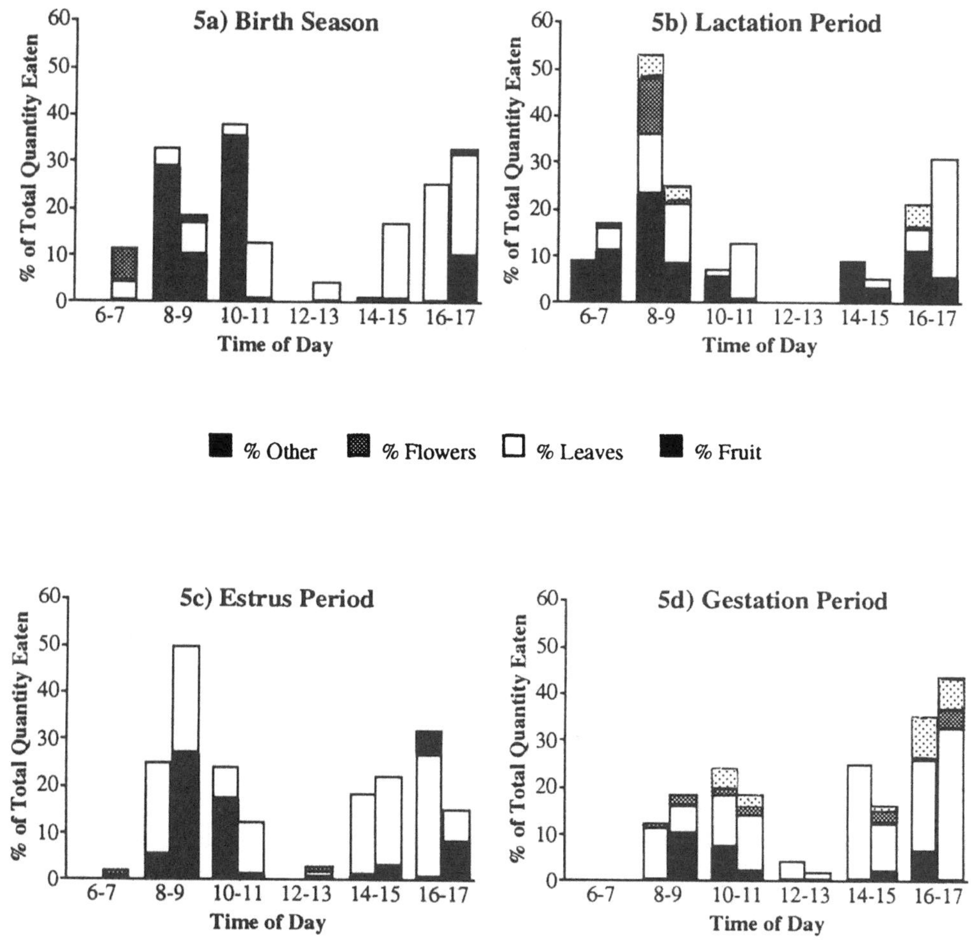


**Figure 5.** Percent of feeding by hour of day.

During gestation, females in Troop H also ate less than during the wet season and spent much time resting. Their diet was more varied than Troop F's, about the same in quantity, and also primarily folivorous. The main difference between F and H was H's consumption of flowers. Twelve percent of H's diet consisted of flowers, of which 93% were from introduced *Eucalyptus* species. Curiously, females in troop H also spent much time licking dry branches.

Finally, during the gestation period females of both troops chased other individuals who were near them, for reasons that were not evident. This happened at any time of the day, and not only during foraging. Even during feeding, however, females frequently did not occupy the victim's feeding site.

**Female Behavior During The Birth Season**

During the birth season, the actual food quantity consumed by Troop F was minimal (Fig. 1). Both total and actual feeding were lowest at this time of the year (Fig. 4). Females of troop F spent 60% of their feeding time on fruit. Of this proportion, 30% consisted of *Celtis phillipensis* and 63% *Tamarindus indica.* Leaves accounted for 36% feeding time, of which 59% were new leaves of *Tamarindus indica* and 41% new leaves of *Acacia rovumae*. The number of mouthfuls of each category of food was proportional to the time spent eating it. From both measures, we conclude that Troop F adopted a more frugivorous than folivorous diet during the birth season. As for Troop H, it spent only 30% of its feeding time on fruit, of which 45% consisted of *Tamarindus indica* and 55% were consecrated to various sweet fruits, including 10% on bananas offered by tourists. Leaf-eating occupied 68% of feeding time, of which 70% were *Tamarindus indica* leaves. Thus, Troop H continued in this season with a predominantly folivorous diet, unlike Troop F in natural forest, and in spite of the contribution of tourists' bananas. The very low food consumption of both troops during the birth season was striking (Fig. 1). Pauses occupied 57% and 38% of feeding sessions for Troop F and Troop H, respectively.

**Female Behavior During Lactation**

During lactation food consumption increased two-fold for Troop F and more than four-fold for Troop H. Species composition also increased, two-fold for Troop F and three-fold for Troop H. Troop F remained more frugivorous than folivorous, with 62% fruit-eating. Of this fruit 35% was ripe *Rinorea greveana*, 25% *Celtis bifida*, 20% *Azima tetracantha*, and the remaining 20% were made up by four other plant species. The remaining 38% of the diet consisted of tamarind leaves (74%) and leaves from only 10 other species. Tamarind was thus the chief leaf source, but it seems that Troop F was more opportunistic in its choice of fruit, eating every ripe fruit available.

Troop H in contrast remained more folivorous than frugivorous. Of the 49% time spent eating leaves, 52% corresponded to tamarind leaves; the other 48% were distributed among 11 other species. Females of troop H ate 40% fruits, of which 65% was *Cordia rothii* and the rest distributed among 5 species. Troop H fed on 22 plant species, as opposed to Troop F's 12, reflecting a greater diversification of diet.

**Female Behavior During The Mating Season**

Differences in feeding behavior between females from troop F and H were most pronounced during the mating season. The quantity of food ingested by F females fell, whereas the quantity eaten by H females increased in this late period of the wet season. Troop F fed on 10 plant species, as opposed to 12 during lactation, and 5 or 6 in the two dry periods. Percent time eating fruit was 49%, compared to 47% for leaves. Tamarind took up 48% and 95% of fruit and leaf-eating time, respectively. Because the number of mouthfulls of leaves was much larger than that of fruit, Troop F was predominantly folivorous. Travelling and resting took up most of the troop's time at this season. Troop H retained its primarily folivorous diet, but it was dominated by introduced *Melia azedarach*, and not by tamarind.

The mating season in both troops was characterized by the frenzy of mating activities (Jolly, 1966; Koyama, 1988; Sauther, 1991) and a high level of aggression between females of both the same and different groups.

**Some General Troop Differences And Similarities**

The diet of *L. catta* females consisted mainly of fruit and leaves. Flowers and buds did not figure largely in either troop's diet. Soil eating was seen nearly all year in both troops, except during the birth season. It occupied about 1% of time in both troops. One habit common to both troops was alternation of food categories. That is, when there has been a session on leaves, buds or flowers, this was nearly always followed by a session on fruits, and *vice versa*. The largest quantity of fruit was eaten in the morning, 1 or 2 hours before the mid-day, while the largest quantity of leaves was eaten at the end of the day. Finally, Troop H always consumed more food than troop F, except for the birth season. H's territory included introduced plant species, such as *Caesalpina pulcherrima* and *Melia azedarach*. Troop H also received daily banana handouts, though these formed at most 10% of the diet. Throughout the year, F was more folivorous than frugivorous. Troop F was primarily folivorous during mating and gestation season, but shifted to a more frugivorous diet during birth and lactation.

## DISCUSSION

### Diverse Strategies to Maximize Energy Gain

**Dietary selectivity.** Primates tend to become less selective in their diet or spend more time seeking food in the harsh season (Schoener, 1971; Clutton-Brock, 1977; Milton, 1978), although this does not apply to all species (Jolly, 1985). It certainly did not apply to Troop F, which varied its dietary habits from one season to another according to the availability of foods. It was an opportunistic frugivore during the wet season and became a limited folivore during the dry season. This is also a contrast to the *Lemur catta* of the nearby forest of Bealoka (O'Connor, 1987). Troop F seemed unselective in its fruit-eating, even in periods of abundance. During the dry season, for example, females of troop F even ate green tamarind fruits. However, since the bulk of their diet in the lean period consisted of leaves of a very limited number of species, overall they became more selective.

**Economizing locomotion.** Reiss (1985) defines energy available for reproduction as equal to energy gained from the environment minus energy required to gain it. Troop F minimized movement in the dry season by: (1) Rising late in the morning. The sun rises late during the gestation period, and temperatures are low. In the early morning, the environment furnishes little heat, and animals would probably lose energy rather than gain it, if they moved around to feed. (2) Sleeping earier in the evening, as the temperature drops more than in other seasons. (3) Travelling less to forage. During this period the group fed mainly in six individual trees, all in virtually the same place. (4) Increasing energy gained by increasing the amount of feeding time and the number of leaves eaten. The variety of the diet being more limited than usual, it is probably of lower quality, which is in part compensated by quantity.

This strategy seems efficient enough to support an elevated metabolic rate, and a high rate of maternal investment in the embryo (Young et al., 1990). It is an old and untested argument that the epitheliochorial placenta of lemurs is a relatively ineffecient means of transferring nutrients and oxygen to the embryo, and that this adds to the energy needs of the mother. It may be that the growth rate of lemur embryos puts extra demands on the mother (Pereira, 1993). However watching the social behavior of pregnant females, it seems clear that they are not exclusively investing in the embryo. The females have quite enough energy to chase and supplant targeted individuals in their entourage.

**"Snacking" in the birth season.** The only notable change in the feeding behavior of Troop H was that it "snacked" during the birth season instead of eating normally. This snacking consisted, on the one hand, of an increase in the number of pauses during feeding sessions, and, on the other hand, of a decrease in the number of mouthfuls ingested during this period. When births are taking place, fruits are only beginning to ripen, flowers to open, and new leaves to appear. Therefore, there is little available nutrition, although the quality promises to improve soon. The female still needs to economize the energy accumulated during pregnancy, gaining the maximum possible from her environment, while spending as little as possible. She rises early to profit by the warming rays of the morning sun, and eats until total darkness. During this extended day, however, she can find little available food on most trees. Pauses to forage and to find the few uneaten tamarind pods, or simply pauses, eke out her remaining energy.

### Dietary Composition

**Folivory-frugivory.** Troop H, in contrast to Troop F, was predominantly folivorous throughout the year. Its diet varied, however, in relation to the availability of fruit, from 1 fruit/1.5 leaves to 1 fruit/3.5 leaves. We speculate that H remained so folivorous because it was always fed some fruit, even when these were not naturally available. It may be necessary for lemurs to maintain a balanced intake of ascorbic acid and tannin to avoid the risk of hemosiderosis (Spelman et al., 1989). It seems that the leaves of tropical species contain ascorbic acid: provisioning by bananas and other introduced fruits may demand an even greater quantity and variety of leaves as counterweight.

*Tamarindus indica* (Leguminosae) constituted the essential nutritional basis for *Lemur catta*. It is well known as containing many tannins, although it is also rich in protein. There is evidence that the detrimental effects of tannins may be resisted in many folivores that regularly consume tannins (Cork and Foley, 1992). Certain species of the family Leguminosaceae and forbs of the family Nyctiginaceae compliment this diet. Concerning *Tamarindus indica*, Troops H and F doubled their consumtion of its leaves during pregnancy. Plants of the family Leguminosaceae are generally rich in nitrogen. This could explain the high preference for them over other available leaves.

One can conclude that the availability and abundance of essential foods has more influence on feeding habits than the nature of the foods themselves. Troops F and H live in very different micro-habitats. F is not provisioned in any way, neither with fruit nor water during the dry season, so it adapts its feeding to what is available, limiting its choice of leaves. Bananas and water are available to H throughout the year, and it eats at will. Ornamental plants in its territory continually give leaves and fruit, even in the off-season. H is visibly less restrictive than F in its food selection.

### Reproductive Success

Comparison of the feeding behavior of the two troops show how adaptable *Lemur catta* is to varying environments, without much affecting its reproductive capacity, since the growth rate of the two groups in two years was 50% and 17.5%, respectively. O'Connor's (1987) data on the number of individuals in the two troops in 1985 corroborate these rates of growth for the two populations. In fact, at that period group size was 15 for F and 13 for H. In the meantime, therefore, Troop F must have fissioned into two subgroups. If the rate of growth of the troop continued, it is not surprising that the troop seemed to have divided again in 1990 (A. Jolly, unpubl. data).

No one doubts the female-centered nature of *Lemur catta* troops. Only the females are full members of the group. Males are visitors who can change every year. It would not be mistaken to consider that the actual growth rate of a *Lemur catta* troop should be based on the number of adult females in a troop. In that case Troop F grew by 100% in two years, while H did not grow at all. How can this be reconciled with the overall growth rates of 50% and 100% that we cited? It is the result of the higher infant mortality in H than in F, of the low juvenile survival in both troops, and of the higher proportion of males born in Troop H. It even seems that the limited, frugivorous diet of F in the primary forest could be more effective than the broader folivorous diet of H in the semi-planted forest. One could argue that the *Lemur catta* adapted more readily to the harsh conditions of their original milieu than to a habitat which differs too much from the original one. But where are the limits to adaptation?

It should be noted that the numbers of animals in any two troops are small and these differences may simply be the result of random variation. Further, there may be a large effect of social conflict, which depends only indirectly in the immediate nutritional situation. Thus, troop H's diet may have had no necessary relation to its relative lack of success. However, the relative success of troop F is indeed an indication that a troop can prosper on a very limited, natural diet.

## REFERENCES


Clutton-Brock, T.H., 1977, Some aspects of intraspecific variation in feeding and ranging behaviour in primates, *in*: "Primate Ecology," Clutton-Brock, T.H., ed., Academic Press, New York.

Cork, S.J., and Foley, W.J., 1991, Digestive and metabolic strategies of arboreal mammalian folivores in relation to chemical defenses in temperate and tropical forests, *in*: "Plant Defenses Against Mammalian Herbivory," R.T. Palo and C.T. Robbins, eds., C.R.C. Press, Boca Raton.

Jolly, A., 1966, "Lemur Behavior," Chicago University Press, Chicago.

Jolly, A., 1985, "The Evolution of Primate Behavior," Macmillan, New York.

Jolly, A., Oliver, W.L.R. and O'Connor, S.M., 1982, Population and troop ranges of *Lemur catta* and *Lemur fulvus* at Berenty, Madagascar: 1980 census, *Folia Primatol.* 39:115.

Mertl-Millhollen, A.S., Gustafson, H.L., Budnitz, N., Dainis, K., and Jolly, A., 1979, Population and territory stability of the *Lemur catta* at Berenty, Madagascar, *Folia Primatol.* 31:106.

Milton, K., 1980, "The Foraging Strategy of Howler Monkeys," Columbia University Press, New York.

O'Connor, S.M, 1987, "The Effect of Human Impact on Vegetation and the Consequences to Primates in Two Riverine Forests in Southern Madagascar," PhD Dissertation, Darwin College, Cambridge.

Sauther, M.L., 1991, Reproductive behavior in free-ranging *Lemur catta* at Beza Mahafaly Special Reserve, Madagascar, *Amer J. Phys. Anthropol.* 84:463.

Schoener, T.W., 1971, Theory of feeding strategies, *Ann. Rev. Ecol. Syst.* 2:369.

Spelman, L.H., Osborn, K.G. and Anderson, M.P., 1989, Pathogenesis of hemosiderosis in lemurs: role of dietary iron, tannin, and ascorbic acid, *Zoo Biol.* 8:239.

Reiss, M.J., 1985, The allometry of reproduction: why larger species invest less in their offspring, *J. Theor. Biol.* 113:529.

Young, A.L., Richard, A.F. and Aiello, L.C, 1990, Female dominance and maternal investment in strepsirhine primates, *Amer. Nat.* 135:473.

# RESOURCE COMPETITION IN WILD POPULATIONS OF RINGTAILED LEMURS (*LEMUR CATTA* ): IMPLICATIONS FOR FEMALE DOMINANCE

Michelle L. Sauther

Department of Anthropology
Washington University
St. Louis, MO 63130
U.S.A.

## ABSTRACT

Stresses imposed on female lemurs by adapting their reproductive events to seasonal resource availability may be exacerbated by high pre- and post-natal maternal investment and by group living. Within such a context female dominance can be seen as a critical behavior, enabling females to coexist with more than one non-natal male within a highly seasonal environment. Females may tolerate year-round male membership as males provide low-cost sentinels for predator detection and defense. Furthermore, male membership may be a viable reproductive tactic for both females and males, especially because females mate first with group males. A combination of female choice of small males and ecological and reproductive constraints on male size may have led to the current system of female dominance in ringtailed lemurs.

## INTRODUCTION

Although females form coalitions against males in many primate species (Smuts, 1987), true female dominance wherein an adult female consistently evokes submissive behavior from an adult male is rare. Among anthropoid primates there is some evidence that female *Cercopithecus talapoin* supplant males without female coalitions (Wolfheim, 1977), and among *Cebus olivaceus* the highest ranking female is dominant to all individuals, male and female, with the exception of the top-ranking male (Robinson, 1981).

Among lemuroid primates, true female dominance involving both feeding and nonfeeding contexts (Kappeler, 1990a) has been observed in captive and wild populations of *Lemur catta* (Taylor, 1986; Kappeler, 1990a; Sauther, 1992), but the existence of female dominance in other lemur species is less clear. It has also been reported for *Indri indri* (Pollock, 1979), *Propithecus verreauxi* (Richard and Heimbuch, 1975), *Phaner furcifer* (Charles-Dominique and Petter, 1980), *Microcebus murinus* (Perret, 1982; Pagès-Feuillade, 1988), *Varecia variegata* (Kaufman, 1991), and *Daubentonia madagascariensis* (Rendall, 1993). However, of these lemur species, only *I. indri*, *V. variegata* and *D. madagascariensis* exhibit some consistent directionality of agonism (see Pereira et al., 1990). None of these studies provide data for female dominance in nonfeeding contexts (e.g., over grooming partners, resting sites etc.),

*Lemur Social Systems and Their Ecological Basis*, Edited by
P.M. Kappeler and J.U. Ganzhorn, Plenum Press, New York, 1993

making it difficult to differentiate female feeding priority from true female dominance (see Kappeler, 1990a). Furthermore, it is becoming clear that the expression of female dominance is variable among lemurs. For example, Sussman (1972) found no indication of a dominance hierarchy, and low levels of aggression in red-fronted lemurs, *Eulemur fulvus rufus*, a species similar in body size and proportions to *L. catta* (Tattersall, 1982), and Pereira et al. (1990) found no evidence of female dominance in any context for semi-captive or wild red-fronted lemurs.

Given the current uncertainty of the nature of female dominance among lemur species, it is important to determine the actual patterns of resource competition in the one lemur species which exhibits undeniable female dominance. This paper presents such information on free-ranging ringtailed lemurs.

## METHODS

### Study Site

Research was conducted at the Beza Mahafaly Special Reserve, which is located approximately 35km northeast of the town of Betioky. The area is dominated by *Tamarindus indica* in the eastern portion of the reserve, with dry-adapted species such as *Salvadora augustifolia* and *Euphorbia tirucalli* becoming more common as one moves to the west. The habitat is very seasonal, with a hot/wet season (December-April) associated with greater food availability as measured by the phenology of ringtailed lemur food resources, a cool/dry season (May-September) when food availability is dramatically reduced, and a transitional period (October-November), which is associated with increased availability of certain resources such as flowers. Rainfall is also seasonal with 506mm falling during the wet season. More details about the study site have been presented elsewhere (Sauther, 1989; Sussman, 1991).

The focus of the study was to document the feeding ecology of two groups of ringtailed lemurs living in a riverine forest within the reserve. A total of 16 *L. catta* (Black troop: 4 females, 2 males; Green troop: 5 males and 5 females) were studied, and 1800h of observations were collected and entered directly into hand-held portable computers.

### Behavior Patterns Sampled

The focal animal sampling method was used (Altmann, 1974). Behavioral categories were: feeding, resting, grooming, sunning, travel (movement as a group), moving (specified as to type, e.g., walk, run, etc.), defecating, urinating, and a category called standing, when the animal was between activities and was simply standing. All these behavior patterns were sampled at five minute intervals and the animal's location was noted. If the animal was feeding or foraging, the plant species and part used was also recorded. At 15min intervals the behavior and location of both the focal animal and its nearest neighbor were noted, and the distance of the nearest neighbor was recorded. Below, I will only consider nearest neighbor data from cases where both individuals were feeding. There were no consistent gender differences in nearest neighbor identity (see Sauther, 1992).

Agonistic behavior was recorded *ad libitum*, even if it did not involve the focal animal. Aggressive behavior included approaches, cuffing, biting, lunging, and chasing; submissive behavior involved retreating and running away with or without spat vocalizations. The behavioral context of all agonistic interactions was also recorded. Feeding agonism involved displacements from a food resource. Nonfeeding agonism included displacements over water, resting sites, or grooming partners. It included stinkfights (see Jolly, 1966) between males, but did not include this behavior when directed by males towards females, as this may be sexual in nature (Sauther, 1991a,b). Nonfeeding agonism also included social spacing, which were agonistic events designed to increase the distance between two individuals, and which were not related to other contexts. Aggressive feeding agonism involved chasing, biting and cuffing, whereas non-aggressive feeding agonism included only approaches and retreats.

The behavior of each adult member was sampled one day per month for at least 7h. Continuous daily observations of both troops were made during the mating season, (May) and during the birth season (from late September through early November). Expected frequencies of agonistic interactions were calculated by multiplying the total number of agonistic episodes for each member for a given period (e.g., wet or dry season) by the expected proportions based on the number of male or female partners potentially available. Once expected frequencies were determined, a $X^2$-value was calculated for each pair of dyads (female-female (FF) versus female-male (FM); FM versus male-male (MM); FF versus MM). The significance of the $X^2$-value was determined by randomization and Monte Carlo procedures, which make no assumption about their distribution, using SAS and Quickbasic (Edington, 1980; Noreen, 1989; Manly, 1991).

## RESULTS

### Resource Competition

**Feeding Agonism.** Of the total agonistic events observed (2301), 86% occurred over access to food resources. The level of feeding agonism differed by gender. Females exhibited significantly greater mean frequencies of feeding agonism for most months than did males (Fig. 1). Feeding agonism between males and females peaked during the late lactation/weaning period (Fig. 2).

To determine how this agonism was distributed by gender, feeding agonism was separated into three dyads of FF, FM and MM for both Green and Black troop (Table 1). During the wet season, in both groups females exhibited more feeding agonism with each other than expected by chance. Males, on the other hand, suffered less feeding agonism from other males than expected in the larger Green troop, whereas the frequencies were as expected in the smaller Black troop. The two troops also differed with respect to FM agonism. In the larger group, FM agonism differed little from expectation, but in the smaller group, FM agonism was considerably less than expected. In the dry season FF agonism did not differ from expectation in the larger troop but was more than expected in the smaller troop. MM agonism was less than expected in the larger group, but more than expected in the smaller group. FM agonism was as expected in the larger troop but less than expected in the smaller troop.

Feeding agonism was split into aggressive (interactions involving chasing, biting and cuffing) and nonaggressive (interactions involving only approaches and retreats) (Fig. 3). Females showed greater mean frequencies of aggressive agonism than did males throughout the wet season, a period of relative food abundance, whereas they showed greater nonfeeding agonism during the dry period of reduced food availability. FM agonistic dyads plotted by month relative to the type of agonism indicate that for most months females aggressively displaced males from feeding patches (Fig. 4).

Agonism was compared by dyads relative to aggressive and nonaggressive feeding agonism. For all seasons, and in both troops aggressive FF feeding agonism was greater than expected by chance based on the number of potential female partners (Table 2). MM was reduced for both seasons in Green troop, but MM agonism was less than expected in the smaller Black troop during the wet season and greater than expected in the dry season. For the dry season in Black troop, and both seasons in Green troop, FM agonism differed little from expected. However, FM was less than expected in the smaller Black troop during the wet season.

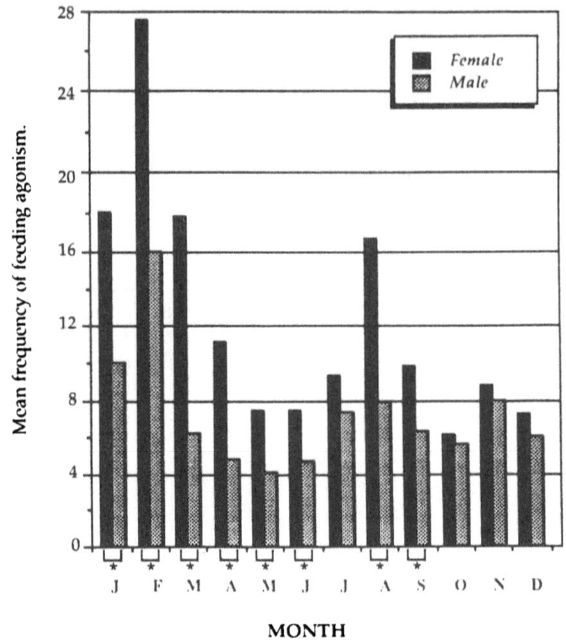


**Figure 1.** Sex differences in feeding agonism among ringtailed lemurs. * indicates $P < 0.05$. Total feeding agonism: N = 1985 bouts.

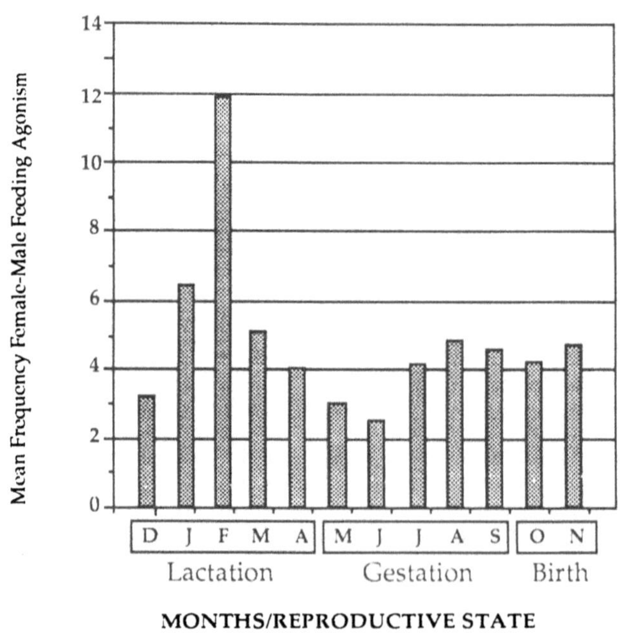


**Figure 2.** Female-male feeding agonism in ringtailed lemurs. Total female-male feeding agonism, N = 942 bouts.

**Table 1.** Dyadic feeding agonism during the wet and dry season by troop. The number of observed agonistic interactions is presented for each dyad-class. FF = female-female dyads, FM = female-male dyads, MM = male-male dyads. *P<0.05, **P<0.01.

| | WET SEASON | | | DRY SEASON | |
|---|---|---|---|---|---|
| Dyads | # Observed | $X^2$ | Dyads | # Observed | $X^2$ |
| Green Troop | | | | | |
| FF/FM | 320/405 | 24.94** | FF/FM | 160/225 | 0.28 |
| FF/MM | 320/67 | 35.28** | FF/MM | 160/41 | 10.73* |
| FM/MM | 405/67 | 11.04* | FM/MM | 225/41 | 10.79* |
| Black Troop | | | | | |
| FF/FM | 162/97 | 21.07** | FF/FM | 138/84 | 17.81** |
| FF/MM | 162/19 | 9.65* | FF/MM | 138/24 | 25.73** |
| FM/MM | 97/19 | 14.98* | FM/MM | 84/24 | 22.70** |

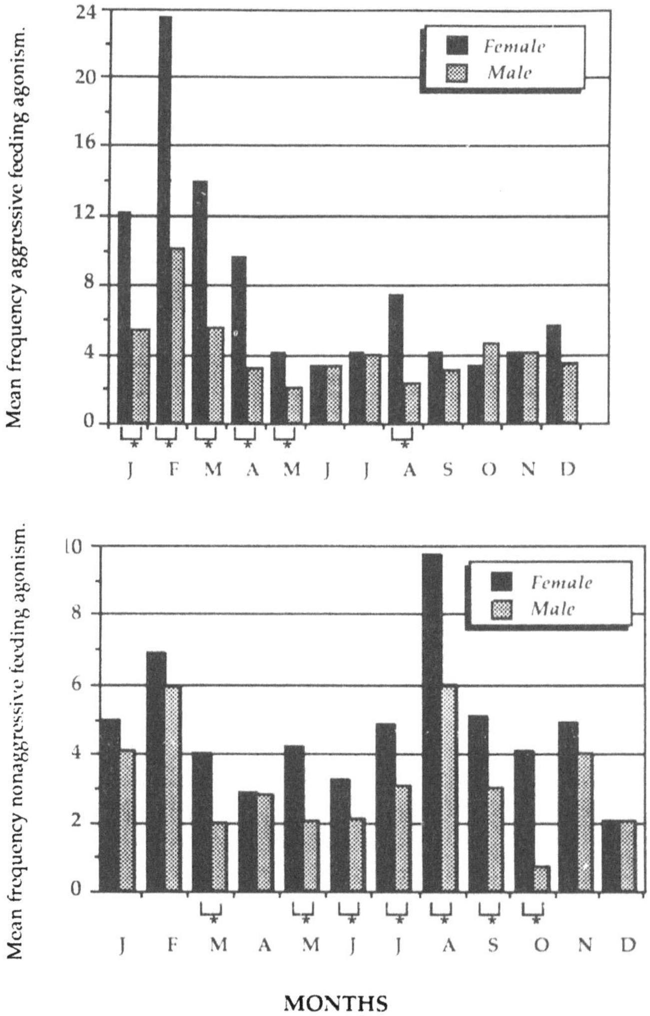


**Figure 3.** Sex differences in aggressive and nonaggressive feeding agonism among ringtailed lemurs. * indicates P<0.05. Total aggressive feeding agonism: N=1208 bouts. Total nonaggressive feeding agonism, N=777 bouts.

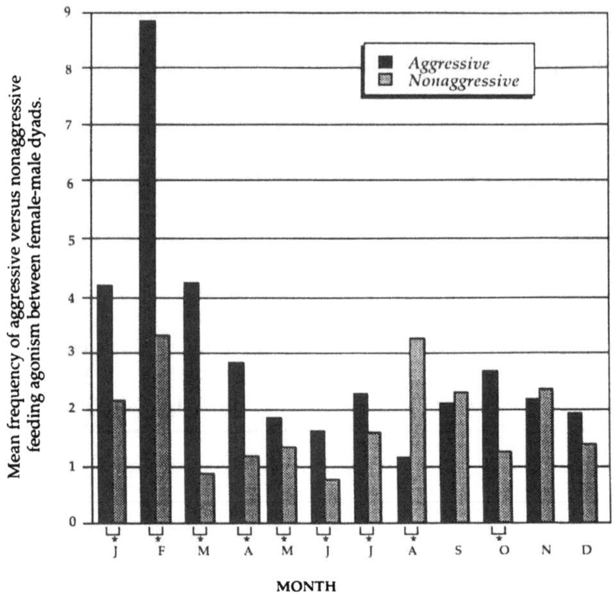


**Figure 4.** Aggressive *vs* nonaggressive feeding agonism in female-male dyads. *indicates $P<0.05$. Total aggressive feeding agonism: $N=577$ bouts. Total nonaggessive feeding agonism, $N=335$ bouts.

**Table 2.** Dyadic aggressive feeding agonism during the wet and dry season by troop. See Table 1 for details.

| | WET SEASON | | | DRY SEASON | |
|---|---|---|---|---|---|
| Dyads | # Observed | $X^2$ | Dyads | # Observed | $X^2$ |
| Green Troop | | | | | |
| FF/FM | 264/304 | 30.55** | FF/FM | 75/122 | 2.72* |
| FF/MM | 264/38 | 34.15** | FF/MM | 75/22 | 8.80** |
| FM/MM | 304/38 | 34.15** | FM/MM | 122/22 | 6.08** |
| Black Troop | | | | | |
| FF/FM | 109/57 | 13.09* | FF/FM | 55/34 | 6.67** |
| FF/MM | 109/2 | 10.10* | FF/MM | 55/6 | 9.56** |
| FM/MM | 57/2 | 8.31** | FM/MM | 34/6 | 3.11** |

Nonaggressive feeding agonism among females was near expected frequencies for the wet season, but greater than expected during the dry season (Table 3). MM was as expected in the wet season, but varied between the two groups for the dry season. MM was less than expected in Green troop and greater than expected in Black troop. FM was as expected in Green troop but was less than expected in the smaller Black troop.

**Nonfeeding Agonism.** Only 14% of the total agonism observed could be categorized as nonfeeding agonism. Sexually mature, reproductive females decidedly won all such encounters with males (but see below). Comparing males and females, nonfeeding agonism was more variable than feeding agonism, with males actually surpassing females during April and the mating period in May (Fig. 5). When broken down by context, most nonfeeding agonism (52%) was over access to water, which collected in tree hollows. Some nonfeeding agonism also occurred in the context of social spacing (23%), over access to favored resting sites (10%), and stinkfights between males (7%).

**Table 3.** Dyadic nonaggressive feeding agonism during the wet and dry season by troop. See Table 1 for details.

| Dyads | WET SEASON # Observed | $X^2$ | Dyads | DRY SEASON # Observed | $X^2$ |
|---|---|---|---|---|---|
| Green Troop | | | | | |
| FF/FM | 56/101 | 0.49 | FF/FM | 85/103 | 10.69** |
| FF/MM | 56/29 | 1.20 | FF/MM | 85/19 | 15.37** |
| FM/MM | 101/29 | 0.49 | FM/MM | 103/19 | 7.18** |
| Black Troop | | | | | |
| FF/FM | 53/40 | 5.57* | FF/FM | 83/50 | 15.03** |
| FF/MM | 53/17 | 1.45 | FF/MM | 83/18 | 17.48** |
| FM/MM | 40/17 | 5.02 | FM/MM | 50/18 | 22.55** |

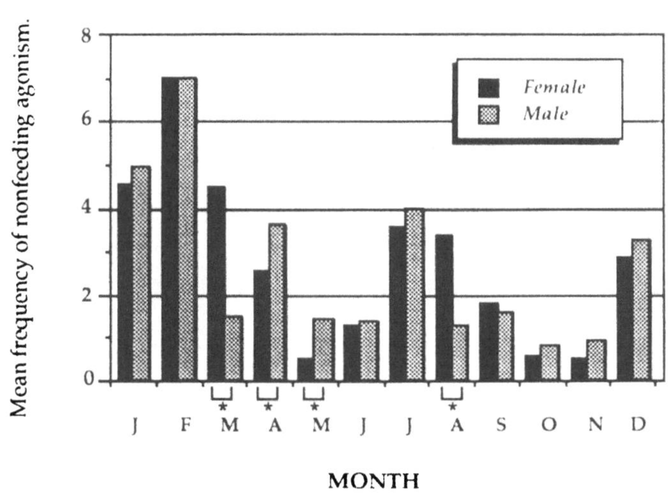


**Figure 5.** Sex differences in nonfeeding agonism among ringtailed lemurs. * indicates $P<0.05$. Total nonfeeding agonism, N=313 bouts.

**Displacements Of Females By Males.** During the course of the study a total of 35 displacements, or 3% of the total agonism between females and males, involved males displacing young adult, nulliparous females who were two years old at the start of the study (November, 1987) (Table 4). All but one of these were nonaggressive in nature; either a male approached a female and she retreated, a male took a food item (e.g., a *Tamarindus indica* fruit) from her, or a male displaced a female from a drinking site by moving her aside with his body. Most of these displacements were over food or water resources. Only one interaction could be characterized as aggressive. During the mating season, males would continuously approach females to monitor their receptivity, and if they were not cuffed or otherwise discouraged by the female, they would try to mate with her (Sauther, 1991b). On one occasion the top-ranked male of Green troop approached one of the young females, and when she did not cuff him away, he attempted to mount her. A brief jumpfight ensued, the female cuffed the male, he cuffed her, and he jumped away. There were no cases where a young female approached a male and he refused to leave the feeding site. All of these displacements occurred between November, 1987 and the mating season in May, 1988. Males were no longer able to displace these females after their first mating season.

**Table 4.** Context of male wins over females for ringtailed lemurs.

| Context of Male's Win | Frequency | Percentage of Total Male Wins | Percentage of Total Agonism between Females and Males |
|---|---|---|---|
| Feeding | 16 | 45% | 2% |
| Drink/Lick | 10 | 29% | 1% |
| Social Spacing | 5 | 14% | 0.005% |
| Resting Sites | 3 | 9% | 0.003% |
| Sexual | 1 | 3% | 0.001% |
| TOTAL | 35 | | 3.09% |

**Nearest Neighbor Distances During Feeding.** One potential source of feeding competition is having close neighbors during feeding bouts. Spatial relationships during feeding can presumably affect feeding efficiency, and close neighbors could limit resource availability. Males differed from females with regard to how close they fed to others (Fig. 6). Males tended to spread out and feed farther away from other individuals. During the wet season, males had greater percentages of "distant" nearest neighbors (>6m), whereas females had higher percentages of "close" nearest neighbors (<3m).

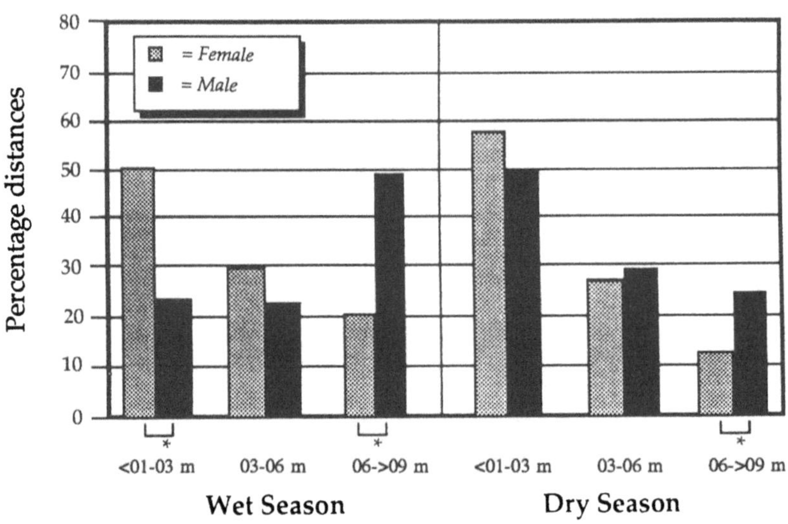


**Figure 6.** Proximity of nearest neighbors while both feed. * denotes $P<0.05$.

During the dry season males and females had similar percentages of close nearest neighbors during feeding although males still had greater percentages of distant nearest neighbors. Comparing the wet and dry season, males more than doubled their time spent feeding near others. This tendency to feed closer in the dry season was most likely the result of the availability of resources, which tended to be limited to small patches of herbs or single fruit trees.

To establish whether all females were equally impacted by the closer proximity of foraging males during the dry season, the percentage of time males spent near individual females was determined (Fig. 7). In both groups males tended to feed most often near low-ranking females. Relative to the wet season, this resulted in low-ranking females maintaining or increasing agonistic bouts with males during the dry season, whereas high ranking females decreased such agonism (Fig. 8).

Another measure of direct feeding competition by males was the frequency that the nearest neighbor fed on the same food item and part. Figure 9 shows the mean frequency per month that females had males as nearest neighbor who were also feeding on the same species and plant part. Two peaks occurred, one during late lactation/early weaning, and the other during late pregnancy-birth-early lactation.

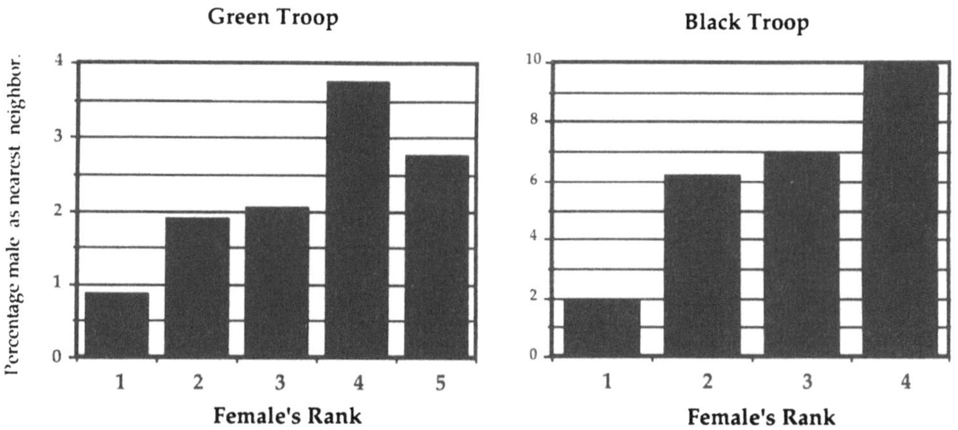


**Figure 7.** Percentage of time females and males were nearest neighbors during foraging and feeding in the dry season.

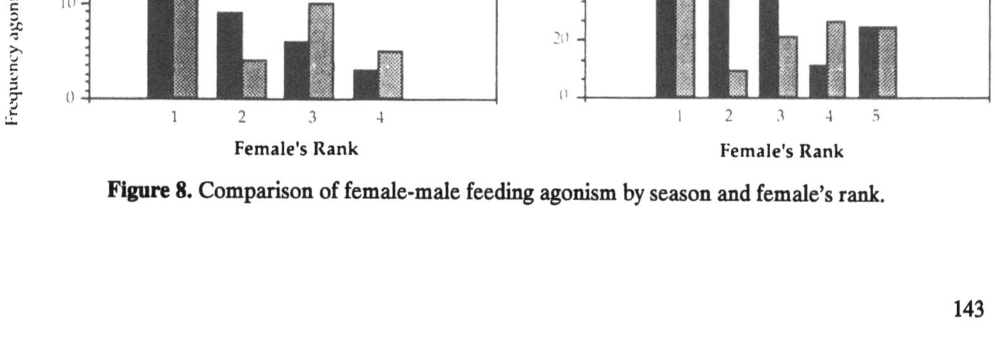


**Figure 8.** Comparison of female-male feeding agonism by season and female's rank.

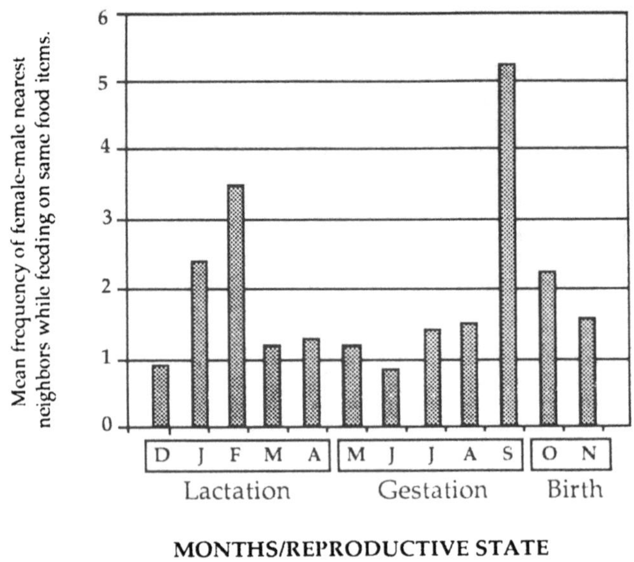


**Figure 9.** Feeding bouts where a female had a male as a nearest neighbor, and both were feeding on the same food item, N=381 bouts.

## DISCUSSION

### Function and Evolution of Female Social Dominance In Ringtailed Lemurs

As already noted, female dominance among lemur species is not as ubiquitous as once thought (e.g., Pereira et al., 1990). For most lemur species, we currently lack the appropriate ecological and behavioral data which would clarify the nature of female dominance in free-ranging groups, e.g., how synchronous is female reproduction, what is the degree of resource seasonality, and what is the frequency and context of male displacements by females. Nevertheless, there are currently three arguments to explain social dominance among lemurs: paternal investment (Pollock, 1979), male deference as a reproductive strategy (Hrdy, 1981; Pereira et al., 1990), and high reproductive costs (Jolly, 1984; Young et al., 1990). As already noted by Young et al. (1990), paternal investment is a possibility only for monogamous species. Is there any support for the latter two hypotheses with regards to free-ranging *L. catta* behavior and ecology?

### Male Deference

With regard to male foraging strategies, Hrdy (1981) has suggested that in female-dominant species males may defer to females in feeding contexts to save energy for mate competition. Males should therefore concentrate on conserving energy and engage in intense male-male competition only during the brief breeding season. This makes sense only if rank-related advantages are limited to the mating season, and seems to assume that feeding agonism among males is not important.

*L. catta* has been presented as a species to fit Hrdy's model (Jolly, 1984). Under such a scenario, male ringtailed lemurs would maintain themselves at minimal levels of feeding competition in the dry season, when food is limited, and then compete for food and status when food availability increases just prior to the mating season. This model is not tenable for a number of reasons.

First, rank-related advantages were not limited solely to the breeding season. High-ranking, central males used less expensive forms of locomotion, they had greater access to drinking sites than lower-ranking males, they won more agonistic bouts over food, and they fed more, by weight, on "valuable" but limited foods such as some fruits (Sauther, 1992). Second, male-male agonism does not support the contention that males were "saving themselves" during the lean period. Males maintained similar mean frequencies of feeding agonism with each other during each season (lactation season = 15; gestation season = 13; birth season = 15; $P > 0.05$ for all pairwise comparisons). Third, there is no indication of male "chivalry" (sensu Jolly, 1984) with regards to lessening female feeding pressure. The frequency of female-male feeding agonism remained similar throughout most of gestation and the birth season. Furthermore, it dramatically increased at a time when females were most stressed, i.e., during the late lactation/weaning period.

Pereira et al. (1990), found some support for Hrdy's argument among captive ringtailed lemurs. They found that most agonistic "wins" (75%) by females over males involved no aggression, simply approach-retreat interactions, and they suggested that females might choose to mate with defering males. However, in the present study on free-ranging populations, there was no indication that males were simply defering to females in feeding contexts. Females had feeding priority over males because they actively usurped males from feeding sites. Females would not hesitate to chase males from a feeding site, and retribution for a slow responding male could be quick and violent. For example, on one occasion a female feeding in a tree grabbed the shoulders of a lingering male, held him in place, and bit him soundly on the top of his head. Over 63% of all feeding agonism between males and females involved a female aggressively displacing the male. In addition, during most months females continued to employ more aggressive, rather than nonaggressive forms of feeding agonism when displacing males from feeding sites. It is possible that the pattern of greater active, aggressive feeding agonism observed in the wild populations was absent in the captive population due to provisioning.

Males could defer to females indirectly, by simply feeding further away from them. However, in this study males fed more closely to females, and directly competed with them by focusing on the same foods during two critical reproductive stages, late lactation/early weaning and late pregnancy/early lactation. Furthermore, all individuals, regardless of gender, fed more closely to one another during the dry season, a period of low food availability. Thus, evidence from the current study provides no indication of male deference in ringtailed lemurs.

**High Reproductive Costs**

Both Jolly (1984) and Young et al. (1990) have made the case that especially high reproductive costs among lemurs might explain the evolution of female dominance in these species. Specifically, a number of lemurs combine low basal metabolic rates with high prenatal maternal investment (Young et al., 1990), and they produce altricial neonates which must be supported by the mother during their infant's rapid postnatal growth period (Jolly, 1984). In addition, many lemur species exhibit seasonal reproduction (Jolly, 1984). Ringtailed lemurs exhibit one of the most highly constrained mating seasons, approximately 24h (Evans and Goy, 1968), with all females within any one troop breeding during a short period of 1-3 weeks (Jolly, 1966; Sussman, 1977; Sauther, 1991b). At Beza Mahafaly this strict breeding seasonality is tied to the availability of resources (Sauther, 1993). Females lactate during the period of food abundance (the wet season, December-April), gestate during the period of relatively low food availability (May-September), and give birth during peaks of important food resources such as flowers (October-November). Given the close reproductive synchrony in this species (Jolly, 1966; Pereira, 1991; Sauther, 1991b) reproducing females will experience identical reproductive events and undergo similar reproductive stresses leading to high levels of interindividual resource competition (Sauther, 1993). Under such conditions, extra feeding competition from group males would be a distinct disadvantage not only to females, but also to their infants, unless adult females had feeding priority. Female dominance might develop under such conditions to mitigate male resource competition. Results from this study support this argument.

**Female Feeding Investment.** Gender differences in resource competition should reflect greater feeding investment by females, i.e., if food is a more limited resource for females (sensu Wrangham, 1980) females should expend more energy in its acquisition. This was seen for *L. catta* at Beza Mahafaly. Not only were females involved in more feeding agonism than males throughout most of the year, but they also engaged in more expensive (i.e., aggressive) forms of contest competition such as chasing, cuffing and biting during the critical lactation period.

**Female Feeding Stress.** Given their reproductive role, females should be under greater feeding stress than males. During lactation, which is the most costly reproductive state for placental mammals (Sadleir, 1969; Robbins, 1983; Clutton-Brock et al., 1989), female ringtailed lemurs were displacing each other at higher frequencies than would be expected, and more of these displacements were aggressive in nature. Males, on the other hand, either received expected or less than expected feeding agonism from other males. During gestation, high frequencies of aggressive and nonaggressive displacements continued between females. In addition, females had greater percentages of close neighbors during feeding throughout the year, whereas males tended to spread out and feed further from others during the wet season.

**Feeding Competition From Males.** Males were costly competitors for females. They did not simply retreat from feeding patches, and if the food item in question was portable (e.g., *Tamarindus indica* fruit), males could run off with the fruit in their mouth. In most cases females had to aggressively displace males from a feeding site, albeit males never contested such displacements. Due to their tendency to feed on identical food items near females, males provided females with both direct and indirect feeding competition especially during the two periods of expected maximal stress, late gestation/early birth, and late lactation/early weaning. Feeding near other individuals can be costly as it increases the chances of aggressive displacements (Mori, 1977; Furuichi, 1983) and can limit foraging success (Robinson, 1981). In addition, male feeding competition was especially acute for lower-ranking pregnant females who had to contend with close male feeding proximity during the period of lowest food availability. This resulted in low ranking females increasing feeding agonism directed at males, unlike high ranking females who actually decreased male-directed feeding agonism relative to the wet season.

There is also evidence that females living in groups with fewer males have an advantage in terms of feeding agonism. Black troop had only two resident males, whereas Green troop had five. Female-male feeding agonism, both aggressive and nonaggressive, was consistently less than expected for most seasons in the group with fewer males, whereas it was at expected levels in the group with more males.

**Nonfeeding Agonism.** It has been suggested that this model, which focuses on high reproductive costs, does not account for female dominance in nonfeeding contexts, which have been observed among captive populations of ringtailed lemurs (Kappeler, 1990a). In this study while female dominance in nonfeeding contexts was also observed, the majority of these were over important resources such as drinking sites, and thus fits with the reproductive energetics model. The second most frequent context was social spacing. If, as is argued here, males compete with females for resources, maintaining social distance from other individuals (including males) may help alleviate direct resource competition for females. It is not possible to currently reject the premise that female dominance in nonfeeding context may also relate to female reproductive costs.

### Lemurs Not Exhibiting Female Dominance: The Case of *Eulemur fulvus rufus*

As already noted, not all lemurs exhibit female dominance. Pereira et al. (1990) found no evidence of female dominance in either captive or free-ranging groups of *E.*

*fulvus rufus*. However, overt expression of female dominance may be tied to the availability of resources. Although red-fronted lemurs are sympatric with ringtailed lemurs in limited areas, field studies on both free-ranging and semi-free-ranging populations indicate that *E. fulvus rufus* is adapted to continuous canopy forests, but that *L. catta* can exist alone in drier brush and scrub forests, and is thus adapted to a more variable environment (Sussman 1972; Ganzhorn, 1985). These two species overlap only in mixed forest which contain both types of forest (Sussman, 1972). Furthermore, distributions of ringtailed lemurs are restricted to the drier southwestern and southern portion of Madagascar, whereas red-fronted lemurs are found in the more moist western and eastern portions of the island (Tattersall, 1982). Both Sussman (1974) and Ganzhorn (1985) have suggested that red-fronted lemurs are adapted to more stable environments (i.e., foraging in continuous canopy where resources are more abundant and evenly distributed), whereas ringtailed lemurs are adapted to a more variable environment (i.e., drier, more seasonal forests where foods are sparsely distributed). Within this highly seasonal environment, *L. catta* females may also suffer greater post-natal maternal costs than *E. fulvus rufus* females, because ringtailed lemur infants exhibit a more precocial rate of development (Sussman, 1977). While red-fronted lemurs may encounter less feeding pressure, leading to low levels of agonism and more egalitarian social relationships, ringtailed lemur females incur high reproductive costs which are exacerbated by foraging on discrete patches of seasonal resources. This may lead to greater feeding agonism and a more overt expression of female dominance in this species.

**Male Displacement of Females**

Female dominance in ringtailed lemurs is a developmental process in which females in a troop must eventually alter their relationship with adult males. Female infants may be buffered from male feeding competition by feeding near their mothers. During this period they can take food from males with impunity, although males will not hesitate to steal food from weaned infants if the mother is not in sight. Once weaned, males can and do displace females, but these displacements are not aggressive in nature. As the female reaches sexual maturity, her relationship with males goes through a transitional process which appears to be mitigated by male sexual advances. As noted elsewhere (Sauther, 1991b), during the mating season males continuously approach females to monitor their receptivity. Young females are thus "forced" to interact aggressively with males in order to thwart mating attempts prior to their receptive period. After the mating season, this new relationship expands to other, non-sexual contexts, i.e., feeding. Once this transitional period is over, males appear unable (or unwilling) to attempt such displacements, and females begin to displace males on a regular basis. One of the consequences of this process is that young females may incur higher levels of male feeding competition just prior to their first mating season.

**Multimale Membership in a Female Dominant Species**

Why do males live together in multi-female groups wherein they suffer feeding agonism from all adult females, instead of forming all-male groups or foraging as solitary males? Not surprisingly, multi-male membership is most likely tied, in part, to tactics for increasing reproductive success. Both Andelman (1986) and Altmann (1990) have noted a positive relationship between the number of females in a group and the number of group males. This is also seen at Beza Mahafaly, where the number of males in a troop is significantly positively correlated with the number of adult females ($r=0.78$, $P<0.01$; $N=70$). This is most likely a reproductive strategy, i.e., more adult females mean more potential mates. Among ringtailed lemurs, there is evidence that establishing close relationships with troop females may provide a reproductive advantage. It is likely that males use olfactory cues from females to determine the onset of estrus (Jolly, 1967; Schilling, 1979) since behavioral cues from females only occur when she is receptive (Evans and Goy, 1968; Sauther, pers. obs.). Males occupying a high-ranking, central position are able to maintain a closer relationship with females to monitor their reproductive state, they can limit monitoring by other males, and they are the first to mate (Sauther, 1991b). Being a female's first mating partner may be critical, as there is evidence from captive studies of ringtailed lemurs that mating with ejaculation leads to a loss of receptivity (Evans and Goy, 1968; van Horn and Resko, 1977). Furthermore, although females will mate with nongroup males, group males are able to mate first (Sauther, 1991b). In addition, there may be a first mate advantage in this species (Pereira and Weiss, 1991). Therefore, living within a social group may increase a male's chances for successful matings, and males may attempt to transfer into troops where they can acquire the central position.

Male membership in a group may also help buffer males from nutritional stress, and predation. There is evidence that migrating males at Beza Mahafaly may be under higher levels of nutritional stress (Sussman, 1991). Three males of an all-male group (which included both old and young adults) weighed significantly less than other group-living males. In addition, during the study one adult male was found dead and observations of the remains showed signs of predation (Sauther, 1989). It is possible that this male was in the process of migrating, as the remains were found outside the normal home range of his group.

Because troop males are food competitors, especially for lower-ranking females, why do females tolerate more than one non-natal male? Three suggested advantages of a multi-male membership are intergroup resource defense (Wrangham, 1980), predation detection and defense (Leutenegger and Kelly 1977; Busse, 1976; but see Cheney and Wrangham, 1987) and female reproductive success. As already noted, reproductive events appear to be tied to the availability of critical resources between late pregnancy and weaning in ringtailed lemurs. Under such conditions, successful intergroup agonism over access to resources may provide females with a reproductive advantage, and males could improve intergroup competitive abilities. However, among the ringtailed lemur groups observed, males were less involved in intergroup encounters, and often fed on resources while females fought (Sauther, 1992).

*L. catta* are semi-terrestrial and are therefore exposed to a number of terrestrial as well as arboreal predators (Sauther, 1989). Antipredator defense in this species involves vigilance and mobbing behaviors (for aerial predators) once potential predators are sighted. Throughout the study when predators such as raptors were encountered males took an equal or greater role in approaching, and even climbing into trees where such predators were perched. Furthermore, male attention is not focused on infants, allowing them to be more aware of potential dangers in the environment, especially since they tend to feed away from the main core of females. In this sense, males may provide lower-cost sentinels, because increasing the number of females would increase female-female competition for resources due to reproductive synchrony (c.f. Terborgh and Janson, 1986).

A third possibility may be tied to female reproductive success. The common view is that reproductive females are limited resources for which males compete (Trivers 1972; Wrangham, 1980), and that females should primarily be concerned with access to resources which may more directly affect their reproductive success. However, in a species with a highly constrained breeding period such as *L. catta*, female reproductive success may converge with that of males. Among free-ranging *L. catta* all females within a group experience estrus within 1-3 weeks of each other. There is also strong evidence that ringtailed lemur females living within the same forest enter estrus during similar periods. At Berenty Reserve females in four troops of ringtailed lemurs were either observed mating or showed physical signs of estrus (i.e., flushed and swollen genitalia) between April 16 and May 1 (Jolly, 1966). At Beza Mahafaly Reserve mating by females in four separate groups were observed between May 7 and May 26 (Sauther, unpubl. data). With such a large number of females entering estrus during the same short period, males could potentially become swamped and some females might not conceive during their first estrus. Ringtailed lemur females who do not conceive during the first estrous period will not have a second estrous cycle till 40 days later (Evans and Goy, 1968; van Horn and Resko, 1977) and a third cycle may occur 80 days after the second estrous period (van Horn and Resko, 1977). Late births will occur if females are fertilized during this second cycle, which can result in early weaning or weaning during food scarcity which creates undue stress on the infant and the mother. In this study even infants conceived during the first estrous period, but born two weeks later than others were at a distinct disadvantage compared to other infants due to rapid infant development. In both troops these were infants born to lower-ranking multiparous females. For example, during troop progressions such infants attempted to be carried at a time when other infants were moving independently. On numerous occasions these infants also lagged behind the rest of the group, and lost contact with their mothers, who had to come back and retrieve them. Such infants were also required to compete with other infants whose foraging skills were more advanced. Both of these infants disappeared by the beginning of the dry season.

Male group membership may therefore increase reproductive success for both females and males. From the male's perspective, living within a group both increases a male's chance to successfully mate, and to mate earlier than non-group males due to female choice. Furthermore, close male contact may facilitate the onset of estrus in female ringtailed lemurs (Evans and Goy, 1968). Because *L. catta* females have such a narrow reproductive "window", mating with many group males who are immediately available during their restricted estrous period may lead to successful fertilization, avoidance of secondary estrus, and increase their chances of producing viable offspring. Thus females may tolerate a number of non-natal males, but the higher levels of aggression shown by lower-ranking females toward group males indicates that there may be a disparity in the number of males that females of differing ranks may tolerate.

## Sexual Selection and the Evolution of Female Dominance in *Lemur catta*

Among most sexually dimorphic species, larger male size confers feeding priority over females (Smuts, 1987). However, female dominance in ringtailed lemurs is accompanied by a lack of sexual size dimorphism (Kappeler, 1991; Sussman, unpubl. data). If, as is argued here, females face high reproductive costs, and males do not defer to females in feeding contexts, then females must be able to aggressively displace males from feeding sites. It is thus feasible that female choice of small males may have occurred. In addition to female choice, body size in male ringtailed lemurs may also be constrained by reproductive energetics associated with seasonal reproduction. Kappeler (1990b) has made the important point that for smaller species such as lemurs, large body size in males might not be feasible in seasonal environments. Males do appear to be under constant feeding stress throughout the year, either from females or from other males. Furthermore, Clutton-Brock (1985) has suggested that if size dimorphism is not important in intra-male competition, other characteristics will be selected for. Kappeler (1990a) has proposed that agility may be more important than large body size for reproductive competition among male lemurs. Another essential characteristic is endurance. Sauther (1991b) observed the mating season among free-ranging *L. catta* at Beza Mahafaly, and noted that male-male competition for females involved great physical exertion, including climbing up and down trees, and protracted chases involving spectacular leaps. Furthermore, males were repeatedly interrupted during coitus, requiring the male to dismount and chase away other males as many as 25 times prior to ejaculation. In addition, this vigorous behavior occurred during a period of high temperatures (average = 37°C.). After the mating season males had visibly lost weight and appeared nutritionally stressed. Smaller, but physically agile and endurant males may therefore have a reproductive advantage in direct mate competition with other males, as well as being more attractive to females.

## Conclusions

These results support the contention that female dominance in ringtailed lemurs is a response to high reproductive costs which are exacerbated by a stressful, seasonal environment (Jolly, 1984; Young et al., 1990). The patchy distribution, and seasonal nature of food resources, have led to highly synchronized reproduction in this species. While this allows all females to lactate and wean their infants during a period of relative food abundance, it also results in greater feeding competition among females. Females not only compete with other troop females for access to resources, they are also the main participants in intergroup encounters over important seasonal resources (Sauther, 1992). Although males are able to alleviate some feeding pressure to themselves by feeding further away from others, and by seeking out more displaced foods (Sauther, 1993), for females, reproductive costs make such a strategy less viable. Instead, ringtailed lemur females respond to greater feeding pressures, as well as to reproduction during periods of fluctuating resource availability, by exhibiting greater feeding investment which takes the form of direct resource (contest) competition. In this species, male feeding competition is mediated by female dominance. Females actively and aggressively maintain priority of access to resources. Males provide females with direct and indirect feeding competition which coincides with periods of costly reproductive states such as lactation, and periods of low food availability, such as the dry season. All females are affected, but lower-ranking females face dual competition from both males and higher-ranking females. For such females, and their just weaned infants, female dominance may be especially critical.

## ACKNOWLEDGEMENTS

A number of people deserve recognition for their help during the course of this study. I wish to thank Jeff Kaufmann, Behaligno, and the reserve guards for their assistance at Beza Mahafaly. In addition I wish to thank Mme Berthe Rakotosamimanana, Benjamin Andriamihaja, Joseph Andriamampianina, Pothin Rakotomanga, Mark Pidgeon and Sheila O'Connor, all of whom shared their expertise and help. My gratitude also goes to the Ministry of Higher Education and the School of Agronomy of the University of Madagascar. I would also like to thank R.W. Sussman, B.A.Hayes, P. Kappeler, J. Ganzhorn, D. Overdorff, C. Hildebolt, and an anonymous reviewer for helpful comments on on earlier draft, and J. Cheverud for his help on the randomization procedure. This project was funded by NSF Grant #BNS-8619240, a Fulbright Collaborative Grant, a grant from the National Geographic Society #3619-87, the Leakey Fund, and a fellowship from the American Association of University Women.

## REFERENCES

Altmann, J., 1974, Observational study of behavior: sampling methods, *Behaviour* 49:227.

Altmann, J., 1990, Primate males go where the females are. *Anim. Behav.* 39:193.

Andelman, S.J., 1986, Ecological and social determinants of cercopithecine mating patterns, *in*: "Ecological Aspects of Social Evolution," D.I. Rubenstein, and R.W. Wrangham, eds., Princeton University Press, Princeton.

Busse, C.D., 1976, Chimpanzee predation as a possible factor in the evolution of red colobus monkey social organization, *Evolution* 31:907.

Cheney, D.L., and Wrangham, R.W., 1987, Predation *in*: "Primate Societies," B.B. Smuts, D.L., Cheney, R.M Seyfarth, R.W. Wrangham, and T.T. Struhsaker, eds., The University of Chicago Press, Chicago.

Charles-Dominique, P., and Petter, J.J., 1980, Ecology and social life of *Phaner furcifer*, *in*: "Nocturnal Malagasy Primates: Ecology, Physiology and Behavior," P. Charles-Dominique, H.M.Cooper, A. Hladik, E. Pages, G.F. Pariente, A. Petter-Rousseaux, J.J. Petter, and A. Schilling, eds., Academic Press, New York.

Clutton-Brock, T.H., 1985, Size, sexual dimorphism, and phylogeny in primates, *in*: "Size and Scaling in Primate Biology" W.L. Jungers, ed., Plenum Press, New York.

Clutton-Brock, T.H., Albon, S.D. and Guiness, F.E., 1989, Fitness costs of gestation and lactation in wild mammals, *Nature* 337:360.

Edington, E.S., 1980, "Randomization Tests," Marcel Dekker, Inc., New York.

Evans, C.S., and Goy, R.W., 1968, Social behaviour and reproductive cycles in captive ring-tailed lemurs (*Lemur catta*), *J. Zool.* 156:181.

Furuichi, T., 1983, Interindividual distance and influence of dominance on feeding in a natural Japanese macaque troop, *Primates* 24:445.

Ganzhorn, J.U., 1985, Habitat separation of semifree-ranging *Lemur catta* and *Lemur fulvus*, *Folia Primatol.* 45:76.

Hrdy, S.B., 1981, "The Woman That Never Evolved," Harvard University Press, Cambridge.
Jolly, A., 1966, "Lemur Behavior," University Press, Chicago.
Jolly, A., 1967, Breeding synchrony in lemurs, *in*: "Social Communication among Primates," S.A. Altmann, ed., The University of Chicago Press, Chicago.
Jolly, A. 1984, The puzzle of female feeding priority. *in*: "Female Primates: Studies by Women Primatologists," M.F. Small, ed., Alan R. Liss, Inc., New York.
Kappeler, P.M., 1990a, Female dominance in *Lemur catta*: more than just female feeding priority?, *Folia Primatol.* 55:92.
Kappeler, P.M., 1990b, The evolution of sexual size dimorphism in prosimian primates, *Am. J. Primatol.* 21:201.
Kappeler, P.M., 1991, Patterns of sexual dimorphism in body weight among prosimian primates, *Folia Primatol.* 57:132.
Kaufman, R., 1991, Female dominance in semifree-ranging black and white ruffed lemurs, *Varecia variegata variegata*, *Folia Primatol.* 57:39.
Leutenegger, W, and Kelly J.T., 1977, Relationship of sexual dimorphism in canine size and body size to social, behavioral, and ecological correlates in anthropoid primates, *Primates* 18:117.
Manly, B.F.J., 1991, "Randomization and Monte Carlo Methods in Biology," Chapman and Hall, New York.
Mori, A., 1977, Intra-troop spacing mechanisms among Japanese monkeys, *Primates* 14:113.
Noreen, E.W., 1989, "Computer-intensive Methods for Testing Hypotheses: An Introduction," John Wiley and Sons, New York.
Pagès-Feuillade, E., 1988, Modalités de l'occupation de l'espace et relations interindividuelles chez un prosimien nocturne malagache (*Microcebus murinus*), *Folia Primatol.* 50:204.
Pereira, M.E., 1991, Asynchrony within estrous synchrony among ringtailed lemurs (Primates: Lemuridae), *Physiol. Behav.* 49: 47.
Pereira, M.E., and Weiss, M.L., 1991, Female mate choice, male migration, and the threat of infanticide in ringtailed lemurs, *Behav. Ecol. Sociobiol.* 28:141.
Pereira, M.E., Kaufman, R., Kappeler, P.M., and Overdorff, D.J., 1990, Female dominance does not characterize all of the Lemuridae, *Folia Primatol.* 55:96.
Perret, M., 1982, Influence du groupement social sur la reproduction de la femelle de *Microcebus murinus*, *Z. Tierpsychol.* 60:47.
Pollock, J.I., 1979, Female dominance in *Indri indri*, *Folia Primatol.* 30:143.
Rendall, D., 1993, Does female social precedence characterize captive aye-ayes (*Daubentonia madagascariensis*)?, *Int. J. Primatol.* 14: 125.
Richard, A.F., and Heimbuch, R., 1975, An analysis of the social behavior of three groups of *Propithecus verreauxi*, *in*: "Lemur Biology," I. Tattersall, and R.W. Sussman, eds., Plenum Press, New York.
Robbins, C.T., 1983, "Wildlife Feeding and Nutrition," Academic Press, London.
Robinson, J.G., 1981, Spatial structure in foraging groups of wedge-capped capuchin monkeys *Cebus nigrivittatus*, *Anim. Behav.* 29:1036.
Sadleir, R.M.F.S., 1969, "The Ecology of Reproduction in Wild and Domestic Mammals," Methuen and Co., London.
Sauther, M.L., 1989, Antipredator behavior in troops of free-ranging *Lemur catta* at Beza Mahafaly Special Reserve, Madagascar, *Int. J. Primatol.* 10:595.
Sauther, M.L., 1991a, Reproductive costs in free-ranging ringtailed lemurs, *Am. J. Primatol.* 24:133.
Sauther, M.L., 1991b, Reproductive behavior of free-ranging *Lemur catta* at Beza Mahafaly Special Reserve, Madagascar, *Am. J. Phys. Anthropol.* 84:463.
Sauther, M.L., 1992, The Effect of Reproductive State, Social Rank, and Group Size on Resource Use Among Free-ranging Ringtailed Lemurs (*Lemur catta*) of Madagascar, Ph.D. thesis, Washington University, St. Louis.
Sauther, M.L., 1993, Changes in the use of wild plant foods in free-ranging ringtailed lemurs during lactation and pregnancy: Some implications for early hominid foraging strategies, *in*: "Eating on the Wild Side: The Pharmacologic, Ecologic, and Social Implications of Using Noncultigens," N.L. Etkin, ed., University of Arizona Press, Tucson.
Schilling, A., 1979, Olfactory communication in prosimians, *in*: "The Study of Prosimian Behavior," G.A. Doyle, and R.D. Martin, eds., Academic Press, New York.
Smuts, B.B., 1987, Gender, aggression, and influence, *in*: "Primate Societies," B.B. Smuts, D.L. Cheney, R.M. Seyfarth, R.W. Wrangham, and T.T. Struhsaker, eds., The University of Chicago Press, Chicago.
Sussman, R.W., 1972, An ecological study of two Madagascan Primates: *Lemur fulvus rufus* (Audebert) and *Lemur catta* (Linnaeus), Ph.D. thesis, Duke University, Durham.

Sussman, R.W., 1974, Ecological distinctions of sympatric species of Lemur, *in*: "Prosimian Biology," R.D.Martin, G.A. Doyle, and A.C. Walker, Duckworth, London.
Sussman, R.W., 1977, Socialization, social structure, and ecology of two sympatric species of *Lemur*, *in*: "Primate Bio-social Development: Biological, Social, and Ecological Determinants," S. Chevalier-Skolnikoff, and F.E. Poirier, eds., Garland Publishing, Inc., New York.
Sussman, R.W., 1991, Demography and social organization of free-ranging *Lemur catta* in the Beza Mahafaly Reserve, Madagascar, *Am. J. Phys. Anthropol.* 84:43.
Tattersall, I., 1982, "The Primates of Madagascar," Columbia University Press, New York.
Taylor, L., 1986, Kinship, Dominance, and Social Organization in a Semi Free-ranging Group of Ringtailed Lemurs (Lemur catta), Unpublished Ph.D. thesis, Washington University, St. Louis.
Terborgh J., and Janson, C.H., 1986, The socioecology of primate groups, *Ann. Rev. Ecol. Syst.* 17:111.
Trivers, R.L., 1972, Parental investment and sexual selection, *in*: "Sexual Selection and the Descent of Man," B. Cambell, ed., Aldine, Chicago.
van Horn, R.N., and Resko, J.A., 1977, Reproductive cycle of the ring-tailed lemur (*Lemur catta*): Sex steroid levels and sexual receptivity under controlled photoperiods, *Endocrinology* 101:1579.
Wolfheim, J.H., 1977, A quantitative analysis of the organization of a group of captive talapoin monkeys (*Miopithecus talapoin*), *Folia Primatol.* 27:1.
Wrangham, R.W., 1980, An ecological model of female-bonded primate groups, *Behaviour* 75: 262.
Young, A.L., Richard, A.F., and Aiello, L.C. , 1990, Female dominance and maternal investment in strepsirhine primates, *Am. Nat.* 135:473.

# FLEXIBILITY AND CONSTRAINTS OF *LEPILEMUR* ECOLOGY

Jörg U. Ganzhorn

Abteilung Verhaltensphysiologie
Universität Tübingen
Beim Kupferhammer 8
7400 Tübingen

Present address: Deutsches Primatenzentrum
Kellnerweg 4
3400 Göttingen
Germany

## ABSTRACT

Using *Lepilemur mustelinus* as an example of the folivorous guild of lemurs, I address the questions of how potential interspecific competition with *Avahi laniger* and declining habitat suitability affect its leaf selection and habitat utilization, respectively. The quality of leaves eaten by *Lepilemur* has been measured in four different forests of Madagascar. Leaf quality was expressed as the ratio of protein to fiber content. In forests where *Avahi* was absent *Lepilemur* ate high quality leaves. Where *Avahi* was present *Lepilemur* fed on low quality leaves which were inferior to the items eaten by *Avahi*. Unidentified site effects had no effect on leaf choice of *Lepilemur*, once the effect of *Avahi* had been accounted for. This pattern can be considered as behavioral character displacement. Microhabitats used by *Lepilemur mustelinus ruficaudatus* are characterized by dense stands of trees, interpreted as a prerequisite for locomotion. Up to a certain point *Lepilemur* tolerates thinning of trees in the wake of logging and uses microhabitats with greater distances between trees. But tree distances in microhabitats used by *Lepilemur* remain constant when average tree distances are further increased by more intensive forms of logging. This indicates an upper limit due to energetic constraints for distances between trees needed for travel by *Lepilemur*. Clumping in the distribution of food trees increases with increasing group size from the solitary *Lepilemur* to the pair-living *Avahi* to the group-living *Propithecus verreauxi*. All three results are predicted by deterministic models of optimization theory. This suggests that leaf quality and habitat structure are limiting resources for *Lepilemur* and that interspecific competition over high quality leaves is an active force in the folivorous guild of Malagasy lemurs.

## INTRODUCTION

Many primate species can adapt to a wide range of environmental situations (Box, 1991; Johns, 1991; Lee, 1991). Yet, there are limits to a species' behavioral, morphological and physiological flexibility, inhibiting its persistence in modified habitats (Johns and Skorupa, 1987). Features associated with food acquisition and

*Lemur Social Systems and Their Ecological Basis*, Edited by
P.M. Kappeler and J.U. Ganzhorn, Plenum Press, New York, 1993

processing play a central role in this process (Chivers et al., 1984; Chivers, 1991). Food acquisition can be restricted by inter- and intraspecific competition or by limited access to food resources without competitive constraints due to habitat characteristics. Any of these factors pose different boundaries and have different consequences for the primate niche. Using the folivorous sportive lemur (*Lepilemur mustelinus*) as an example, I address the questions how the availability of high quality leaves, potential interspecific competition with the woolly lemur (*Avahi laniger*), and declining habitat suitability affect the shape of the niche of *Lepilemur*. *Lepilemur's* social system will be related to the spatial distribution of food trees in comparison to *Avahi* and Verreaux's sifaka (*Propithecus verreauxi*), two folivorous species with different social systems (Harcourt and Thornback, 1990).

Interspecific competition can take two major forms. Species can either interfere directly, the stronger species excluding other species from preferred resources. Or, species can be restricted by exploitation competition without direct interference (Richard, 1985; Keddy, 1989). While interference competition can be observed directly, exploitation competition is far more difficult to demonstrate (Connell 1983; Schoener, 1983). However, behavioral character displacement, i.e., niche shift and compression in relation to the presence of supposedly competing species are among the best indicators that exploitation competition is acting in a given assemblage of species (Brown and Wilson, 1956; Hutchinson, 1957; MacArthur and Levins, 1967; Diamond, 1978; Dayan et al., 1990; for evidence from field experiments: Belovsky, 1984; Alatalo et al., 1987). While increased interspecific competition should result in niche shift and compression of the potential interspecific competitors, limited access to food resources due to unfavorable habitat characteristics should stretch a species' niche up to the limits where parts of the population can no longer persist. This is what happens to many species in disturbed habitats (Johns and Skorupa, 1987).

A review of leaf selection of all folivorous Malagasy lemur species in relation to leaf chemistry revealed that all species favored leaves with high protein and/or low fiber content. Comparison of various sites in Madagascar also depicted a strong positive correlation between the biomass of folivorous lemurs in a given forest and its leaf quality in representative samples of leaves, when leaf quality was expressed as the ratio of protein to fiber concentrations (Ganzhorn, 1992). This correlation matches the relation between leaf chemistry and the biomass of folivorous colobine monkeys in Africa and Asia (Waterman et al., 1988; Oates et al., 1990). Therefore it is likely that this ratio actually is one of the factors limiting populations of folivorous primates.

Microhabitat structures are poorly defined in terms of components limiting primate populations. Yet, clingers and leapers use microhabitats with vertical supports more frequently than expected by chance (Fleagle, 1984; Ganzhorn, 1989a). As small increments in the energetic demands of locomotion are considered to be biologically relevant (Demes et al., 1991), the availability of these vertical structures for travel may actually constrain the distribution and thus overall population densities of lemurs. Energetic considerations of the cost of locomotion in *Lepilemur mustelinus leucopus* revealed that the vertical component of a leap (i.e. the loss of height and thus energetically the most demanding component of locomotion [Schmidt-Nielsen, 1984]) scales to the square of leap distance (Charles-Dominique and Hladik, 1971). Any increase in distances between trees will therefore impose a disproportionally higher energy demand on this vertical clinger and leaper up to the point where the species reaches its energetic limits. Thus, for folivorous lemurs, and for *Lepilemur* in particular, leaf quality and microhabitat structure may represent two vital resources.

In the present paper, I first discuss effects of interspecific competition on leaf selection of *Lepilemur*. This will be done in relation to food availability in four different forests in Madagascar and in relation to the presence of *Avahi* as a potential competitor. Food availability was measured in terms of the chemical composition of mature leaves available in a given forest. This yields a chemical profile of leaves for each of the four forests. *Avahi* was chosen as a potential competitor because, based on morphological and behavioral characteristics, the two species seem to have the potential to occupy the same niche with respect to feeding behavior and habitat structure. Both species are nocturnal folivores, both weigh about 1kg, both are classified as vertical clingers and leapers, and both are solitary (*Lepilemur*) or occur in pairs or family groups (*Avahi*) (Petter et al., 1977; Tattersall, 1982). I therefore consider them potential competitors even though the two species differ in gut and tooth morphology, possibly allowing niche differentiation on a scale not considered here (Chivers and Hladik, 1980; Maier, 1980; Razanahoera, 1988).

Second, I use selective logging of different intensity to study the effects of experimentally increasing the distance of suitable supports for travel on microhabitat utilization of *Lepilemur*. Even though these structural components of the habitat may not qualify as resources in a sense that animals consume and compete over them (Wiens, 1989), they certainly pose constraints and thus impose selection pressure upon animals with specialized locomotion. If tree distance would in fact constrain *Lepilemur*, then this niche dimension should expand with increasing tree distance up to the point where the species reaches its morphological and energetic limits. Increasing tree spacing beyond this point should not alter the width of this niche dimension any more. Instead the population density should decline (MacArthur and Pianka, 1966; Schoener, 1971; Charnov, 1976).

Third, in view of leaf chemistry and tree distribution as limiting resources I shall discuss links between these ecological variables and the social system of the solitary *Lepilemur* and of *Avahi*, which live in pairs or family groups. This comparison will be extended by including *Propithecus verreauxi*, another folivorous lemur species which lives in larger groups than *Avahi*.

## METHODS

### Taxonomic Problems

The taxonomy of *Lepilemur* is unclear. The formerly six subspecies (Tattersall, 1982) are now considered seven full species (Petter et al., 1977; Tattersall, 1986). Yet, given their parapatric distribution, *Lepilemur* could also represent a superspecies (Haffer, 1986; Richard and Dewar, 1991). Adopting the different taxonomic classifications would change the argument but would not affect the conclusions. For the present article I therefore combined all species or subspecies of *Lepilemur* considered here (*L. mustelinus edwardsi* [at Ampijoroa], *L.m.microdon* [at Analamazaotra], *L.m.ruficaudatus* [at Morondava], and *L.m.septentrionalis* [at Ankarana]) under the name of *Lepilemur mustelinus*.

### Leaf Quality and Interspecific Competition: Behavioral Character Displacement and Ecological Release in Food Quality

I studied leaf selection of *Lepilemur mustelinus* at four different sites: Analamazaotra, Ampijoroa, Ankarana and Morondava (Fig. 1). At Analamazaotra and Ampijoroa *Avahi laniger* was present, at the study sites of Ankarana and Morondava it was absent. Leaves eaten by the two species, as well as representative samples of leaves available in the four forests were collected, dried and analyzed for concentrations of extractable protein, acid detergent fiber, procyanidin tannins, alkaloids, and other components, not considered here. Leaf quality was expressed as the ratio of extractable protein to acid detergent fiber. More information on the sites and methodological details on leaf sampling, chemical plant analysis, evaluation of lemur food selection and statistical analyses are given in Ganzhorn (1992 and quoted lit.).

### Tree Spacing as a Limiting Constraint for *Lepilemur*

To investigate the flexibility in the structural niche I chose tree distance as a potentially constraining variable for *Lepilemur*. This choice was based on the study of Charles-Dominique and Hladik (1971) and the energetic consideration discussed above. Since tree distance increases systematically with increasing logging intensity, the structural suitability of a forest for *Lepilemur* should decline with increasing logging activities. This should allow to test the flexibility of *Lepilemur* towards this niche component and to determine its limits. This study has been carried out in the

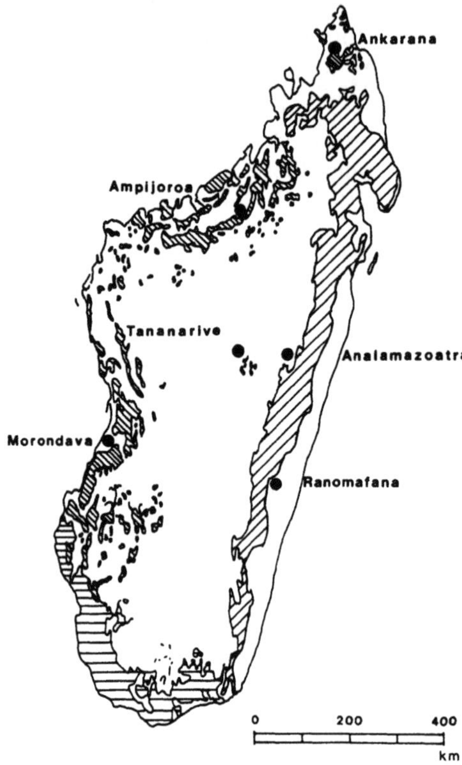


**Figure 1.** Vegetation formations (eastern rain forest, western dry deciduous forest, and southern dry thorn scrub) and study sites in Madagascar (after Nicoll and Langrand, 1989). North is up.

forestry concession of the Centre de Formation Professionnelle Forestière (CFPF) de Morondava, some 40 km north of Morondava (Covi, 1988). Lemur populations were estimated from repeated transect counts (Ganzhorn, 1992).

### Resource Distribution

Food composition of *Lepilemur* and *Propithecus* has been studied in the forestry concession of the CFPF (Morondava) since 1988 (Ganzhorn, unpubl. data). The distribution of food tree species for these two species has been mapped in a one hectare quadrat in the same forest by J.-P. Abraham and H.J. Andrianaina. The distribution of food trees in this forest will be compared with the distribution of food trees at Ampijoroa for the same two lemur species and for *Avahi*. Food tree species used by these lemur species around Ampijoroa have been described by Richard (1978), Albignac (1981), Razanahoera-Rakotomalala (1981), and Ganzhorn (1988). Here, the distribution of some of the food tree species has been mapped by Andriantsoa-Rahelinirina (1985) in an area of 7.1 ha.

To assess patterns in the spatial distribution of trees the mean number of food trees and its variance was determined for different sized quadrats. The variance-to-mean ratio represents an index of dispersion. The resulting frequency distribution per quadrat size were compared to a Poisson distribution. If the variance in the number of trees per quadrat is about equal to the mean and the frequency distribution of trees per quadrat does not differ from a Poisson distribution, the spatial distribution of trees is considered random. If the variance is greater than the mean, the frequency distribution differs from the Poisson distribution, but is not different from a negative binomial, the spatial distribution of trees is considered to be clumped. Procedures and programs follow Ludwig and Reynolds (1988).

## RESULTS

### Leaf Quality and Interspecific Competition: Behavioral Character Displacement and Ecological Release in Food Quality

**Quality of Leaves Available in Different Forests.** The mean ratio between protein and fiber of leaves available in a given forest does not differ either between Analamazaotra and Ankarana, or between Ampijoroa and Morondava. But the two former sites have significantly lower ratios of protein to fiber than the two latter sites (Table 1; unbalanced ANOVA: $F_{3,81}=5.33$; $P=0.002$). Subsequently low or high ratios of protein to fiber will be termed "low or high quality" leaves, even though this measurement does not take into account effects of secondary plant components. Accordingly, the representative samples of leaves at Analamazaotra and Ankarana are classified as being of low quality, whereas leaves available at Ampijoroa and Morondava are classified as high quality.

**Table 1.** Ratio of extractable protein (%) to acid detergent fiber (%) in leaves eaten by *Avahi laniger*, *Lepilemur mustelinus* and in representative samples of mature leaves from four different forests in Madagascar ("Random sample"). Values are means ± standard deviation and sample size.

| | Analamazaotra | Ampijoroa | Ankarana | Morondava |
|---|---|---|---|---|
| *Avahi laniger* | 0.20±0.09<br>N=18 | 0.24±0.14<br>N=9 | absent | absent |
| *Lepilemur mustelinus* | 0.15±0.12<br>N=14 | 0.13±0.08<br>N=13 | 0.45±0.28<br>N=13 | 0.38±0.26<br>N=20 |
| Random sample | 0.11±0.06<br>N=13 | 0.21±0.13<br>N=14 | 0.13±0.09<br>N=24 | 0.19±0.07<br>N=34 |

**Interspecific Competition: Behavioral Character Displacement and Ecological Release.** The ratio of protein to fiber in leaves eaten by *Lepilemur* in the four different forests varies significantly between sites. Where *Avahi* is present (Analamazaotra and Ampijoroa) *Lepilemur* feeds on low quality items. Where *Avahi* is absent (Ankarana and Morondava), *Lepilemur* eats high quality leaves (Table 1; Fig. 2). The presence of *Avahi* explains some 29.4% of the variance in the chemical composition of leaves eaten by *Lepilemur*. According to analysis of variance the differences in the chemical composition of leaves eaten by *Lepilemur* is not influenced by the quality of leaves available in the forest (Table 2). Also, once the presence of *Avahi* is accounted for, unspecified site effects do not contribute significantly to the variance in leaf quality of *Lepilemur* food items ($P>0.5$ according to type I SS in unbalanced ANOVA).

This result is not only due to a shift in the mean quality of consumed leaves, but also due to an increase in niche width, defined as the standard deviation of the mean. In the absence of *Avahi* the variance of the quality of leaves eaten by *Lepilemur* increases significantly (comparison of sites with low quality of the random sample of leaves: Ankarana-Analamazaotra: $F=4.95$, $df=12/13$, $P<0.01$; comparison of sites with high quality of the random sample of leaves: Morondava-Ampijoroa: $F=9.73$, $df=19/12$, $P<0.001$). This increase in variance is independent of the variance in the chemical composition of leaves available in the forest.

**Table 2.** Factors affecting the quality of leaves eaten by *Lepilemur* in four different forests according to unbalanced ANOVA, type III SS. The quality of leaves available in a forest (leaf quality of random sample) was classified as either low (Analamazaotra and Ankarana) or high (Ampijoroa and Morondava).

| | df | F | P |
|---|---|---|---|
| Model | 3/56 | 8.14 | <0.001 |
| Presence of *Avahi laniger* | 1 | 24.39 | <0.001 |
| Leaf quality of random sample | 1 | 0.63 | 0.430 |
| Interaction *Avahi**Leaf quality | 1 | 0.12 | 0.729 |

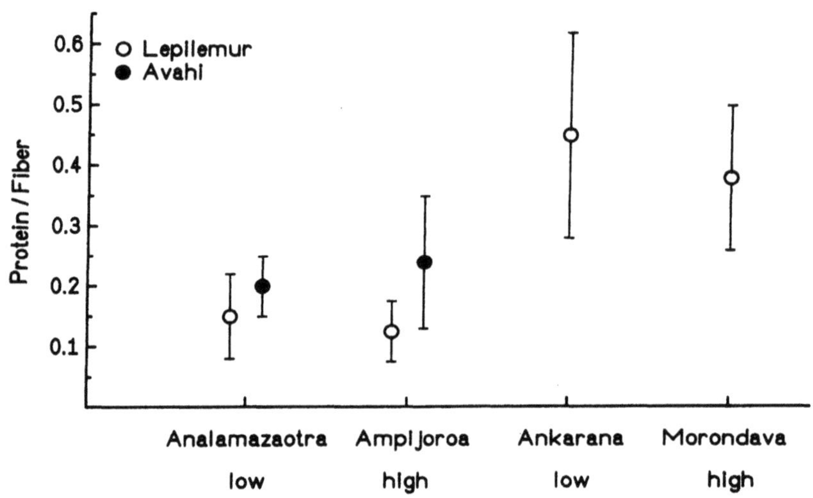


**Figure 2.** Quality of leaves, expressed as the ratio of extractable protein to acid detergent fiber, eaten by *Lepilemur mustelinus* and *Avahi laniger* in four different forests of Madagascar. Values are means and 95% confidence intervals. "low" and "high" mark the average quality of leaves availabe at a given site.

## Leaf Selectivity in Relation to *Avahi* and to Average Leaf Quality of a Forest

**Primary Leaf Chemicals.** For the subsequent analysis all items per site were standardized by the mean quality of the random sample of leaves at that site. This procedure makes it possible to assess the selectivity of *Lepilemur* under different circumstances. Positive deviations of the standardized quality of leaves eaten by *Lepilemur* from 1 (=standardized mean value of leaves available in a forest) indicate selectivity towards high quality items.

According to analysis of variance, using the standardized ratios of protein to fiber as the dependent variable, the presence of *Avahi* still contributes most significantly to the variance in *Lepilemur* leaf selection. But now there is also a significant effect of the quality of leaves available in the forest (Table 3).

Comparing the selectivity of *Lepilemur* within the two pairs of sites with and without *Avahi*, the selectivity of *Lepilemur* is significantly higher at sites characterized by low quality leaves available in the forest than at sites characterized by high quality leaves.

**Table 3.** Factors affecting the selectivity of *Lepilemur* in four different forests according to unbalanced ANOVA, type III SS. To assess selectivity the chemical composition of leaves eaten by *Lepilemur* has been standardized by the mean ratio of protein to fiber of the random sample of leaves per site. For more details see Table 2.

| | df/error | F | P |
|---|---|---|---|
| Model | 3/56 | 10.21 | <0.001 |
| Presence of *Avahi laniger* | 1 | 23.17 | <0.001 |
| Leaf quality of random sample | 1 | 9.57 | 0.003 |
| Interaction *Avahi**Leaf quality | 1 | 1.30 | 0.260 |

**Table 4.** Values of procyanidin tannins standardized by the mean of the random sample of mature leaves per forest and percentages of leaves containing alkaloids in leaves eaten by *Avahi laniger* and *Lepilemur mustelinus*. For tannins values are means ± standard deviations of relative units. For alkaloids the percentage of leaves containing alkaloids is given.

| | Analamazaotra | Ampijoroa | Ankarana | Morondava |
|---|---|---|---|---|
| ***Avahi*** | | | | |
| Tannin | 0.98±0.40<br>N=34 | 1.08±0.71<br>N=9 | absent | absent |
| Alkaloids | 0%<br>N=36 | 0%<br>N=11 | | |
| ***Lepilemur*** | | | | |
| Tannin | 0.73±0.39<br>N=25 | 0.81±0.43<br>N=16 | 0.86±0.44<br>N=13 | 1.04±0.54<br>N=24 |
| Alkaloids | 43%<br>N=28 | 12.5%<br>N=16 | 36%<br>N=14 | 31%<br>N=26 |

**Secondary Leaf Chemicals.** In the areas of sympatry *Lepilemur* shows high selectivity against condensed tannins but includes large numbers of leaves containing alkaloids in its diet. Here *Avahi* avoids alkaloids, but does not discriminate against condensed tannins. Where *Avahi* is absent, the variance of the concentrations of condensed tannin in *Lepilemur's* diet increase due to inclusion of high tannin leaves and the proportion of its food items containing alkaloids declines (Table 4). With respect to procyanidin tannins the selectivity of *Lepilemur* varies in relation to the presence of *Avahi* (analysis of variance: $F_{1,76}=4.23$, $P=0.04$). But there was no effect of *Avahi* on *Lepilemur's* selectivity of leaves with alkaloids ($\chi^2=2.93$, df=1, $P>0.05$). Both tests have been performed after standardizing *Lepilemur* food items by the tannin concentrations and alkaloid presence in the random samples of leaves of the site in question.

## Tree Spacing as a Limiting Constraint for *Lepilemur*

**Tree Spacing.** Among the three forest plots considered here, mean tree distance increased consistently from unlogged plots to areas of low and high logging intensity (Fig. 3). Low logging intensity involves loss of 5-10% of trees >10cm diameter at breast height (DBH) (Deleporte et al., in prep.). High logging intensity is not normal for the study site. This was due to over-zealous logging efforts and approximately doubled the loss of trees.

In the unlogged parts of the forest, *Lepilemur* was found more often in plots characterized by dense stands of trees >10cm DBH compared to random samples of microhabitats described along transects (t=3.09, df=58, $P<0.01$; Fig. 3). In the areas with moderate logging there was no difference in mean tree distances between microhabitats used by *Lepilemur* and the random sample, indicating that *Lepilemur* used all of the available habitat. In the most heavily logged area, *Lepilemur* again used microhabitats with smaller tree distances than found in the representative sample (t=3.45, df=78, $P<0.001$), but there was no difference between microhabitats used by *Lepilemur* in lightly and heavily logged areas (Fig. 3). The variance in tree distance in microhabitats used by *Lepilemur* increased significantly between unlogged and lightly logged parts of the forest. In the transition from light to more intensive logging, the variability in *Lepilemur* microhabitat characteristics remained constant with respect to tree distance. This indicates that *Lepilemur* reached its tolerance with respect to tree distances between the situation at low and high logging intensity as depicted in Figure 3.

**Possible Consequences for *Lepilemur* Population Densities.** The populations of *Lepilemur* remained constant between 1990 and 1992 in the unlogged control of the forest and also in the part of the forest which had been unlogged in 1990 and logged with low intensity by the end of 1990 and in 1991 (Mann-Whitney U-test: $P>0.05$). But populations declined significantly between 1990 and 1992 in the 1990 still unlogged forest plot subject to intensive logging in 1990/91 ($P<0.05$; Fig. 3).

## Resource Distribution

In the unlogged parts of the CFPF forestry concession, the spatial distribution of trees used for leaf eating by *Lepilemur* do not differ from a random distribution at quadrat sizes of $100m^2$, $400m^2$, $900m^2$ and $2500m^2$ ($P>0.05$ for differences to a Poisson distribution; pooled distribution of 11 food tree species used for leaf eating; Fig. 4).

At Ampijoroa, the distribution of *Noronhia oblanceolata*, an occasional food tree for *Lepilemur* (Razanaohera-Rakotomalala, 1981; Ganzhorn, 1988) appears clumped at a scale of $625m^2$ ($P<0.001$ with respect to a Poisson distribution; $P>0.05$ compared to a negative binomial; variance-to-mean ratio [VMR] as index of dispersion = 8.45), but is not different from random, even approaching uniform distribution at a scale of $2500m^2$ ($0.51<\text{VMR}<1.00$, value depending on frequency classes). *Poupartia gymmifera*, another food tree of *Lepilemur* and a food tree for *Avahi* occurs clumped at quadrat sizes equal or greater than $625m^2$ (VMR=1.40).

All trees used for leaf eating by *Propithecus verreauxi* at Morondava (pooled distribution of 11 food tree species; Fig. 4) show significant clumping at quadrat sizes equal or greater than $400m^2$. Due to small sample size statistical tests were not possible at quadrat sizes of $2500m^2$ . But the high variance to mean ratio also indicates clumped distribution at this scale.

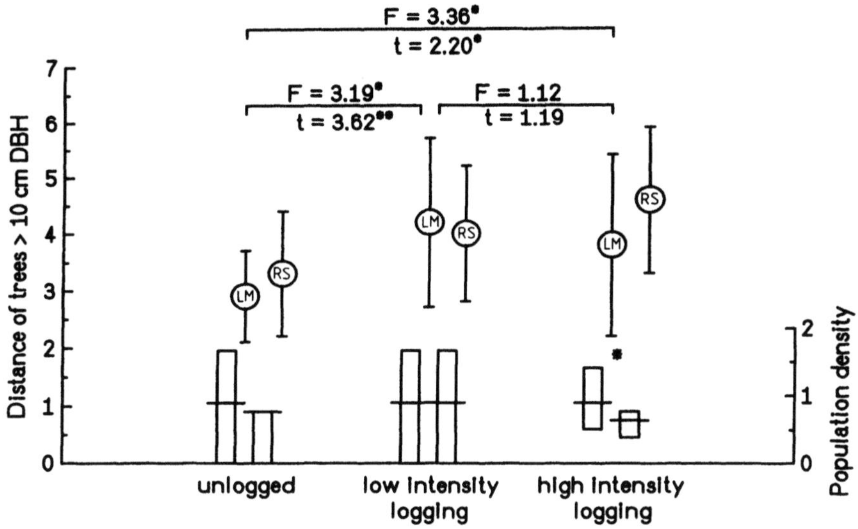


**Figure 3.** Distance (m) of the 4 nearest trees > 10 cm diameter at breast height (DBH) to the center of microhabitats used by *Lepilemur mustelinus* (LM) or to representative samples (RS) of the forest along transect lines. Values are means and standard deviation; N = 10, 35 and 30 for LM and N = 50 for each of the RS. The F and t values refer to differences in variance and means, respectively, between microhabitats used by LM in areas of different logging intensity. The bars at the bottom of the figure represent medians and quartiles of relative *Lepilemur* population densities in 1990 and 1992 (left and right bar of each pair of bars, respectively). * indicates significant changes (P < 0.05) between 1990 (all sites unlogged) and 1992. Population densities are relative units standardized with respect to the median population density of 1990.

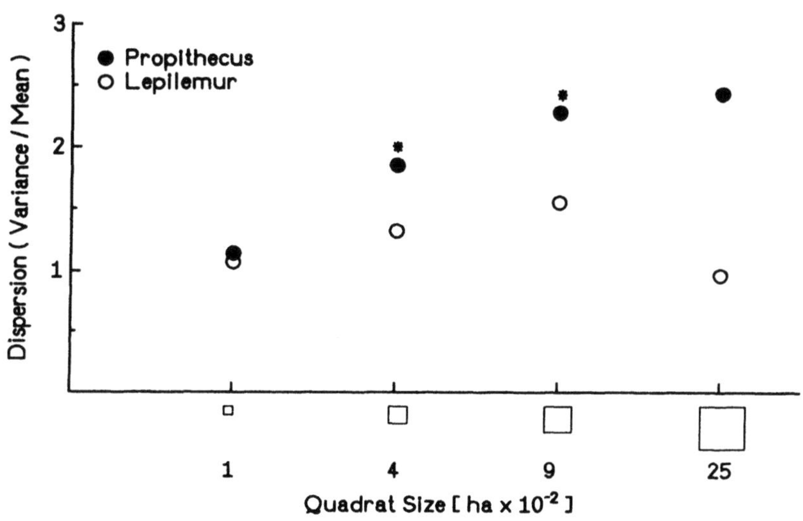


**Figure 4.** Dispersion of trees > 10 cm DBH used by *Lepilemur* or *Propithecus* for leaf eating in the forestry concession of the CFPF, Morondava. Dispersion indices of food trees were calculated for different sized quadrats. Significant deviations (P < 0.05) from a Poisson distribution are indicated by *. For 0.25ha quadrats no statistical test could be applied.

At Ampijoroa variance-to-mean ratios of two food trees of *Propithecus verreauxi* (*Poupartia gymmifera* and *Protorhus deflexa*) range between 1.40 and 4.63. This indicates clumped distributions at all scales considered here. Thus, on average, food trees are most clumped for *Propithecus* and least clumped for *Lepilemur*. The spatial distribution of the one food tree species of *Avahi* is somewhere in between.

## DISCUSSION

### Ecological Factors

Based on positive correlations between the ratio of protein to fiber in leaves available in a given forest and the biomass of folivorous colobine monkeys and lemurs (Waterman et al., 1988; Oates et al., 1990; Ganzhorn, 1992) it was hypothesized that leaf quality expressed by this ratio, is one of the factors limiting populations of folivorous primates. Using *Lepilemur mustelinus* as an example, the present study addressed the question to what extent this resource is in fact limiting and how the availability of high quality leaves and potential interspecific competition with *Avahi laniger* affect the shape of the niche of *Lepilemur*.

Both questions were investigated by applying deterministic optimization models as put forward by MacArthur and Levins (1967). With respect to chemical characteristics of food items, there is a clear separation between *Lepilemur* and *Avahi* in sympatry (Ganzhorn, 1988). In areas without *Avahi*, the quality of leaves consumed by *Lepilemur* is higher than in the areas where both species are present. This feature is best interpreted as behavioral character displacement in the wake of interspecific competition. The results also agree with the prediction that niche width should be reduced under increased competition. These findings suggest strongly that leaf quality is a limiting resource for the guild of folivorous lemurs and that interspecific competition is an important component acting within this guild.

So far the argument has been based on the ratio of protein to fiber concentrations, i.e. on primary plant components. However, there has been increasing awareness that secondary plant chemicals affect primate food selection and species separation (reviews: Glander, 1982; Waterman, 1984; Ganzhorn, 1989b; Glander et al., 1989). The question arises whether or not these components also vary predictably with changes in the lemur community. This ought to be expected if these secondary plant chemicals play a significant role in the evolution of primate communities. Changes in *Lepilemur's* selectivity in relation to the presence of *Avahi* are barely significant in the case of condensed tannins, explaining only some 5.3% of the variance in the data, and not significant in the case of changes in the proportion of leaves containing alkaloids (Table 4). Therefore, with respect to limiting resources secondary plant chemicals play a secondary role compared to primary components at least for the the folivorous guild of lemurs.

As predicted by basic models in community ecology considering a decline of limited resources, the width of the structural microhabitat niche increases with increasing logging intensity. *Lepilemur* is characterized as "slow clinger and leaper" (Richard and Dewar, 1991), which preferentially uses vertical supports for travelling. As tree distances increase with logging, locomotion and travel costs increase and make the logged habitat energetically more demanding for *Lepilemur*. Since tree distances in microhabitats used by *Lepilemur* in the heavily logged area did not increase beyond the tree distances in microhabitats used in moderately logged areas, the flexibility of *Lepilemur* with respect to acceptable inter-tree distances seems to have reached its limits in the moderately logged areas. The population decline in forest logged at high intensity may then be the result of reduced suitability of structural habitat components along with a reduction in food availability.

The present analysis suggests that *Lepilemur* populations are filling the habitat to its carrying capacity set by leaf quality and habitat structure, and that interspecific competition is strongly acting upon the shape of their niche. Starting from the assumption that leaf quality is a prime ecological force acting upon the folivorous guild of lemurs and primates in general, current models in optimization theory and community ecology assuming equilibrium predict all the observed changes in the niche of *Lepilemur*.

### Ecology and Social systems

The spatial distributions of food trees used by the solitary *Lepilemur* do not differ from a random distribution. In contrast, the food trees used by the group-living *Propithecus* appear clumped and the only known dispersion of a food tree of the pair living *Avahi* is intermediate between the dispersion of *Lepilemur* and *Propithecus* food trees. The distributions of trees used for leaf eating by the three folivorous lemur species thus follow the pattern known from other groups of animals, i.e., solitary species feed on food sources distributed evenly or randomly while increased group sizes are linked to increased clumping of food. So far this pattern has been interpreted in relation to maximizing individual energy gain and minimizing the risk of predation (review: Terborgh and Janson, 1986). Yet this can not be all. All three lemur species considered are folivores. Without interspecific interference, all three select leaves based on similar criteria, and all three are preyed upon by the same predator, *Cryptoprocta ferox* (Goodman et al., 1993). They therefore ought to be subject to similar evolutionary selection criteria. But their social systems are different. Thus interactions between plants and animals as well as between different animal species alone are insufficient to explain the evolution of the various social systems. The role of intraspecific processes such as outlined by the "infanticide prevention hypothesis" (van Schaik and Dunbar, 1990; van Schaik and Kappeler, 1993) might have been underestimated in the evolution of social systems so far. Yet, hypotheses ultimately favoring ecological or social causes are linked by circular arguments and mutual dependencies. This may not be due to flimpsy argumentation but rather represent the course of evolution. This should not be described by mutually exclusive polarities of cause and effect favoring social over ecological factors or *vice versa*, but rather by "hypercycles" (Eigen and Schuster, 1977, 1978a,b) and synergetic processes (Haken, 1990) between and within animal and plant species.

## ACKNOWLEDGEMENTS

I am grateful to the Government of Madagascar for permission to conduct this research. I appreciate the help and advice of J.P. Abraham, H. Andrianaina, K. Böhning-Gaese, P.M. Kappeler, O. Langrand, M. Nicoll, J. Pollock, J. Rabesoa, B. Rakotosamimanana, V. and B. Randrianasolo, K. Schmidt-Koenig, E.L. Simons, L. Wilmé, and P.C. Wright. Comments by S. Bearder, R. Crompton, and J.F. Fleagle improved the manuscript. The research was supported by the Fondation Cordama, Deutsche Forschungsgemeinschaft, Fritz Thyssen Stiftung, and the World Wide Fund for Nature.

## REFERENCES

Alatalo, R.V., Eriksson, D., Gustafsson, L., and Larsson, K., 1987, Exploitation competition influences the use of foraging sites by tits: experimental evidence, *Ecology* 68:284.

Albignac, R., 1981, Variabilité dans l'organisation territoriale et l'écologie de *Avahi laniger* (Lémurien nocturne de Madagascar), *Compte Rendus Academie Science Paris 292 Série III*:331.

Andriantsoa-Rahelinirina, L., 1985, Distribution de quelques espèces végétales consommées par les lémuriens dans la forêt d'Ampijoroa, *Mémoire de Diplome d'Etudes Approfondies de Sciences Biologique Appliquées*, Antananarivo: Université de Madagascar.

Belovsky, G.E., 1984, Moose and snowshoe hare competition and a mechanistic explanation from foraging theory, *Oecologia* 61:150.

Box, H.O., 1991,"Primate Responses to Environmental Change," Chapman and Hall, London.

Brown, W.L.Jr., and Wilson, E.O., 1956, Character displacement, *Systematic Zoology* 7:49.

Charles-Dominique, P., and Hladik, C.M., 1971, Le Lépilemur du sud de Madagascar: écologie, alimentation et vie sociale, *Terre Vie* 25:3.

Charnov, E.L., 1976, Optimal foraging: the marginal value theorem, *Theoret. Pop. Biol.* 9:129.

Chivers, D.J., 1991, Species differences in tolerance to environmental change, *in*: "Primate Responses to Environmental Change," H.O. Box, ed., Chapman and Hall, London.

Chivers, D.J., and Hladik, C.M., 1980, Morphology of the gastrointestinal tract in primates: comparisons with other mammals in relation to diet, *J. Morphol.* 166:337.

Chivers, D.J., Wood, B.A., and Bilsborough, A., 1984, "Food Aquisition and Processing in Primates," Plenum Press, New York.

Connell, J.H., 1983, On the prevalence and relative importance of interspecific competition: evidence from field experiments, *Amer. Nat.* 122:661.
Covi, S., 1988, Mise en valeur équilibrée d'un écosystème forestier. La reconstition d'un forêt dense sèche après exploitation, *in*: "L'Equilibre des Ecosystèmes Forestiers à Madagascar: Actes d'un Séminaire International," L. Rakotovao, V. Barre, and J. Sayer, eds., Gland: UICN.
Dayan, T., Simberloff, D., Tchernov, E., and Yom-Tov, Y., 1990, Feline canines: community-wide character displacement among the small cats of Israel, *Amer. Nat.* 136:39.
Deleporte, P., Randrianasolo, J., and Rakotonirina, in prep., Forestry in the dense tropical dry forest, *in*: "Economy and Ecology of a Tropical dry Forest," J.U. Ganzhorn and J.-P. Sorg, eds., Springer, Berlin.
Demes, B., Forchap, E., and Herwig, H., 1991, They seem to glide. Are there aerodynamic effects in prosimian primates? *Z. Morph. Anthrop.* 78:373.
Diamond, J.M., 1978, Niche shifts and the rediscovery of interspecific competition, *Amer. Sci.* 66:322.
Eigen, M., and Schuster, P., 1977, The hypercycle: a principle of natural self-organization, Part A: Emergence of the hypercycle, *Naturwiss.* 64:541.
Eigen, M., and Schuster, P., 1978a, The hypercycle: a principle of natural self-organization, Part B: The abstract hypercycle, *Naturwiss.* 65:7.
Eigen, M., and Schuster, P, 1978b, The hypercycle: a principle of natural self-organization, Part C: The realistic hypercycle, *Naturwiss.* 65:341.
Fleagle J.G., 1984, Primate locomotion and diet, *in*: "Food Aquisition and Processing in Primates," D.J. Chivers, B.A. Wood, and A. Bilsborough, eds., Plenum Press, New York.
Ganzhorn, J.U., 1988, Food partitioning among Malagasy primates, *Oecologia* 75:436.
Ganzhorn, J.U., 1989a, Niche separation of seven lemur species in the eastern rainforest of Madagascar, *Oecologia* 79:279.
Ganzhorn, J.U., 1989b, Primate species separation in relation to secondary plant chemicals, *Hum. Evol.* 4:125.
Ganzhorn, J.U., 1992, Leaf chemistry and the biomass of folivorous primates in tropical forests: Test of a hypothesis, *Oecologia* 91: 540.
Glander, K.E., 1982, The impact of plant secondary compounds on primate feeding behavior, *Ybk Phys. Anthropol.* 25:1.
Glander, K.E., Wright, P.C., Seigler, D.S., Randrianasolo, V., and Randrianasolo, B., 1989, Consumption of cyanogenic bamboo by a newly discovered species of bamboo lemur, *Am. J. Primatol.* 19:119.
Goodman, S.M., O'Connor, S, and Langrand, O., 1993, A review of predation on lemurs: implications for the evolution of social behavior in small, nocturnal primates, *in*: "Lemur Social Systems and Their Ecological Basis," P.M. Kappeler and J.U. Ganzhorn, eds., Plenum Press, New York.
Haffer, J., 1986, Superspecies and species limits in vertebrates, *Z. Zool. Syst. Evolutionsforschung* 24:169.
Haken, H., 1990, "Synergetik. Eine Einführung," Springer, Berlin Heidelberg.
Harcourt, C., and Thornback, J., 1990, "Lemurs of Madagascar and the Comoros", Gland: IUCN.
Hutchinson, G.E., 1957, Concluding remarks, *Cold Spring Harbor Symposia on Quantitative Biology* 22:415.
Johns, A.D., 1991, Forest disturbance and Amazonian primates, *in*: "Primate Responses to Environmental Change," H.O. Box, ed., Chapman and Hall, London.
Johns, A.D., and Skorupa, J.P., 1987, Responses of rain-forest primates to habitat disturbance: a review, *Int. J. Primatol.* 8:157.
Keddy, P.A., 1989, "Competition," Chapman and Hall, London.
Lee, P.C., 1991, Adaptations to environmental change: an evolutionary perspective, *in*: "Primate Responses to Environmental Change," H.O. Box, ed., Chapman and Hall, London.
Ludwig, J.A., and Reynolds, J.F., 1988, "Statistical Ecology," John Wiley & Sons, New York.
MacArthur, R.H., and Levins, R., 1967, The limiting similarity, convergence, and divergence of species, *Amer. Nat.* 101:377.
MacArthur, R.H., and Pianka, E.R., 1966, On optimal use of a patchy environment, *Amer. Nat.* 100:603.
Maier, W., 1980, Konstruktionsmorphologische Untersuchungen am Gebiß der rezenten Prosimiae (Primates), *Abh. Senckenb. Naturforsch. Ges.* 538:1.
Nicoll, M.L., and Langrand, O., 1989, *Madagascar: Revue de la Conservation et des Aires Protégées*, Gland: WWF.
Oates, J.F., Whitesides, G.H., Davies, A.G., Waterman, P.G., Green, S.M., Dasilva G.L., and Mole, S., 1990, Determinants of variation in tropical forest primate biomass: new evidence from West Africa, *Ecology* 71:328.

Petter J.J., Albignac R., and Rumpler Y., 1977, "Faune de Madagascar 44: Mammifères Lémuriens (Primates Prosimiens)," ORSTOM/CNRS, Paris.
Razanaohera, M. R., 1988, Comportement alimentaire de deux espèces sympatriques dans la forêt d'Ankarafantsika (nord-ouest de Madagascar): *Lepilemur edwardsi* et *Avahi laniger* (Lémuriens nocturnes), *in*: "L'Equilibre des Ecosystèmes Forestiers à Madagascar: Actes d'un Séminaire International," L. Rakotovao, V. Barre, and J. Sayer, eds., Gland: UICN.
Razanaohera-Rakotomalala, M., 1981, Les adaptations alimentaires comparées de deux lémuriens folivores sympatriques: Avahi JOURDAN, 1834, Lepilemur I. GEOFFROY, 1851, Thèse de 3ème cycle, Laboratoire de Zoologie-Biologie Générale, Antananarivo: Université de Madagascar.
Richard, A.F., 1978, "Behavioral Variation," Bucknell University Press, Lewisburg, Pa.
Richard, A.F., 1985, "Primates in Nature," Freeman, New York.
Richard, A.F., and Dewar, R.E., 1991, Lemur ecology, *Ann. Rev. Ecol. Syst.* 22:145.
Richard, A.F., Rakotomanga, P., and Schwartz, M., 1991, Demography of *Propithecus verreauxi* at Beza Mahafaly: sex ratio, survival and fertility, *Am. J. Phys. Anthropl.* 84:307.
Schmidt-Nielsen, K., 1984, "Scaling: Why is Animal Size so important?" Cambridge University Press, Cambridge.
Schoener, T.W., 1971, Theory of feeding strategies, *Ann. Rev. Ecol. Syst.* 2:369.
Schoener, T.W., 1983, Field experiments on interspecific competition, *Amer. Nat.* 122:240.
Tattersall, I., 1982, "The Primates of Madagascar," Columbia University Press, New York.
Tattersall, I., 1986, Systematics of the Malagasy strepsirhine primates, *in*: "Comparative Primate Biology, Vol. 1: Systematics, Evolution and Anatomy," D.R. Swindler, and J. Erwin, eds., Liss, New York.
Terborgh, J., 1983, "Five New World Primates," Princeton University Press, Princeton.
Terborgh, J., and Janson, C.H., 1986, The socioecology of primate groups, *Ann. Rev. Ecol. Syst.* 17:111.
van Schaik, C.P., and Dunbar, R.I.M., 1990, The evolution of monogamy in large primates: a new hypothesis and some crucial tests, *Behaviour* 115:30.
van Schaik, C.P., and Kappeler, P.M., 1993, Life history, activity period and lemur social systems, *in*: "Lemur Social Systems and Their Ecological Basis," P.M. Kappeler and J.U. Ganzhorn, eds., Plenum Press, New York.
Waterman, P.G., 1984, Food acquisition and processing as a function of plant chemistry, *in*: "Food Aquisition and Processing in Primates," D.J. Chivers, B.A. Wood, and A. Bilsborough, eds., Plenum Press, New York.
Waterman, P.G., Ross, J.A.M., Bennett, E.L., and Davies, A.G., 1988, A comparison of the floristics and leaf chemistry of the tree flora in two Malaysian rain forests and the influence of leaf chemistry on populations of colobine monkeys in the Old World, *Biol. J. Linn. Soc.* 34:1.
Wiens, J.A., 1989, "The Ecology of Bird Communities," Vol 1 and 2, Cambridge University Press, Cambridge.

## ECOLOGICAL AND REPRODUCTIVE CORRELATES TO RANGE USE IN RED-BELLIED LEMURS (*EULEMUR RUBRIVENTER*) AND RUFOUS LEMURS (*EULEMUR FULVUS RUFUS*)

Deborah J. Overdorff

Duke University
Department of Biological Anthropology and Anatomy
3705-B Erwin Road, The Wheeler Building
Durham, NC 27705
U.S.A.

### ABSTRACT

In this paper, I describe how ecological variables (diet, food availability, patch characteristics) and variables unrelated to food availability such as reproduction and territoriality, affect daily and long-term ranging patterns in two primate species in southeastern Madagascar: the rufous lemur (*Eulemur fulvus rufus*) and the red-bellied lemur (*Eulemur rubriventer*). Daily path lengths (DPL), home range size, and the distance traveled between patches were compared and contrasted between lemur species. Rufous lemurs ranged further on a daily and long-term basis than red-bellied lemurs. Group ranges overlapped extensively and rufous lemurs did not defend home range borders. In contrast, red-bellied lemurs maintained exclusive use of their home range and actively defended boundaries. Additionally, rufous lemurs used more scattered patches and traveled further between these patches than red-bellied lemurs. Ranging patterns in both lemur species were not correlated with food availability but were influenced by three seasonal variations in diet: the number of food patches visited, dietary diversity, and the number of feeding bouts. Rufous lemurs traveled furthest when each of these variables was highest and coincided to the same time period when females were lactating. A similar seasonal peak in red-bellied lemurs ranging patterns was not observed although they traveled further when daily diet was more diverse. It is suggested that the presence of several reproductively active females in rufous lemur groups may influence food choice and consequently, ranging patterns. In contrast, red-bellied lemur's territoriality may limit group movements and constrain daily and seasonal ranging.

### INTRODUCTION

This paper will examine the possible effects of ecological variables (diet, food availability, patch characteristics) and variables unrelated to food availability such as reproductive patterns and territoriality, on the ranging patterns of two congeneric, sympatric species of primate in Madagascar: the rufous lemur (*Eulemur fulvus rufus*) and the red-bellied lemur (*Eulemur rubriventer*). These two species are ecologically and morphologically similar in many ways, yet they have different social structures (group size and composition). Both lemur species are the same body size (~2kg,

*Lemur Social Systems and Their Ecological Basis*, Edited by
P.M. Kappeler and J.U. Ganzhorn, Plenum Press, New York, 1993

Glander et al., 1992), are highly frugivorous, and they eat fruit from many of the same plant species (Overdorff, in press). Due to the consistency in body size and diet between these two lemur species, their ranging patterns also should be similar, particularly if food preferences predispose a species to a particular ranging pattern. Folivores, for example, generally have small ranges compared to frugivores (Milton and May, 1976; Clutton-Brock, 1977; Richard, 1978). In addition, many species feeding on highly patchy food, such as fruits or new leaf flushes, travel further because their range use and travel routes are influenced by the timing, availability, quality, and distribution of their preferred food (Waser, 1975; Pollock, 1979; Terborgh, 1983; Strier, 1987; Meyers and Wright, 1993). Consequently, both lemur species' ranging patterns should change seasonally and vary similarly in response to seasonal food availability.

Despite the similarities in body size and ecology described above, these two lemur species have different social structures (*E. fulvus rufus*: mean group size = 8, range 6-12, N=8; *E. rubriventer*: mean group size = 3, range 3-5, N=12). As a result, this difference in group size may influence ranging differences between the two species. Many authors have shown that daily and long-term ranging patterns vary between large and small groups, based on the assumption that groups must travel further searching for food as group size increases (within-species comparisons: Waser, 1978; Robinson, 1981; van Schaik et al., 1983; between-species comparison: Milton and May, 1976; Terborgh and Janson, 1986; Wright, 1986; Chapman, 1990). Based on these authors' assumptions that larger groups will use patches that differ in size and distribution from those used by small groups, rufous lemurs should use larger, rarer patches that are more widely distributed in the environment than those used by red-bellied lemurs. Therefore, rufous lemurs will have longer daily path lengths, larger home ranges, and travel further between patches than red-bellied lemurs.

Finally, ranging patterns are not always affected by food availability and food preferences but can be limited by other variables such as territoriality or reproductive patterns. Territorial species, for example, defend borders routinely and spend time crossing from side to side to either maintain exclusive access to resources and/or mates (Mitani and Rodman, 1979). Seasonal patterns in reproduction and social organization also can influence the direction of group movements, how far animals travel, and how often groups come into contact with one another (Rasmussen, 1979; 1983). In addition, individual ranges can be affected by reproductive state. Female chimpanzees in estrus will disperse to a neighboring group (Pusey, 1979) and male *Propithecus* will often leave their social group when females in neighboring groups cycle (Richard, 1974, 1985). Because most lemurs are highly seasonal breeders (Rasmussen, 1985; Richard and Dewar, 1992), the timing of reproductive events such as gestation and lactation, may influence a group's ranging patterns as these events require a high energy investment on the part of the female (Oftedal, 1978; Young et al., 1990). Therefore, there are two possible hypotheses to consider. First, groups containing lactating females may travel further searching for higher quality, larger food sources when food is scarce. Alternatively, the birth season may be coordinated with peak food availability so that groups travel less. Therefore, this paper will examine how the ecological, reproductive state, and territoriality may influence daily and long-term ranging patterns and the timing of changes in range use in these two lemur species.

## METHODS

### Study Site

The study site was located in the Ranomafana National Park (RNP) region, a large (40,000ha) southeastern rain forest in Madagascar. RNP is located between 47°18'-47°37' and 21°02'-21°25' S and ranges from montane cloud forest (1500m) to lowland rain forest (500m). The 3.5km$^2$ site for this study, Vatoharanana, is approximately 5km south from the Duke Research station and is a high montane rain forest (altitude: 1125m) with an annual rainfall of 2300mm during this study. Ten sympatric lemur species are found in the area in addition to the study species.

**Study Groups**

Two groups of red-bellied lemurs (RB, group size = 3) and one group of rufous lemurs (RL, group size = 9) were followed from dawn to dusk at least eight days a month from July 1988 through August 1989 with the help of a field assistant. A second study group of rufous lemurs (group size = 7) was observed throughout the study on an opportunistic basis. Over 3000h of data were collected and data are presented from August 1988 through August 1989. Data, however, are unavailable on rufous lemurs for April 1989 as they migrated from the study site (Overdorff, in press). Differences were tested between study groups, within each species, and male and female focal animals. Results were combined if no differences were found.

**Phenology and Botanical Samples**

In July 1988, a sample of 104 trees, representing four mature trees of 26 plant species, were selected that were known to have fruits, flowers, or leaves consumed by primates at RNP. All trees in the sample were checked for the presence of fruit, flowers, and new leaves twice a month (on the fifteenth and thirtieth) throughout the study. The availability of fruit flowers, and new leaves was measured by ranking each phase on a scale from 0 (not present) to 5 (full fruit, flowers, and/or new leaves). This ranking scale reflected the proportion of the tree's crown containing a particular phenological phase and was called an abundance score. Food availability was determined using the following formula:

$$FA = \sum_{i=1}^{s} CV*(A/5)$$

where FA = food availability, i = number of trees with fruit in each sample, CV = crown volume, A = amplitude score. Crown volume was calculated for every tree over 5cm diameter at breast height (DBH) fed in by the focal animal by visually estimating crown radius (m), crown depth (m), crown shape (circular, cylindrical, triangular, dome) to calculate three dimensional volume. Because more than 50% of both lemur species' diets consisted of fruit (Overdorff, 1991, in press), it was assumed that patterns of fruit availability would affect seasonal patch use most critically.

The word "patch" is used to describe a single tree or liana used by a lemur for food. Patch size was calculated for every tree used by the focal animal over five centimeters DBH by using the same techniques described previously (crown volume). When animals fed on liana species, the area of the tree crown in which the liana was growing was considered as the patch size.

Patch density was defined as the number of patches per ha of each plant species eaten by either lemur species. Density estimates were based on ten botanical plots that were laid out randomly in ten areas of the study site (two 25m x 25m plots, and eight 50m x 5m plots). Data were divided into quartiles to categorize plant species as common, moderately common, moderately rare, or rare. The percentage of time an animal spent feeding on common plant species could then be calculated using the data described in the following section.

**Diet**

A combination of focal animal sampling and all occurrence data were used to collect data on feeding (Altmann, 1974). During all day follows, each observer followed a different adult male and female focal animal and focal samples were balanced between all adult individuals in each study group. Different animals were followed on consecutive days. Observers recorded five types of data for each feeding bout: duration, patch size, number of individuals feeding with the focal animal (subgroup size), number and type of food removed by the focal animal, and location of the patch. A feeding bout was defined as a period of time greater than 15 seconds in which the focal animal was eating. Using these data, the following aspects of both lemur species' diets were quantified daily: total amount of time devoted to feeding, number of feeding bouts initiated, number and type of plant parts eaten, number of patches visited, length of feeding bouts, and dietary diversity. The methods used to quantify each of these variables are described in detail elsewhere (Overdorff, in press). Dietary diversity was determined daily by calculating the proportion of time feeding in each food category using the following formula:

$$D = \left(\sum_{i=0}^{i} p_i^2\right)^{-1}$$

where D = dietary diversity index, $p_i$ the proportion of time spent feeding in food category i of each food item within the total sample (see Gautier-Hion, 1980).

## Ranging

Trails were mapped in 50m intervals using a 50m measuring tape and a compass (Overdorff, 1991). All patches greater than 5cm DBH used by either lemur species for food or resting sites were mapped onto a two-dimensional map of the study site using Pathfinder 1.5 (Winslett, unpubl. program). Each new patch visited was labeled with a numbered tree tag so that subsequent visits to the same patch could be recorded. Each number patch was mapped onto the study site map by measuring distance from previously mapped landmarks and taking a compass bearing.

These data were used to calculate daily path lengths (DPL), home range, the location of patches within each home range, and the distance traveled between patches. Daily path lengths were calculated for every sample in which a focal animal was followed at least 12h (usually from dawn to dusk). Days were longest in December but animals were not followed significantly longer than during any other month. Daily path lengths were reconstructed using Pathfinder 1.5 by connecting patches and/or trail markers in the order they were visited or crossed each day. Pathfinder assumes paths traveled between patches were linear. Because both of the study animals fed in almost all the same patches daily, the DPL estimates for males and females were identical.

Home range was calculated by fitting a minimum convex polygon around all DPLs collected for each study group and calculating the area of the polygon. The trail map was divided into hectare quadrats and the number of times a DPL passed through each quadrat was counted to determine if the study group frequented some areas of the range more often than others.

The distance traveled between each mapped food tree was also calculated, using the data described previously. The distance between patches was calculated only for mapped patches and only for patches eaten in consecutively. As with DPL, the distance traveled between patches was the same for males and females as the group typically used the same patches and did not fission while traveling.

## Statistics

Data were compared on a daily basis except when data were compared to seasonal food availability. In this case, data were averaged in semi-monthly intervals (N=24) to coincide with phenological sampling. The Mann-Whitney U-test or Kruskal-Wallis test was used for differences between groups and for differences between study groups of the same species. Spearman's rank correlation coefficient was used to test for correlations between daily ranging patterns, dietary variables, patch size (averaged daily), and distribution of patches.

## RESULTS

### Phenology

Seasonal fruiting, new leafing, and flowering patterns were observed during the study (Fig. 1). Fruit availability peaked from August through October and again in February, but was lowest from April through August 1989. New leaf availability peaked in January and May. Two distinct peaks in flower availability occurred in August 1988, followed by a smaller peak in November-December 1988.

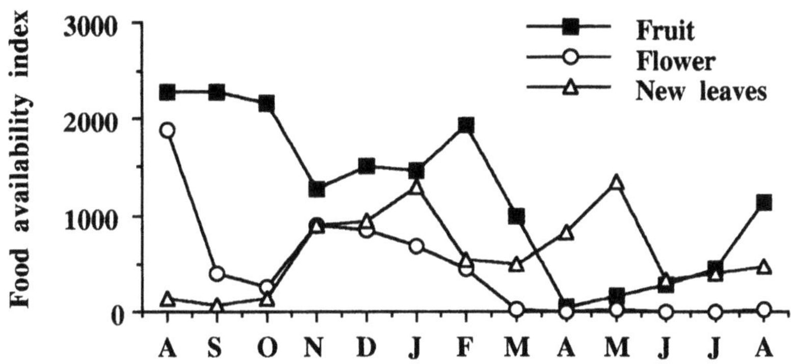


**Figure 1.** Mean fruit, flower,and new leaf availability index from August 1988-August 1989.

### Ranging Behavior

**Daily path length (DPL).** DPL did not differ between the two red-bellied lemur study groups (Group I: N=69; Group II: N=51, z=-0.9, N.S.) or between the two rufous lemur study groups (Group I: N=99, Group II: N=27, z=-1.0, N.S.). Red-bellied lemurs mean monthly DPL varied (H=17.9, df=12, P<0.03) and peaked slightly in October, December, and February (range 70-1020m, mean=444.0m, SE=28.9, N=120, Fig 2.). In comparison, rufous lemurs traveled significantly further than red-bellied lemurs (range 132-2114m, mean=961.0m, SE=52.5, N=126, z=-6.7, P<0.0001). Rufous lemur's mean monthly DPL also varied (H=26.9, df=11, P<0.007) and peaked distinctively from October through January (Fig. 2).

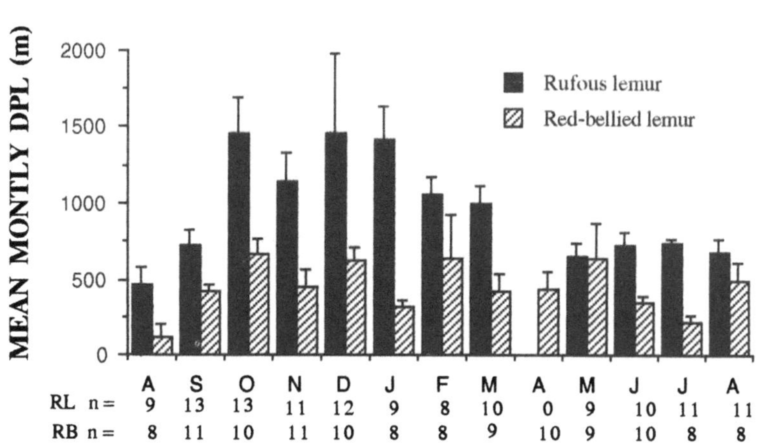


**Figure 2.** Mean monthly DPL and standard error for red-bellied lemurs and rufous lemurs from August 1988 through August 1989 (n = sample sizes for each month).

**Home range.** Both red-bellied lemur study groups had the same size home range (19ha) and groups used their home ranges similarly. Red-bellied lemurs moved evenly through their home range, crossing through the center (Fig. 3). Borders were approached daily and were actively defended by the resident group against intrusion by neighboring groups. Rufous lemur home ranges were approximately ten times larger than red-bellied lemur ranges (Group I = 100ha, Group II = 95ha). In addition, rufous lemurs used their range in different ways compared to red-bellied lemurs. Over 50% of the total number of Group I's DPLs crossed through a 40ha core area in the center of their range (Fig. 4). In addition, study groups ranges completely overlapped and were undefended. In April, I was unable to locate either rufous lemur study group until groups were relocated again in mid-May. It was assumed that the study groups had traveled outside of their known home range.

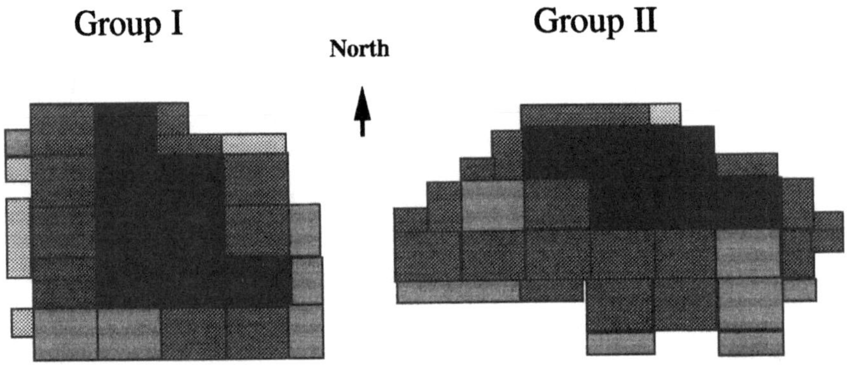


**Figure 3.** Home range and use by red-bellied lemur groups I and II. See Fig. 4 for legend.

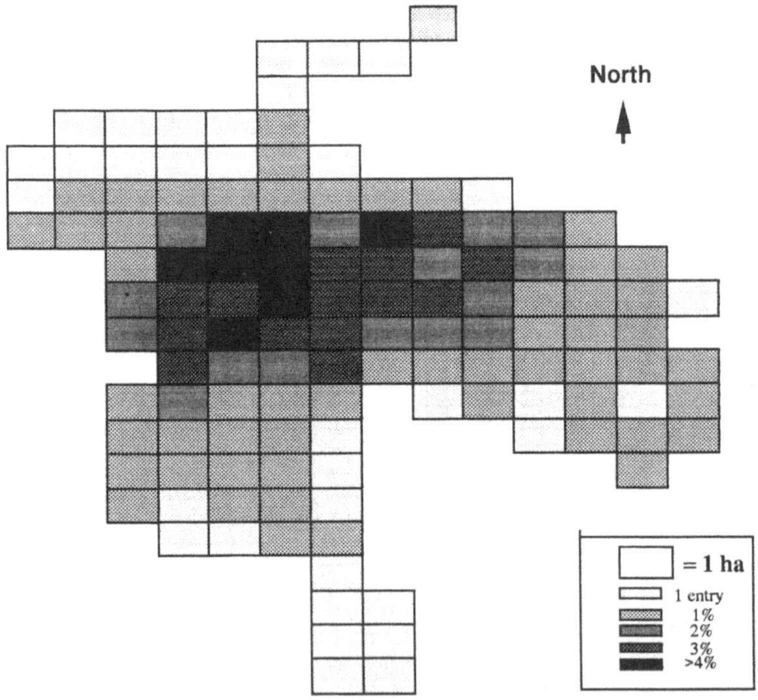


**Figure 4.** Total home range and use by rufous lemur group I from August 1988 through August 1989.

**Distance traveled between patches.** It was assumed that groups that used more clumped patches would travel shorter distances between patches than groups that used more scattered patches. Overall, red-bellied lemurs (mean=49.9m, SE=1.6m, range: 21.9-147.6m, N=863) traveled shorter distances between patches than rufous lemurs (mean=62.3m, SE=2.4m, range: 28.1-191.2m, N=998; z=-3.6, P<0.0004). Both lemur species traveled furthest between patches on days when DPL was longest (RL: $r_s=0.33$, P<0.002; RB: $r_s=0.41$; P<0.0001; Fig. 5).

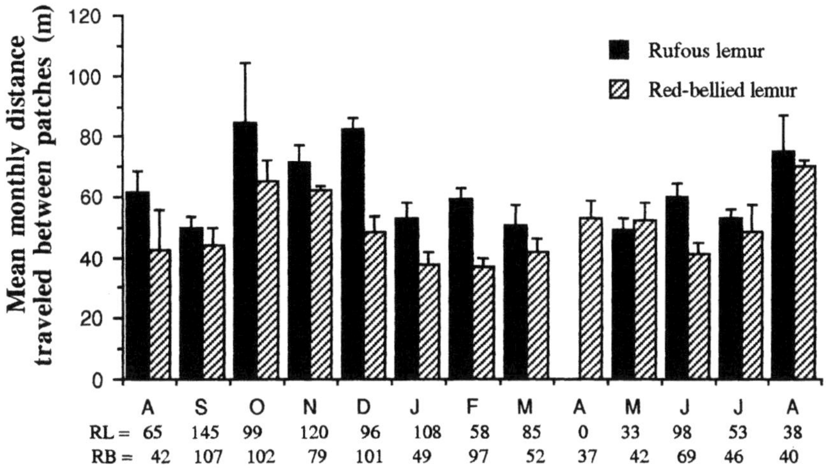


**Figure 5.** Mean distance and standard error traveled each month between patches from August 1988-August 1989. RL = sample size for rufous lemurs; RB = sample size for red-bellied lemurs.

**Climate and Phenology.** Interspecific differences were observed in the relationship of climate with DPL. Rainfall and temperature were not correlated with red-bellied lemur's DPL. In contrast, rufous lemurs traveled further during the warmer, rainy season (Table 1). Neither lemur species' DPL however, was correlated with fruit, flower, or new leaf availability (Table 1) although rufous lemurs tended to travel less when fruit was scarce.

**Table 1.** Correlation statistics ($r_s$) for the relationship of DPL to daily rainfall, maximum temperature, fruit, flower, and new leaf availability for red-bellied lemurs and rufous lemurs.

| | Red-bellied Lemurs | | | Rufous Lemurs | | |
|---|---|---|---|---|---|---|
| | $r_s$ | P | N | $r_s$ | P | N |
| Rainfall | -.19 | NS | 120 | .64 | NS | 126 |
| Temperature | -.24 | NS | 120 | .66 | NS | 126 |
| Fruit | .04 | NS | 24 | .08 | NS | 24 |
| Flowers | -.32 | NS | 24 | .34 | NS | 24 |
| New Leaf | .09 | NS | 24 | .22 | NS | 24 |

**Diet.** The relationship of DPL with the following dietary variables was tested: the amount of time spent feeding daily, the percentage of fruit in the diet, the number of daily feeding bouts, the number of patches visited daily, and dietary diversity. Neither lemur species traveled further on days when they devoted more time to feeding or on days when they ate more fruit (Table 2). However, rufous lemurs traveled further on days when they initiated more feeding bouts, whereas red-bellied lemurs did not (Table 2). Both lemur species also traveled further when they visited more patches and when dietary diversity was highest (Table 2). Both species diets were more diverse from October through February.

**Table 2.** Correlation coefficients ($r_s$) for DPL in relation to the mean number of patches visited, number of bouts initiated daily, and dietary diversity for rufous lemurs and red-bellied lemurs.

| | Rufous Lemur | | | Red-bellied Lemur | | |
|---|---|---|---|---|---|---|
| | $r_s$ | P | N | $r_s$ | P | N |
| # of Patches | .45 | <0.0001 | 126 | .45 | <0.0001 | 120 |
| # of Bouts | .42 | <0.0001 | 126 | .52 | <0.0001 | 120 |
| Dietary Diversity | .52 | <0.0001 | 126 | .51 | <0.0001 | 120 |

## Patch Characteristics

Patch size and the percentage of time spent feeding on common plant species did not correlate with ranging patterns for either species. Neither lemur species traveled further between larger patches and they did not travel further daily when using larger patches (Table 3). Both lemur species, however, used larger patches when fruit was scarce (RB: $r_s = -0.44$, $P < 0.04$; RL: $r_s = -0.42$, $P < 0.04$, $N = 24$). Neither lemur species traveled any more or less when they spent more time feeding on common plant species (Table 3). Rufous lemurs fed more often on common plant species when fruit was most available and rare plant species when fruit was scarce ($r_s = 0.73$, $P < 0.01$). Red-bellied lemurs continued feeding on common plant species throughout the study (Fig. 6).

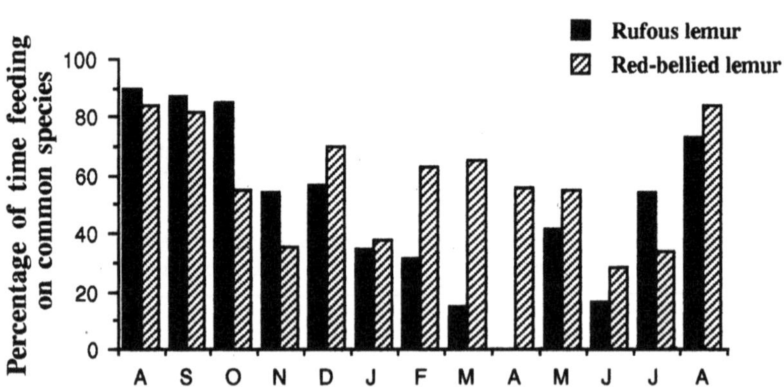


**Figure 6.** Percentage of time spent feeding on common plant species each month by rufous lemurs and red-bellied lemurs.

**Table 3.** Correlation coefficients for the relationship of the distance traveled between patches and DPL to patch size and the percentage of time spent feeding on common plant species for red-bellied lemurs and rufous lemurs.

| | Rufous Lemur | | | Red-bellied Lemur | | |
|---|---|---|---|---|---|---|
| | $r_s$ | P | N | $r_s$ | P | N |
| DPL Patch Size | .05 | NS | 120 | -.05 | NS | 126 |
| DPL % Common | .07 | NS | 120 | -.09 | NS | 126 |
| Distance Patch Size | -.10 | NS | 863 | .11 | NS | 998 |
| Distance Patch Size | .02 | NS | 120 | .01 | NS | 126 |

## DISCUSSION

Although diets were similar, rufous lemurs and red-bellied lemurs did not have similar ranging patterns and did not respond similarly to changes in food availability. Seasonal variations in ranging were observed but did not follow predicted patterns. Variations in ranging patterns (DPL and the distance traveled between patches) between rufous lemurs and red-bellied lemurs were related to dietary changes that were independent of food availability. In fact, rufous lemurs tended to travel less during fruit scarcity, but they fed in larger patches of rare plant species and they initiated fewer bouts that lasted longer. When fruit was most scarce, rufous lemurs disappeared from the study site in April until mid-May. Subsequent annual observations suggest that this is not an unusual behavior when fruit is scarce (Overdorff, in press; Merenlender, pers. comm.). Consequently, it seems that food availability can influence rufous lemur to initially travel long distances (up to 5km) to new food sources when food availability is critical. In contrast, red-bellied lemurs did not show any distinctive peaks in ranging activity and never moved outside the boundaries of their territory. Therefore, food availability does not consistently influence ranging year-round in either lemur species.

Because food availability did not seem to influence ranging, several other dietary variables unrelated to food availability were considered: diet, patch size, dietary diversity, number of bouts and patches visited, and patch distribution. Unlike in other prosimian primates (Pollock, 1979), the percentage of time feeding on fruit was not correlated with ranging patterns. Additionally, neither lemur species traveled further when using larger, rarer patches, as observed in other primates (*Saimiri*: Wright, 1986; *Brachyteles*: Strier, 1987). There were, however, three feeding patterns linked to ranging patterns: dietary diversity, the number of patches visited, and the number of feeding bouts initiated daily. Rufous lemurs traveled further from October through January when diets were more diverse and they initiated more feeding bouts. The plant species eaten during this period, however, were not as common as those exploited earlier in the study (August, September 1988) or as rare as those exploited later (May, June 1989). Patches also were smaller than those used at other times of the year (Overdorff, 1991) and were more scattered based on the mean distance traveled between patches.

Red-bellied lemurs also traveled further on days when dietary diversity was highest and when they visited more patches daily. There was no distinctive peak in travel although travel peaked intermittently from October through February. They also used common plant species year round and exploited patches similar in size to those used by rufous lemurs (Overdorff, 1991). However, they used their range more evenly and had smaller ranges and shorter DPL compared to rufous lemurs.

Although the ecological patterns described above correlated with ranging patterns, these patterns do not, however, explain why rufous lemurs ranging patterns fluctuated out of phase with food availability, or why red-bellied lemur ranging patterns did not peak as distinctively as rufous lemur travel patterns. In addition, the ranging differences between lemur species remains unresolved. One possible explanation is that the changes in rufous lemurs' diet and ranging could be related to seasonal reproductive patterns because reproductive status has been shown to affect ranging patterns in other primates (Rasmussen, 1979, 1983; Richard, 1974, 1985). In this study, infants were born in mid-September (Overdorff, 1991) so females are lactating and nursing from mid-September until infants are weaned in late January or early February. This corresponds with the same time frame as the observed peak in rufous lemur DPL, dietary diversity, and number of feeding bouts. Females may need to diversify diet and initiate more feeding bouts per day than at other times of year due to the presumed energetic burden placed on them while nursing (Crampton and Lloyd, 1959; Portman, 1970; Oftedal, 1978; Young et al., 1990). This may be critical to rufous lemur groups because they are generally larger than red-bellied lemur groups and can contain up to four adult females with infants. In addition, rufous lemur females exclusively carry infants. Because fruit was not as abundant as in previous months, groups may travel further looking for higher quality, more abundant patches. For example, females also eat more insect material which is high in protein (October-November) and more flower nectar which is potentially high in sugar at this time (December and January; Overdorff, in press). Therefore, reproduction may be timed with the availability of high quality foods or essential nutrients rather than strict fruit availability.

An alternative hypothesis, however, is that the presumed costs of travel imposed by larger group size may outweigh the benefits gained from higher quality food (Milton and May, 1976; van Schaik et al., 1983). To further test these hypotheses, comparisons of ranging patterns between similar sized groups with and without lactating females are necessary.

Red-bellied lemur diets tended to follow similar patterns compared to rufous lemurs when females were lactating. Their diets were diverse and gestating females also ate flower nectar and insects. The relationship between reproductive patterns and changes in red-bellied lemur ranging patterns, however, was not as distinctive as the relationship observed in rufous lemurs. This may result from the differences in the number of reproductive females per group and infant care. Red-bellied lemur groups usually contained only one female with an infant each birth season, and males carried infants (Overdorff, 1991; unpubl. data). Therefore, females are alleviated sometimes from the burden of carrying the infant full-time. To test if ranging patterns differ between groups based on the number of nursing infants, a comparison could be made of ranging patterns between groups containing one or two infants (twins have been observed; Overdorff, 1991).

The differences in ranging observed between the two lemur species also may be related to the fact that red-bellied lemurs are territorial. Although red-bellied lemurs did not cover all of their home range daily, they traveled linearly from border to border (Overdorff, 1992). Consequently, they traveled less and used their range more evenly than rufous lemurs based on the number of DPLs that crossed into each hectare quadrant. Red-bellied lemur's DPL therefore, is potentially controlled by the length and width of their territory, especially if they monitor borders on a daily basis (Mitani and Rodman, 1979). Because of their territoriality, red-bellied lemurs are limited to the patches available within their territory that were more clumped and represented more common plant species. Consequently, they did not travel as far between patches. By maintaining territories, however, red-bellied lemurs may be able to maintain exclusive access to these patches. Additionally, the small group size also may limit ranging. Patches used by red-bellied lemurs may not be depleted as quickly as those used by rufous lemurs because group size is smaller. To test if group size and/or territoriality influences ranging behavior, large and small groups of both lemur species (if they can be located) must be studied to determine if ranging patterns are species-typical.

A final consideration in regard to the ranging patterns of these two lemur species is that both of them are cathemeral (active both day and night; see Tattersall, 1987; Overdorff, 1988, 1991). This study only measured daily and long-term ranging patterns based on 12h diurnal samples. Both species have been observed to move at night and the degree to which they range and the seasonal patterns of night time activity and ranging must be further studied to provide a complete picture of how these species interact with their environment and with each other.

## ACKNOWLEDGEMENTS

I would like to acknowledge M. Raymond Rakotoandriana and M. George Rakotoanrivo of the Department of Water and Forests of Madagascar; M. Berthe Rakotsamimanana and M. Benjamin Andriamihaga of the Ministry of Higher Education for permission to work in Madagascar. I would like to thank Missouri Botanical Garden for plant identifications. This paper was improved by comments from K. Glander, D. Doran, J. Ganzhorn, S. Boinski, and an anonymous reviewer. Special thanks to M. E. Winslett for his patience and field assistance and his contribution of the Pathfinder program. The research for this project was funded by National Science Foundation grant BNS-8819559, Wenner-Gren Foundation, Douroucouli Foundation, Duke University Travel Award, and National Sigma Xi Grand in Aid.

## REFERENCES

Altmann, S. A., 1974, Baboons, space, time and energy, *Amer. Zool.* 14:221.

Chapman, C., 1990, Ecological constraints on group size in three species of Neotropical primates, *Folia Primatol.* 55:1.

Clutton-Brock, T.H. and Harvey, P. A., 1977, Primate ecology and social organisation, *J. Zool., Lond.* 183:1.

Crampton, E. and Lloyd, L., 1959, "Fundamentals of Nutrition," W. H. Freeman, San Francisco.

Glander, K. E., Wright, P.C., Daniels, P. S., and Merenlender, A., 1992, Morphometrics and testicle size of rain forest species from southeastern Madagascar, *J. Hum. Evol.* 22:1.

Gautier-Hion, A., 1980, Seasonal variations of diet related to species and sex in a community of *Cercopithecus* monkeys, *J. Anim. Ecol.* 49:237.

Milton, K. and May, M. L., 1976, Body weight, diet, and home range area in primates, *Nature* 259:459.

Mitani, J. C. and Rodman, P. S., 1979, Territoriality: The relation of ranging pattern and home range size to defendability, with an analysis of territoriality among primate species, *Behav. Ecol. Sociobiol.* 5:241.

Oftedal, O., 1980, Milk and mammalian evolution, *in*: "Comparative Physiology: Primitive mammals," K. Schmidt-Nielsen, L. Bolis and C.R. Taylor, eds., Cambridge University Press, Cambridge.

Overdorff, D. J., 1988, A preliminary report on the activity cycle and diet of the red-bellied lemur (*Lemur rubriventer*) in Madagascar, *Am. J. Primatol.* 16:143.

Overdorff, D.J., 1991, Ecological Correlates to Social Structure in Two Prosimian Primates in Madagascar. Ph.D. thesis, Duke University, Durham.

Overdorff, D. J., 1992, Territoriality and home range use by red-bellied lemurs (*Eulemur rubriventer*) in Madagascar, *Am. J. Phys. Anthropol.* Suppl. 14:139.

Overdorff, D. J., in press, Similarities, differences, and seasonal patterns in the diets of *Eulemur rubriventer* and *Eulemur fulvus rufus* in Madagascar, *Int. J. Primatol.*

Pollock, J. I., 1979, Spatial distributions and ranging behavior in lemurs, *in*: "The Study of Prosimian Behavior," G. A. Doyle and R. D. Martin, eds., Academic Press, New York.

Portman, O.W., 1970, Nutritional requirements (NRC) of non human primates. *in*: "Feeding and Nutrition of Nonhuman Primates," R. H. Harris, ed., Academic Press, New York.

Pusey, A., 1979, Intercommunity transfer of chimpanzees in Gombe National park, *in*: "The Great Apes," D. A. Hamburg and E. R. McCown, eds., Benjamin/Cummings, Menlo Park, California.

Rasmussen, D. R., 1979, Correlates of patterns of range use of a troop of yellow baboons (*Papio cynocephalus*): I. Sleeping sites, impregnable females, births, and male emigrations and immigrations, *Anim. Behav.* 27:1098.

Rasmussen, D. R., 1983, Correlates of patterns of range use of a troop of yellow baboons (*Papio cynocephalus*): II. Spatial structure, cover density, food gathering, and individual behavior patterns, *Anim. Behav.* 31:834.

Rasmussen, D. T., 1985, A comparative study of breeding seasonality and litter size in eleven taxa of captive lemurs (*Lemur* and *Varecia*), *Int. J. Primatol.* 5:501.

Richard, A. F., 1974, Patterns of mating in *Propithecus verreauxi verreauxi*. *in*: "Prosimian Biology," R. D. Martin and G. A. Doyle, eds., Duckworth, London.

Richard, A., 1978, "Behavioral Variation: Case Study of a Malagasy Lemur," Bucknell University Press, Lewisburg.

Richard, A. F., 1985, Social boundaries in a Malagasy prosimian, the sifaka (*Propithecus verreauxi*), *Int. J. Primatol.* 6:553.

Richard, A. F. and Dewar, R. E., 1991, Lemur ecology, *Ann. Rev. Ecol. Syst.* 22:145.

Robinson, J. G., 1981, Spatial structure in foraging groups of wedge-capped capuchin monkeys (*Cebus nigrivirgatus*), *Anim. Behav.* 29:1036.

van Schaik, C.P., van Noordwjik, M. A., de Boer, R.J., and den Tonkelaar, I., 1983, The effect of group size on time budgets and social behavior in wild long-tailed macaques (*Macaca fasicularis*), *Behav. Ecol. Sociobiol.* 13:173.

Strier, K. B., 1987, Ranging behavior of woolly spider monkeys, or muriquis (*Brachyteles arachnoides*), *Int. J. Primatol.* 8:575.

Terborgh, J., 1983, "Five New World Primates," Princeton University Press, Princeton.

Terborgh, J. and Janson, C. H., 1986, The socioecology of primate groups, *Ann. Rev. Ecol. Syst.* 17:111.

Waser, P., 1975, Monthly variations in feeding and activity patterns of the mangabey, *Cercocebus albigena* (Lydekker), *E. Afr. Wildl. J.* 13:249.

Waser, P., 1978, Feeding, ranging, and group size in the Mangabey *Cercocebus albigena*, *in*: "Primate Ecology:Studies of Feeding and Ranging Behavior in Lemurs, Monkeys, and Apes," T. H. Clutton-Brock, ed., Academic Press, London.

Wright, P.C., 1986, Ecological correlates to monogamy. *in*: "Primate Ecology and Conservation," J.G. Else and P.C. Lee, eds., Cambridge University Press, Cambridge.

Young, A. L., Richard, A. F., and Aiello, L. C., 1990, Female dominance and maternal investment in strepsirhine primates, *Am. Nat.* 135:473.

## RESOURCE TRACKING: FOOD AVAILABILITY AND *PROPITHECUS* SEASONAL REPRODUCTION

David M. Meyers[1] and Patricia C. Wright[2]

[1]Biological Anthropology and Anatomy
Duke University
3705B Erwin Rd. Wheeler Bldg.
Durham, NC 27705
U.S.A.

[2]Department of Anthropology
State University of New York at Stony Brook
Stony Brook, NY 11794-4364
U.S.A.

## ABSTRACT

In order to examine the ultimate causes of the strict reproductive seasonality found among Malagasy primates, we compared the feeding ecology and reproductive timing of two *Propithecus* species in dry forest and in rain forest. We examined the associations between phenology, feeding behavior, and foraging effort over an annual cycle in golden-crowned sifaka (*P. tattersalli*) and Milne-Edward's sifaka (*P. diadema edwardsi*). Specifically, we described resource tracking and reproductive timing for *P. tattersalli* in very seasonal northern forests and then examined whether *P. diadema edwardsi* in the less seasonal southeastern rain forest follows similar patterns. *Propithecus tattersalli* tracks immature leaves such that when they are available this sifaka will concentrate on this food type regardless of the high availability of staple foods such as seeds and mature leaves. Correlation matrices were used to examine the relationships among food availability, diet and foraging effort. Immature leaf consumption was correlated positively with both immature leaf availability and with DPL in the golden-crowned sifaka. The strongest positive correlation was immature leaf availability with DPL and the strongest negative correlation was mature leaf feeding with DPL. High immature leaf availability and high levels of immature leaf feeding occurred in the early wet season. *Propithecus diadema edwardsi* also increased immature leaf feeding frequency and as well increased total feeding time during periods of increased immature leaf availability. The timing of high immature leaf feeding and availability was similar at both sites. In both species, infants were born in June or July and were weaned in November and December. As *Propithecus* species are anatomical folivores, the increased nutrient demands on lactating females may require higher quality foods as well as higher levels of food intake. In both species, the period of highest immature leaf intake was during late lactation and weaning. Slight differences in the two species' responses to seasonality suggest that, although reproduction may be timed such that late lactation and weaning coincide with high levels of immature leaf availability, resource seasonality alone may not be the ultimate cause of strict breeding seasonality in sifaka.

## INTRODUCTION

### Female Dominance and Strict Breeding Seasonality

The occurrence of female dominance in Malagasy lemurs has been associated with strict breeding seasonality through the idea that breeding seasonality reflects reproductive stress on females, and high reproductive stress selects for female dominance (Hrdy, 1981; Jolly, 1984; Pollock, 1989; Young et al., 1990). Many primate species show seasonal birth peaks (Lancaster and Lee, 1965), some of these species show female dominance (Hrdy, 1981; Jolly, 1984), but no primate fauna which encompasses such a wide range of ecological adaptations as do lemurs shows such a pervasive trend towards strict breeding seasonality.

In general, breeding seasonality in primates has been tied to seasonal patterns of food availability (Charles-Dominique, 1977; Jolly, 1984; van Schaik and van Noordwijk, 1985; Goldizen et al., 1988). One distinction which sets Malagasy lemurs apart from other primates is that reproductive activity (estrous cycle and male gonadal enlargement) in lemurs is likely triggered by day length variation (van Horn, 1974; Rasmussen, 1985) and, in all but a few species, breeding is limited to a short period each year. Although day length is the proximate cause of breeding seasonality in lemurs, the evolutionary cause remains obscure (see Richard and Dewar, 1991).

Two non-exclusive hypotheses concerning the evolution of strict breeding seasonality in lemurs have been suggested. One hypothesis, noted above, suggests that the temporal distribution of primate food resources in Madagascar is unique such that there is relatively stronger seasonal variation (see Jolly, 1984; Richard and Dewar, 1991). Lemurs exist in habitats ranging from semi-desert through rain forests and occur within the latitudinal range of African and South American primates suggesting that stronger resource seasonality is an unlikely evolutionary cause of strict breeding seasonality. Although unlikely, this hypothesis has never been tested due to the lack of adequate phenological data. The second main hypothesis is that lemurs have high metabolic constraints (Richard and Nicoll, 1987; Pollock, 1989) and are thus more sensitive to seasonal changes in food availability or in other environmental conditions. This hypothesis suggests that the relatively low basal metabolic rates found in lemurs (Pollock, 1989; Young et al., 1990) lead to constraints on thermoregulation (Pollock, 1989), on reproductive energetics (Richard and Nicoll, 1987; Nicoll and Thompson, 1987; Young et al., 1990) and on general behavioral flexibility (Pollock, 1989).

### Reproductive Costs and Constraints in Lemurs

For mammals, the most energetically expensive period of reproduction for females is peak lactation (Oftedal, 1984; Nicoll and Thompson, 1987). In some species of hypometabolic mammals, female resting metabolism increases up to or beyond the levels predicted by the Kleiber relationship (Kleiber, 1961) during lactation and drops shortly after infants are weaned (Nicoll and Thompson, 1987). Sifaka females, and other lemur females, may exhibit similar increases in resting metabolism during pregnancy and lactation (Richard and Nicoll, 1987). From an allometric study of reproductive parameters among primates, Young et al. (1990) have shown that relative to available energy (in this case, basal metabolic rate) prenatal maternal-investment rate was higher in lemurs than in Lorisiformes and, along this relative scale, no grade distinction could be made between strepsirhines and haplorhines. Due to a low basal metabolic rate, lemurs have less energy available for reproduction. In sum, reproductive costs seem to be especially high for lemurs and, although some of the cost is due to high prenatal investment (Young et al., 1990), the period of lactation will likely be the most costly for the mother as the cost of infant transport must be borne in addition to the cost of milk production. Although energy is commonly considered the currency of ecology, growth is dependent on protein (Richard, 1985) such that energy may be limiting for reproduction only in its effects on protein acquisition.

Just prior to weaning and infant locomotor independence, the cost of infant transport is at a peak. When the infant begins feeding on high levels of solid food and achieves independent locomotion, the energetic and nutritive burden of growth switches from the mother to the infant. For the infant, the weaning period is obviously stressful due to 1) the high nutrient demands of growth, 2) the need for solid food mastication, 3) the relatively high cost of travel, and 4) the risk of predation (Haring et al., 1988; Wright, 1990). The latter two result from small body size.

### Feeding and Reproduction in Primates

Several studies of primates in the wild and in captivity have shown a strong relationship between the reproductive state of females and various feeding and foraging parameters. In studies of wild and semi-provisioned baboons, pregnant and lactating females fed more and had higher intakes of energy and protein than they did when cycling (Muruthi et al., 1991; Altmann, 1980). Sauther and Nash (1987), in a study of food consumption in captive galagos, found that lactating females significantly increased food intake over intake levels during pregnancy and during the estrous cycle. These lactating females subsequently increased both caloric intake and protein intake (selecting high protein foods) relative to the other reproductive conditions. Furthermore, these authors found that females increased food intake during lactation such that in the later stages of lactation, higher levels of protein and calories were consumed (Sauther and Nash, 1987). A similar increase in energy intake during lactation was reported for cotton-topped tamarins (Kirkwood and Underwood, 1984).

In this chapter, we compare feeding and ranging patterns of two sifaka living in habitats which differ in their phenological cycles. Through an analysis of feeding, ranging, and reproduction of *Propithecus tattersalli* in a strongly seasonal dry northeastern forest, we examine dietary patterns, foraging effort and reproductive timing. We then determine if the larger *Propithecus diadema edwardsi* living in the less seasonal rain forest environment of Ranomafana National Park follows similar patterns.

## METHODS

### Study Sites

**Daraina.** The golden-crowned sifaka (*P. tattersalli*) was studied at three primary dry forest sites in the region of Daraina, northeastern Madagascar (13°14'S latitude, 49°39'E longitude) from July 1990 through June 1991. Regional yearly rainfall averages 1445mm (at Vohemar, Donque 1972). From information recorded at Antsiranana 100km to the north, on average, 86%of the rain falls during the months of December through March, the driest month is September (very little rain falls from May through October), and the greatest number of hours of sunlight occur in October (Griffiths and Ranaivoson, 1972; see Fig. 1). The three sites were located in isolated forest patches located on hills which ranged in peak altitude from 400m to over 1100m (Table 1).

Five botanical transects (two-50x2, two-50x5, and one-50x10m) covering 1200m$^2$ were set up in each of the three sites and all trees over 5cm diameter at breast height (DBH) were identified. Species composition and structure will be presented elsewhere (Meyers, unpubl. data). At each site, all trees in three transects (a total of 499 trees representing over 150 species) were visually monitored monthly for fruit, flowers, immature leaves, and mature leaves. Immature leaves were distinguished from mature leaves by two of the following characters: color (the presence of red), texture, and size. Availability was scored on a scale of 0-3 and presence in this analysis is a score of one or higher. The percentage of trees with an item present was calculated for each month at each site and the sites were averaged to provide regional scores.

**Table 1.** Study sites and group compositions.

| | DARAINA | | | RANOMAFANA | |
|---|---|---|---|---|---|
| | Site Z | Site A | Site M | | |
| Peak Altitude (m) | 1170 | 670 | 400 | 1374 | |
| Peak Deciduousness | 10% | 25% | 50% | 0% | |
| Group Name | PT-VI | PT-II | PT-X | PDE-I | PDE-II |
| Group Size (start-end) | 9-7 | 6-7 | 7-4 | 6-8 | 4-3 |
| # of Adult Females | 2 | 2 | 2-1 | 2 | 2 |
| # of Adult Males | 2 | 1 | 2-1 | 2 | 2-1 |
| # of Successfully Weaned Offspring 1989-1991 | 0 | 2 | 1 | 2 | 0 |

**Ranomafana National Park.** Milne-Edward's sifaka (*P. diadema edwardsi*) was studied at the Talatakely research station (21°16'S latitude, 47°20'E longitude). This site contains degraded primary rain forest on steep slopes and includes over 12km of marked trails over an area of $4km^2$. Yearly rainfall from July, 1988, through June, 1989 was about 2200mm (estimated from Overdorff, 1991). During the four months of December through March, 74% of the yearly rain falls (Griffiths and Ranaivoson, 1972; see Fig. 1).

Phenology was monitored twice monthly for 100 trees, including 25 species, from July 1990 through June 1991. Availability of fruit, immature leaves, and flowers was scored on a scale of 1 to 5. Values of two or more have been scored as present. Since visual phenology scores are by nature subjective and since the methodology differed between Ranomafana and Daraina, we restricted our comparisons to general phenological patterns and not to specific values.

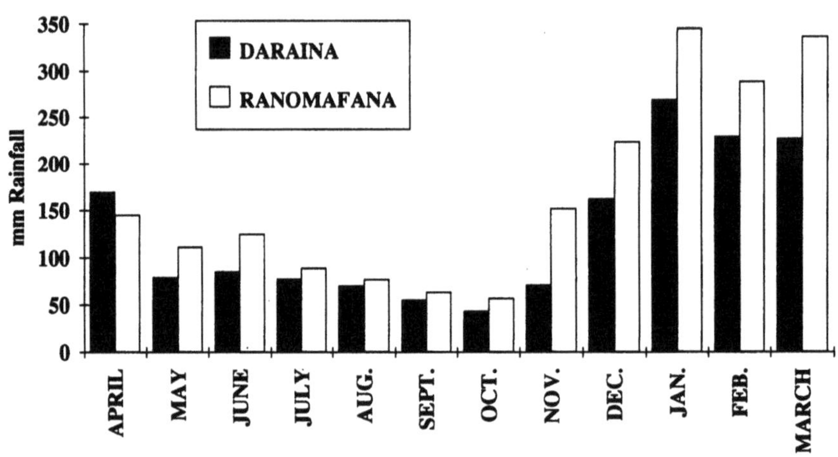


**Figure 1.** Monthly rainfall for the two regions where the study sites are located has been estimated by combining long-term data from the two nearest climatic stations. Antsiranana and Antalaha were used for Daraina; Fiananrantsoa and Mananjary were used for Ranomafana. Data are from Griffiths and Ranaivoson (1972).

### Subjects

**Golden-Crowned Sifaka.** Information on *P. tattersalli* in this chapter was derived from three study groups: one from each site. Group compositions are given in Table 1. Excluding dependent infants, modal group size was 6 animals for groups PT-II and PT-X and was 7 animals for group PT-VI. Individuals were captured and marked for rapid identification using the same methods as Glander et al. (1992).

The golden crowned sifaka has an average adult body weight of 3.5kg. One year old individuals were nearly 70% of adult body weight but are still smaller than similarly aged *P.d. edwardsi* (Ravosa et al., 1993). Different adult body size is the main morphological distinction (aside from coloration) between these two species as they represent scaled versions of each other (Ravosa, 1992; Ravosa et al., 1993).

**Milne-Edward's Sifaka.** Two groups of *P.d. edwardsi* were studied for 10 months from July, 1990 through June, 1991. Individuals had been captured and marked with collars for rapid identification (Glander et al., 1992). Average adult body weight was just under 6kg. By the age of one year, 50% of adult body weight was attained (Ravosa et al., 1993). Group compositions changed during the study and are presented in Table 1.

### Standardized Observations

All groups were followed from waking to sleep tree (or until after dark) for at least three days per month (90% of monthly samples were 4 days or more; one month for *P. d. edwardsi* included only two days). Focal animal observations were conducted on adult and subadult males and females for complete days. At five minute intervals, instantaneous observations (Altmann, 1974) were recorded on activity, height, nearest neighbor, distance to the nearest neighbor, and if the subject was feeding, food species, food type, tree dimensions, and an index of food availability. These observations, hereafter termed individual activity records (IAR), were summed for each sample day then averaged across days for each monthly sample to achieve independent monthly scores expressed as IAR frequencies per day. Opportunistic observations were recorded on activities related to reproduction, infant development, and social behavior.

### Statistical Analysis

All values used for statistical analysis in this chapter were derived from daily frequencies, such as the number of individual activity records (IAR) that were feeding observations, averaged per month. The values for the three study groups of the golden-crowned sifaka were averaged together, as were the values for the two groups of Milne-Edward's sifaka, to get values for each species. Spearman Rank Correlation (Siegel and Castellan, 1988) analysis was used to examine the relationships among indicators of diet, resource availability, and foraging effort (DPL and feeding frequency).

## RESULTS

### Resource Tracking

**Golden-Crowned Sifaka.** In Daraina, the availability of immature leaves varied greatly throughout the study with flowers and fruit varying as well, but to a lesser extent (Fig. 2). All three items had their highest availability during the wet season. The diet of *P. tattersalli*, based on total feeding minutes, consisted of about 17% mature foliage, 22% immature foliage, 13% flowers, 37% unripe entire fruit and seeds, and 9% fruit pulp (Meyers unpubl. data). The seasonal use of main dietary items is presented in Figure 3. Although fruit availability was highest during December through March, seeds were fed upon year-round and were the staple item in the diet of *P. tattersalli* (Fig. 3). When available, flowers and immature leaves were consumed (Fig. 3).

Immature leaf feeding was correlated significantly with immature leaf availability for *P. tattersalli* (Table 2) as was flower feeding, although this is likely due to the predominance of both flowers and immature leaves in the early wet season diet of *P. tattersalli* (note the correlation of immature leaf and flower feeding; Table 2). The strongest positive correlation was that between immature leaf availability and DPL, and the strongest negative correlation was between DPL and mature leaf feeding (Table 2). Seed feeding was not significantly correlated with any measure of availability or foraging effort. The measure of foraging effort, DPL, varied seasonally (Fig. 5). Golden-crowned sifaka traveled further distances when more immature leaves were available ($r_s = 0.860$; $P = 0.001$).

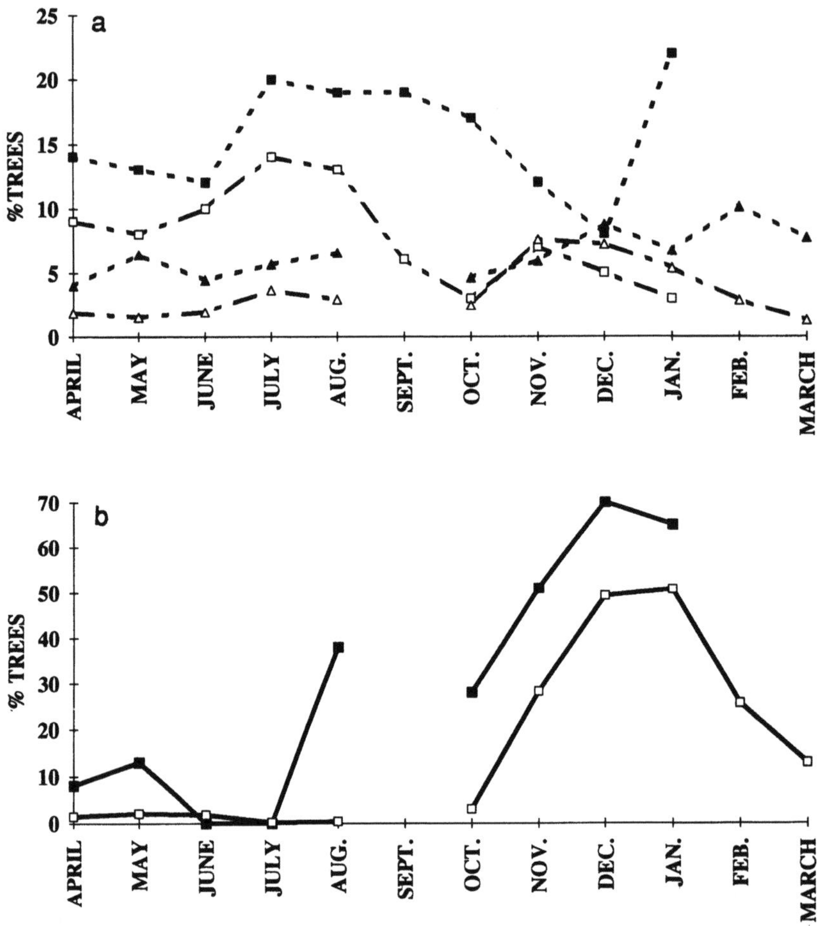


**Figure 2.** Phenological results presented comparatively for the two study sites to show the general similarity in immature leaf phenology (a) and the high variation between sites for flowers and fruit phenology (b). In (a) white and black squares represent data for Daraina and Ranomafanan, respectively. In (b) fruit is presented as short dashes and flowers as short and long dashes. Points for Daraina are triangles and points for Ranomafana are squares. See text for sample sizes and methods.

**Table 2.** Foraging ecology of *P. tattersalli* at Daraina.[1]

| | IMM | MAT | FL | SE | %IMM | %FL | %FR | IAR FEEDING | DPL |
|---|---|---|---|---|---|---|---|---|---|
| IMM | 1 | NS | + | (-) | + + | + | NS | NS | + |
| MAT | -0.420 | 1 | NS | NS | - | NS | - | NS | - |
| FL | 0.685 | -0.378 | 1 | NS | + | NS | NS | (+) | NS |
| SE | -0.504 | -0.482 | -0.329 | 1 | NS | NS | NS | NS | NS |
| %IMM | 0.797 | -0.720 | 0.587 | -0.091 | 1 | NS | (+) | NS | + + + |
| %FL | 0.592 | 0.053 | 0.063 | -0.476 | 0.392 | 1 | NS | NS | NS |
| %FR | 0.259 | -0.706 | 0.259 | 0.315 | 0.573 | 0.158 | 1 | NS | + |
| IAR FEEDING | 0.028 | 0.140 | 0.503 | -0.252 | 0.105 | -0.410 | -0.378 | 1 | NS |
| DPL | 0.685 | -0.895 | 0.496 | 0.161 | 0.860 | 0.245 | 0.587 | -0.084 | 1 |

[1]Spearman Rank Correlation among the following variables: IMM- immature leaf feeding, MAT- mature leaf feeding, FL- flower feeding, SE- seed feeding, %IMM- percent of trees with immature leaves, % FL- flowering trees, % FR- fruiting trees, IAR FEEDING- feeding IAR for all food types, and DPL- daily path length. N=12 in all cases, and significance levels are as follows: (+) $P<0.1$, + $P\leq0.05$, + + $P\leq0.01$, + + + $P\leq0.001$ (two-tailed). See text for further explanations.

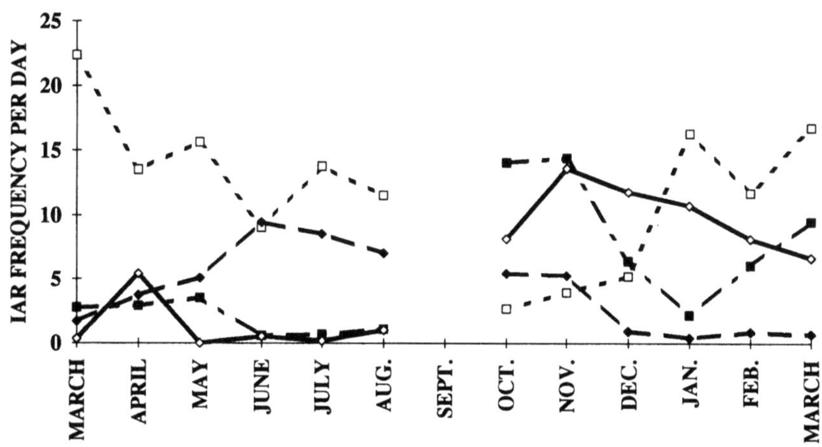


**Figure 3.** Four main dietary items of *P. tattersalli*: seeds (short dashes), mature leaves (long dashes), flowers (short and long dash) and immature leaves (solid line). Data were combined for three study groups and were based on the daily IAR frequencies for each food item.

**Milne-Edward's Sifaka.** Like Daraina, Ranomafana shows seasonal variation in resource availability (Fig. 2) even though rainfall is higher year round (see Table 1). As was the case for Daraina, more trees have immature leaves in the wet season than during the drier months of June, July, and August. The timing of fruit production differed between the two locations (Fig. 2) and may vary yearly at Ranomafana (Overdorff, 1991; Wright, unpubl. data).

Overall, the annual diet of *P. d. edwardsi* consisted of approximately 15% vine leaves, 3% flowers, 26% immature leaves, 20% ripe fruit, and 35% seeds. All fruit (ripe fruit and seeds) accounted for 55% of the diet. The seasonal dietary pattern for *P. d. edwardsi* is presented in Figure 6. For Milne-Edward's sifaka, the only significant positive correlation was between immature leaf feeding and immature leaf availability (Table 3). Unlike the results from Daraina, there is a trend at Ranomafana for an increase in total feeding time associated with increased feeding on immature leaves ($r_s=0.608$, $P<0.10$). For *P. d. edwardsi*, the other measure of foraging effort, DPL, was not correlated to immature leaf availability, nor was DPL correlated with immature leaf feeding (Table 3).

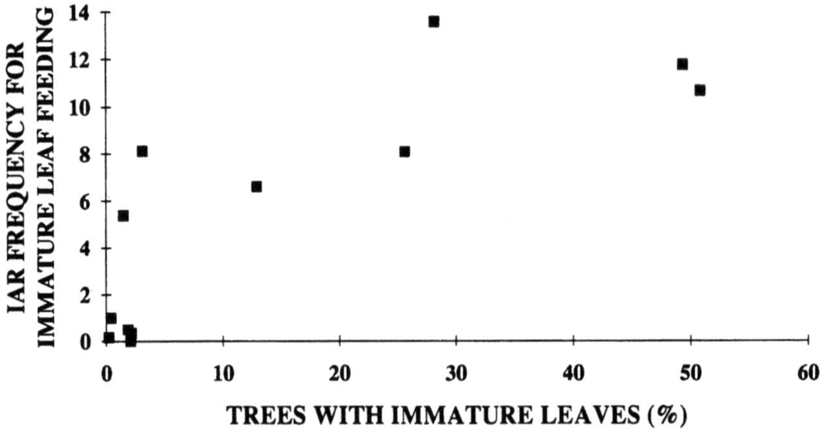


**Figure 4.** Correlation between immature leaf feeding and the percentage of trees with immature leaves ($r_s=0.797$, $P< 0.01$).

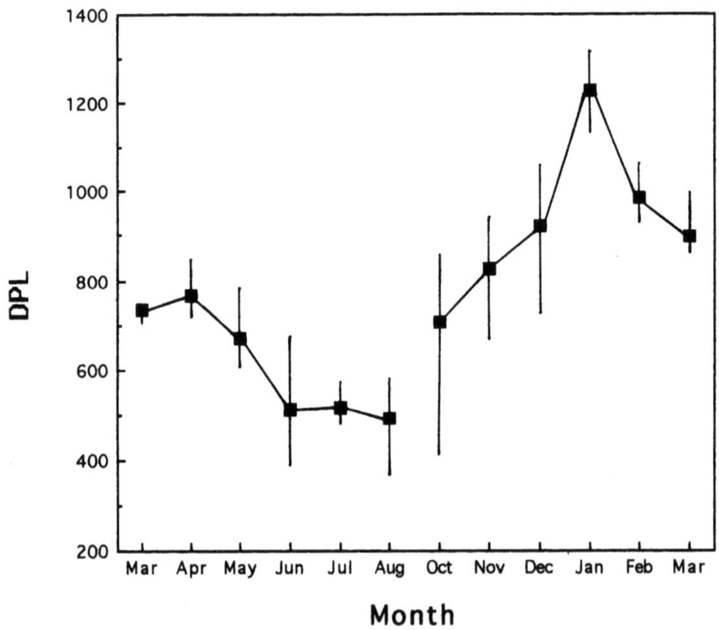


**Figure 5.** Saisonal variation in daily path length of *P. tattersalli* the frequency of immature leaf feeding ($r_s=0.685$, $P<0.05$) and to immature leaf availability ($r_s=0.860$, $P<0.001$).

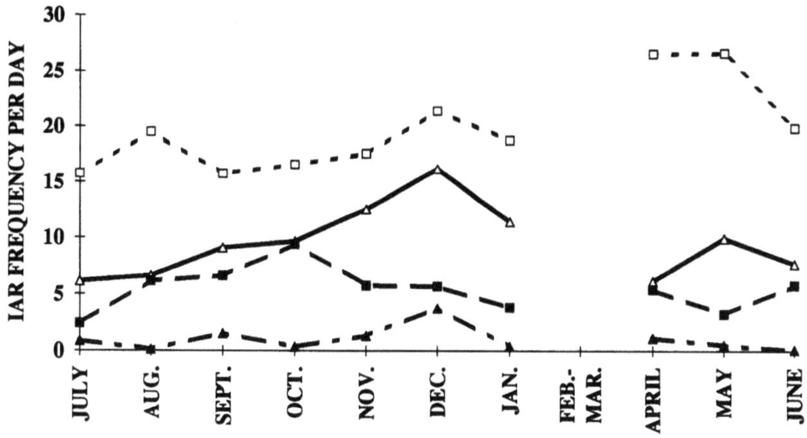


**Figure 6.** Four main dietary items for *P. d. edwardsi*: fruit and seeds (short dashes), immature leaves from trees (solid line), vine leaves (long dashes), and flowers (short and long dashes). Note the different period of study for Ranomafana as compared to Daraina (see Fig. 3).

**Table 3.** Foraging ecology of *P. d. edwardsi* at Ranomafana.[1]

| | IMM | VL | FL | FR | %IMM | %FL | %FR | IAR FEEDING | DPL |
|---|---|---|---|---|---|---|---|---|---|
| IMM | 1 | NS | NS | NS | + | - | NS | (+) | NS |
| VL | 0.049 | 1 | NS | NS | NS | NS | NS | NS | NS |
| FL | 0.035 | -0.176 | 1 | NS | NS | NS | NS | NS | (+) |
| FR | 0.131 | -0.310 | -0.085 | 1 | NS | NS | NS | (+) | NS |
| %IMM | 0.803 | 0.184 | 0.377 | 0.025 | 1 | (-) | NS | NS | NS |
| %FL | -0.732 | -0.292 | -0.255 | 0.092 | -0.681 | 1 | NS | NS | NS |
| %FR | -0.428 | -0.128 | -0.250 | -0.541 | -0.130 | 0.043 | 1 | (-) | NS |
| IAR FEEDING | 0.608 | -0.164 | 0.406 | 0.687 | 0.435 | -0.474 | -0.689 | 1 | NS |
| DPL | 0.255 | 0.418 | 0.576 | -0.353 | 0.226 | -0.274 | 0.384 | 0.212 | 1 |

[1]VL- vine leaves, FR- fruit and seed feeding, all other abbreviations and notations as in Table 2. N=10 for all pairs except those involving %IMM for which N=9.

## Reproductive Timing

**Golden-Crowned Sifaka.** Data concerning the timing of reproduction in *P. tattersalli* are presented in Table 4. The only exact birth date recorded was July 24, 1990, for PT-VI; other birth dates were estimated from the condition of the newborns. Two infants born to group PT-X were found with very sparse thin hair on July 27, 1990 and considered to be at most two days old. On July 30, 1990, another newborn was recorded at the same site and was of similar condition as the infants from PT-X. A fifth infant was estimated to be about two weeks old (from PT-II) when it was first seen on August 7, 1990. The sixth infant born was not observed until about one month after its birth and it was apparently born about the same time as the other infants. In the six groups (including the three main study groups and three additional groups) intensively observed from the birth season through complete infant locomotor and dietary independence, three infants survived, three infants died. In one group of three individual no births were observed. In the one group where two infants were born (PT-X), the oldest of two females was chased from the group shortly after birth and her baby disappeared (considered dead) within three weeks. The other two disappearances (deaths) occurred in September and November.

For the three infants that survived, weaning took place during the four week period of December 20, 1990 through January 18, 1991. Prior to this period, all infants observed were carried by adult females during all movement between trees and they did not feed on large quantities of solid food although they were observed to taste most adult foods. Following this period, suckling was always refused and dorsal riding by the infant was only tolerated for extremely brief periods such as during predator scares.

Matings were observed during the last week in January. All but a few births occurred around July 24th suggesting a gestation period of just under six months. Since lactation lasted five months, there was only one month per year when successfully reproducing females were not gestating or lactating (but were likely cycling).

To summarize the timing of reproductive activities, feeding measures, and forest phenology we note that 1) gestation takes place from the late wet season through the middle of the dry season, 2) lactation occurs from the middle of the dry season through the early wet season, and 3) weaning occurs during the middle of the wet season at a period of high abundance of immature foliage both in the forest and in the diet of *P. tattersalli*.

**Milne-Edward's Sifaka.** The timing of reproduction in *P. d. edwardsi* was very similar to that of *P. tattersalli* (Table 4). Infants were born from May 26 through July 24 and these infants were weaned during December. One female observed to copulate on December 1, gave birth on May 26, suggesting a 176d gestation (Wright, in press). As with the golden crowned sifaka, there was only a short time between weaning and conception. The most notable differences between the two species was that mating occurred almost two months earlier in *P. diadema* whereas weaning was only one month earlier. Thus, although lactation was relatively longer for *P. diadema edwardsi*, weaning still occured earlier than in *P. tattersalli*.

**Table 4.** Comparison of reproductive parameters in two *Propithecus* species.

| | *Propithecus tattersalli* | *Propithecus diadema edwardsi* |
|---|---|---|
| Adult Body Weight | 3.5kg | 5.8kg |
| Mating | late January | early December |
| Births | late July | early June |
| Gestation Length | 6 months | 6 months (176 days) |
| Lactation Period Length | 5 months | 6 months |
| Weaning Period | December | November |
| Age at First Birth | no data | 4 years (N=1) |

## DISCUSSION

### Resource Tracking

At Daraina, the early wet season was characterized by high availability of immature leaves and this food item formed a relatively large part of the golden-crowned sifaka's diet. During this period, seed consumption was at low levels, in spite of the high availability of fruit. In this period of high immature leaf consumption and availability, DPL was also higher. There are two ways to interpret immature leaf consumption. First, the higher foraging effort suggested by the increased DPL might indicate either lower quality resources or widely dispersed patches (Terborgh 1983). Alternatively, a DPL of just over one kilometer is not an absolutely large distance and perhaps a more appropriate question would be why DPL was so short during other times of the year. The months with the shortest DPL (June, July and August) were part of the dry season when mature leaves and petioles, low energy food items, accounted for a large proportion of the diet (Fig. 3; Meyers, 1992; unpubl. data).

Since the phenological pattern at Daraina suggests that the wet season is a good period for fruit and flowers, as well as for immature leaves (Fig. 2), we favor the second interpretation and suggest that the short DPL values during the dry season represent a behavioral mechanism for energy expenditure minimization during periods of low resource abundance (Schoener, 1971; Milton, 1980). The relatively high use of immature leaves during the wet season and the strong correlations of immature leaf feeding and availability indicate that the sifaka are tracking immature leaves. When immature leaves were available, they predominated in the diet; when immature leaves were not available, other items were accepted such as seeds, unripe entire fruit, mature leaves and petioles.

A similar pattern has been found for *Propithecus verreauxi coquereli* in western Madagascar (Richard, 1978) in that immature leaves predominated in the diet when they were available. *P. d. edwardsi* was found to track new leaves and appears to increase foraging effort as well during the period of high immature leaf availability, but there was no increase in DPL with increased immature leaf consumption.

Ganzhorn (1992) has shown that Malagasy rain forests have lower protein to fiber ratios in mature leaves and contain lower folivore biomass than do Malagasy dry forests. Protein to fiber ratios are strong predictors of primate folivore biomass in many areas of the tropics (Waterman et al., 1988; Oates et al., 1990; Ganzhorn, 1992), and immature leaves contain higher ratios than do mature leaves (Davies et al., 1988; Ganzhorn, 1988). These findings together with our results suggest that monitoring protein to fiber ratios in the annual diets of these two sifaka will enable us to better understand foraging and reproductive constraints in this genus.

**Reproductive Seasonality**

The peak in immature leaf availability at Daraina was in December and January, the peak in immature leaf feeding was in November and December, and the peak in DPL was in January. At Ranomafana, the peak in new leaf feeding for *P. d. edwardsi* was from October through December. At this site, immature leaf availability peaked from October through January and again in April; a phenological pattern similar to that found by Overdorff (1991) during 1989. The early rainy season was associated with high availability of immature leaves, and both species in this study responded to the high levels with increased immature leaf intake and increased foraging effort.

Our first conclusion is that weaning was timed as to occur during the period of high immature leaf availability. Furthermore, late lactation occured during the early wet season, which was also within the period of abundant immature leaves. We suggest that successful lactation and weaning in *Propithecus* depends on the high intake of immature foliage for adequate protein intake. Immature leaves tend to have high levels of protein, low levels of fiber and relatively few defensive compounds in comparison to mature leaves (Coley, 1982; Ganzhorn, 1988) and contain higher levels of protein than do seeds (Davies et al., 1988). Increased intake of protein may be an important component of successful lactation in sifaka and in other primates (Sauther and Nash, 1987; Muruthi et al., 1991).

The observation that weaning in *Propithecus* occurred prior to the end of the wet season and during a generally favorable period in terms of immature leaf availability, has two possible implications. First, there is likely a trade-off between the females' needs and the needs of the newly weaned infant. Since infants will continue to grow rapidly until the beginning of the dry season (see Pereira, 1993), upon weaning they are under a high juvenile mortality risk (cf. Janson and van Schaik, 1993). Should locomotor independence occur later in the seasonal cycle, the risk of starvation would be higher. Second, some female primates lose weight during lactation (Goldizen et al., 1988) and female condition impacts on reproductive success. Thus, if lemurs experience relatively high reproductive costs (Jolly, 1984; Richard and Nicoll, 1987; Pollock, 1989; Young et al., 1990) recovering lost condition after weaning could be an important determinant of reproductive success in the following year.

In summary, strict breeding seasonality in *Propithecus* can be considered a mechanism to optimize the timing of nutrient demands on reproducing females and recently weaned offspring such that it coincides with strong availability of high protein food: specifically immature leaves. At most one offspring per female can be produced yearly and successful yearly reproduction is rare in *Propithecus* (Richard et al., 1991; Wright, 1988; Meyers, unpubl. data). Furthermore, *Propithecus* populations have high infant mortality (Richard et al., 1991; Meyers, unpubl. data; Wright, unpubl. data).

Our study has shown that reproductive timing is geared towards late lactation and weaning. Although the hypothesis that resource seasonality is the main determinant of reproductive timing has been supported, it is extremely unlikely that resource seasonality *per se* is the cause of strict breeding seasonality in lemurs. The different phenological patterns and degrees of seasonal variation in resources which we have described in our comparison of Daraina and Ranomafana do not lead to major differences in reproductive timing even though there was large variation in ranging and foraging patterns. In conclusion, resource seasonality is the key to reproductive timing in *Propithecus*, but it is not the ultimate evolutionary cause of strict breeding seasonality. Our results are not incompatible with the hypothesis of metabolic constraints on reproduction as the ultimate cause of strict breeding seasonality in Malagasy primates.

## ACKNOWLEDGEMENTS

The authors greatly acknowledge the permission and assistance of the government of the Democratic Republic of Madagascar, especially the Ministry of Higher Education (MINESUP), the Direction des Eaux et Forôts (MPAEF, DEF), and the Service Provincial des Eaux et Forôts at both Antsiranana and Fianarantsoa. DMM thanks Fred Boltz, Blaid Manantsoa, Bob Razafindrazaka, and Houssein for help at Daraina. Carel van Schaik, Michele Goldsmith, Jörg Ganzhorn, Peter Kappeler, Michelle Sauther, Jukka Jernvall, Teresa Habacker, and an anonymous reviewer are thanked for helpful comments on the manuscript. Special thanks to Jukka Jernvall for data analysis and graphic creations. This work has been supported by Duke University, the Douroucouli Foundation, WWF, and SUNY Stony Brook. This is publication #551 from the Duke University Primate Center.

## REFERENCES


Altmann, J., 1974, Observational study of behavior: Sampling methods, *Behaviour* 49:227.

Altmann, J., 1980, "Baboon Mothers and Infants," Harvard University Press, Cambridge.

Charles-Dominique, P., 1977, "Ecology and Behavior of Nocturnal Prosimians," Columbia University Press, New York.

Coley, P.D., 1982, Rates of herbivory on different tropical trees, *in*: "The Ecology of a Tropical Forest," E.G. Leigh, A.S. Rand, and D.M. Windsor, eds., Smithsonian, Washington.

Davies, A.G., Bennett, E.L., and Waterman, P.G., 1988, Food selection by two Southeast Asian colobine monkeys (*Presbytis rubicunda and Presbytis melalophos*) in relation to plant chemistry, *Biol. J. Linn. Soc.* 34:33.

Donque, G., 1972, The climatology of Madagascar, *in*: "Biogeography and Ecology of Madagascar," R. Battistini and G. Richard-Vindard, eds., W. Junk, The Hague.

Ganzhorn, J.U., 1988, Food partitioning among Malagasy primates, *Oecologia* 75:436.

Ganzhorn, J.U., 1992, Leaf chemistry and the biomass of folivorous primates in tropical forests: test of a hypothesis, *Oecologia* 91:540.

Glander, K.E., Wright, P.C., Daniels, P.S., and Merenlender, A.M., 1992, Morphometrics and testicle size of rainforest lemur species from southeastern Madagascar, *J. Human Evol.* 22:1.

Goldizen, A.W., Terborgh, J., Cornejo, F., Porras, D.T., and Evans, R., 1988, Seasonal food shortage, weight loss, and the timing of births in saddle-back tamarins (*Saguinus fuscicollis*), *J. Animal Ecology* 57:893.

Griffiths, J.F. and Ranaivoson, R., 1972, Madagascar, *in*: "World Survey of Climatology", J.F. Griffiths, ed., W. Junk, The Hague: 10.

Haring, D., Wright, P.C., Izard, M., and Simons, E.L., 1988, Conservation of Madagascar's sifaka (*Propithecus*) in captivity and in the wild, *in*: "Proceedings of Fifth World Conference on Breeding Endangered Species in Captivity," B. Dresser, ed., Cincinnati.

Hrdy, S.B., 1981, "The Woman that Never Evolved," Harvard University Press, Cambridge.

Janson, C.H. and van Schaik, C.P., 1993, Ecological risk aversion in juvenile primates: slow and steady wins the race. *in*: "Juvenile Primates: Life History, Development and Behavior," M.E. Pereira and L.A. Fairbanks, eds., Oxford University Press, New York.

Jolly, A., 1984, The puzzle of female feeding priority, *in*: "Female Primates: Studies by Women Primatologist", M.F. Small, ed., Alan Liss, New York.

Kirkwood, J.K, and Underwood, S.S., 1984, Energy requirements of captive cotton-top tamarins (*Saguinus oedipus oedipus*), *Folia Primatol.* 42:180.
Kleiber, M., 1961, "The Fire of Life," Wiley Press, New York.
Lancaster, J.B., and Lee, R.B., 1965, The annual reproductive cycle in monkeys and apes, *in*: "Primate Behavior: Field Studies of Monkeys and Apes", I. DeVore, ed., Holt, Rinehart and Winston, New York.
Meyers, D.M., 1992, Seasonal and habitat related variation in ranging behavior in the golden-crowned sifaka (*Propithecus tattersalli*), *Am. J. Phys. Anthropol. Suppl.* 14:122.
Milton, K., 1980, "The Foraging Strategy of Howler Monkeys: A Study in Primate Economics," Columbia University Press, New York.
Muruthi, P., Altmann, J., and Altmann, S., 1991, Resource base, parity and reproductive condition affect females feeding time and nutrient intake within and between groups of a baboon population, *Oecologia* 87:467.
Nicoll, M.E., and Thompson, S.D., 1987, Basal metabolic rates and energetics of reproduction in therian mammals: marsupials and placentals compared, *in*: "Reproductive Energetics in Mammals", A.S.I. Loudon and P.A. Racey, eds., *Symp. Zool. Soc. Lond.* 57:7.
Oates, J.F., Whitesides, G.H., Davies, A.G., Waterman, P.G., Green, S.M., Dasilva, G.L., and Mole, S., 1990, Determinants of variation in tropical forest primate biomass: new evidence from West Africa, *Ecology* 71:328.
Oftedal, O.T., 1984, Milk composition, milk yield and energy output at peak lactation: a comparative review, *in*: "Physiological Strategies in Lactation", M. Peaker, G. Vernon, and C.H. Knight, eds., *Symp. Zool. Soc. Lond.* 51:33.
Overdorff, D.J., 1991, Ecological correlates of social structure in two prosimian primates: *Eulemur fulvus rufus* and *Eulemur rubriventer* in Madagascar, Ph.D. Thesis, Duke University, Durham.
Pereira, M.E., 1993, Seasonal adjustment of growth rate and adult body weight in ringtailed lemurs, *in*: "Lemur Social Systems and Their Ecological Basis," Kappeler P.M. and J.U. Ganzhorn, eds., Plenum Press, New York.
Pollock, J.I., 1989, Intersexual relationships amongst prosimians, *Human Evol.* 4:133.
Rasmussen, D.T., 1985, A comparative study of breeding seasonality and litter size in eleven taxa of captive lemurs (*Lemur* and *Varecia*), *Int. J. Primatol.* 6:501.
Ravosa, M.J., 1992, Allometry and heterochrony in extant and extinct Malagasy primates, *J. Human Evol.* 23:197.
Ravosa, M.J., Meyers, D.M., and Glander, K.E., 1993, Relative growth of the limbs and trunk in Sifakas: heterochronic, ecological and functional considerations, *Am. J. Phys. Anthropol.* (in press).
Richard, A.F., 1978, "Behavioral Variation: Case Study of a Malagasy Lemur," Associated University Press, London.
Richard, A.F., 1985, "Primates in Nature," W.H. Freeman and Co., New York.
Richard, A.F. and Dewar, R.E., 1991, Lemur ecology, *Ann. Rev. Ecol. Syst.* 22:145.
Richard, A.F. and Nicoll, M.E., 1987, Female social dominance and basal metabolism in a Malagasy primate, *Propithecus verreauxi*, *Am. J. Primatol.* 12:309.
Richard, A.F., Rakotomanga, P., and Schwartz, M., 1991, Demography of *Propithecus verreauxi* at Beza Mahafaly, Madagascar: sex ratio, survival, and fertility, 1984-1988, *Am. J. Phys. Anthropol.* 84:307.
Sauther, M.L., and L.T., Nash, 1986, The effect of reproductive state and body size on food consumption in captive *Galago senegalensis braccatus*, *Am. J. Phys. Anthropol.* 73:81.
Schoener, T.W., 1971, Theory of feeding strategies, *Ann. Rev. Ecol. Syst.* 2:369.
Siegel, S., and Castellan, N.J., 1988, "Nonparametric Statistics for the Behavioral Sciences," McGraw-Hill, Inc., U.S.A.
Terborgh, J., 1983 "Five New World Primates: a Study in Comparative Ecology," Princeton Univ. Press, New Jersey.
van Horn, R.N., 1975, Primate breeding season: photoperiodic regulation in captive *Lemur catta*, *Folia Primatol.* 24:203.
van Schaik, C.P. and van Noordwijk, M.A., 1985, Interannual variability in fruit abundance and the reproductive seasonality in Sumatran long-tailed macaques (*Macaca fascicularis*), *J. Zool., Lond.* 206:533.
Waterman, P.G., Ross, J.A.M., Bennett, E.L., and Davies, A.G., 1988, A comparison of the floristics and the leaf chemistry of the tree flora in two Malaysian rain forests and the influence of leaf chemistry on populations of colobine monkeys in the Old World, *Biol. J. Linn. Soc.* 34:1.
Wright, P.C., 1987, Diet and ranging patterns of *Propithecus diadema edwardsi* in Madagascar, *Am. J. Phys. Anthropol.* 72 :271.
Wright, P. C., 1990, Patterns of paternal care in primates, *Int. J. Primatol.* 11:89.

Wright, P.C., 1993, Rain forest royalty: the diademed sifaka of Madagascar, *Nat. Hist.*, 14.
Young, A.L., Richard, A.F., and Aiello, L.C., 1990, Female dominance and maternal investment in strepsirhine primates, *Am. Nat.* 135:473.

# SEASONAL BEHAVIORAL VARIATION AND ITS RELATIONSHIP TO THERMOREGULATION IN RUFFED LEMURS (*VARECIA VARIEGATA VARIEGATA*)

Hilary Simons Morland

Department of Anthropology
Yale University
New Haven, CT 06520
U.S.A.

## ABSTRACT

Seasonal behavioral variation has been described in lemur species throughout Madagascar, and has been related to pronounced island-wide environmental seasonality. Although there are regional differences in rainfall patterns, ambient temperature appears to vary in consistent, seasonal cycles throughout Madagascar. Data collected during a field study of black and white ruffed lemurs (*Varecia variegata variegata*) in northeastern Madagascar demonstrated that ruffed lemurs reduced activity levels and travel distances, and increased feeding, sunning, and hunched sitting during the months with the lowest temperatures. Lemurs appear to respond to predictable annual cycles of temperature change with behavioral and social mechanisms rather than purely physiological ones. Ambient temperature is proposed to be a major determinant of seasonal variation in lemur behavior. These hypotheses have implications for the evolution of lemur social organization. The goal of this paper is to focus attention on the relationship between thermoregulation and behavior in lemurs and to suggest a set of issues that merit detailed study.

## INTRODUCTION

Seasonal reproductive cycles and marked seasonal variation in aspects of physiology, behavior, and ecology have been reported in lemur species from all regions of Madagascar (see Tattersall, 1982; Rasmussen, 1985; Richard and Dewar, 1991). Ranging, activity levels, diet, and/or social behavior have been observed to vary seasonally in many species, including *Propithecus verreauxi* (Richard, 1978), *Propithecus tattersalli* (Meyers, 1992), *Indri indri* (Pollock, 1975), *Lemur catta* (Sussman, 1974), and *Eulemur spp.* (Sussman and Richard, 1974; Overdorff, in press). These behavioral changes usually correspond with annual changes in external environmental conditions such as photoperiod, climate, and/or food abundance and availability (e.g., Richard, 1978; Hladik, 1980; Hladik et al., 1980). The most extreme seasonal changes take place in the smallest bodied, nocturnal lemurs, *Cheirogaleus medius* and *Microcebus murinus*. These species undergo distinct annual cycles in body weight, body temperature, tail fat storage, food intake, and activity levels which coincide with annual changes in photoperiod, climate, and food abundance in deciduous forest in western Madagascar (Russell, 1975; Hladik, 1980; Hladik et al., 1980; Petter-Rousseaux, 1980).

Variations in activity levels, diet, habitat use, and/or body weight also have been found to coincide with seasonal changes in one or more environmental conditions in other mammals in temperate, tropical, and subtropical zones (e.g., *Ochotona princeps*: Conner, 1983; *Mustela erminea*: Robitaille and Baron, 1987; rodents: Korn, 1989; *Vulpes vulpes*: Cavallini and Lovari, 1991). Seasonal variation in behavior has been reported in monkeys living in New and Old World tropical rain forests (e.g., Terborgh, 1983; Gautier-Hion, 1980; Raemaekers, 1980; Anderson, 1989) but is by no means universal among primates in tropical forests. For example, Chapman (1988) found few consistent seasonal behavioral changes in three monkey species observed in a highly seasonal Costa Rican forest. Monkeys studied in Kibale forest during dry and wet seasons failed to show strong seasonal trends in ranging and diet, probably because fluctuations in food abundance were not seasonal (Waser, 1975; Struhsaker, 1975).

It has been suggested that many of the traits distinguishing lemurs from other primates can be partially explained as responses to pronounced environmental seasonality in Madagascar (e.g., Jolly, 1984; Rasmussen, 1985; see Richard and Dewar, 1991). However, Richard and Dewar (1991) questioned this argument. Using rainfall as an indicator, they pointed out that climatic seasonality varies widely among regions in Madagascar. For example, the north and west have variable monthly rainfall with prolonged annual dry seasons and the southern region has sparse and irregular rainfall, while the east has no true dry period (Fig. 1). Annual temperature ranges also vary regionally, depending on latitude and coastal proximity, from only 2.8°C in the northwest to 7.7°C in the south (Williams, 1990). Richard and Dewar (1991) also showed that rainfall patterns in Madagascar's eastern forests resemble those in other tropical rain forests which support equally diverse primate communities.

These observations present a paradox: island-wide environmental seasonality has been used to explain seasonal variation in lemur behavior even though behavioral variation has been documented in a wide range of different climatic regimes. In this paper, I illustrate seasonal variation in lemur behavior with data collected during a field study of black and white ruffed lemurs (*Varecia variegata variegata*) in the eastern rain forest in Madagascar. The results suggest that there may be an important relationship between ambient temperature and seasonal behavioral variation in lemurs.

## METHODS

Field work was conducted on the Special Reserve of Nosy Mangabe, a 520ha island located 5km off the coast in the northern end of *V. v. variegata*'s current geographical range. Nosy Mangabe is covered with intact lowland rain forest. Its rugged topography and high rainfall have produced one of the most diverse plant communities in Madagascar, with floral species characteristic of lowland eastern rain forest on laterite (Schatz, unpubl. report).

*V. variegata* is a 3-4kg diurnal, arboreal, frugivorous primate endemic to Madagascar's eastern rain forests (Tattersall, 1982; Morland, 1991). Ruffed lemurs were translocated to Nosy Mangabe in the 1930s (Petter et al., 1977) and current population density is appproximately 29-43/km$^2$ (Morland, 1991). My study population included 26 ruffed lemurs living in two multi-male, multi-female, social groups. Eleven of these ruffed lemurs were captured and marked with color-coded collars while the rest were identified by individual characteristics.

Behavioral data were collected on 14 focal animals using a combination of instantaneous sampling and continuous recording (Altmann, 1974). During dawn-to-dusk observation sessions on one focal animal per day, I sampled behavior category and other variables (e.g., height, location) at 5-min intervals. I collected 1560 hours of observations during 12 months between August 1987 and January 1989.

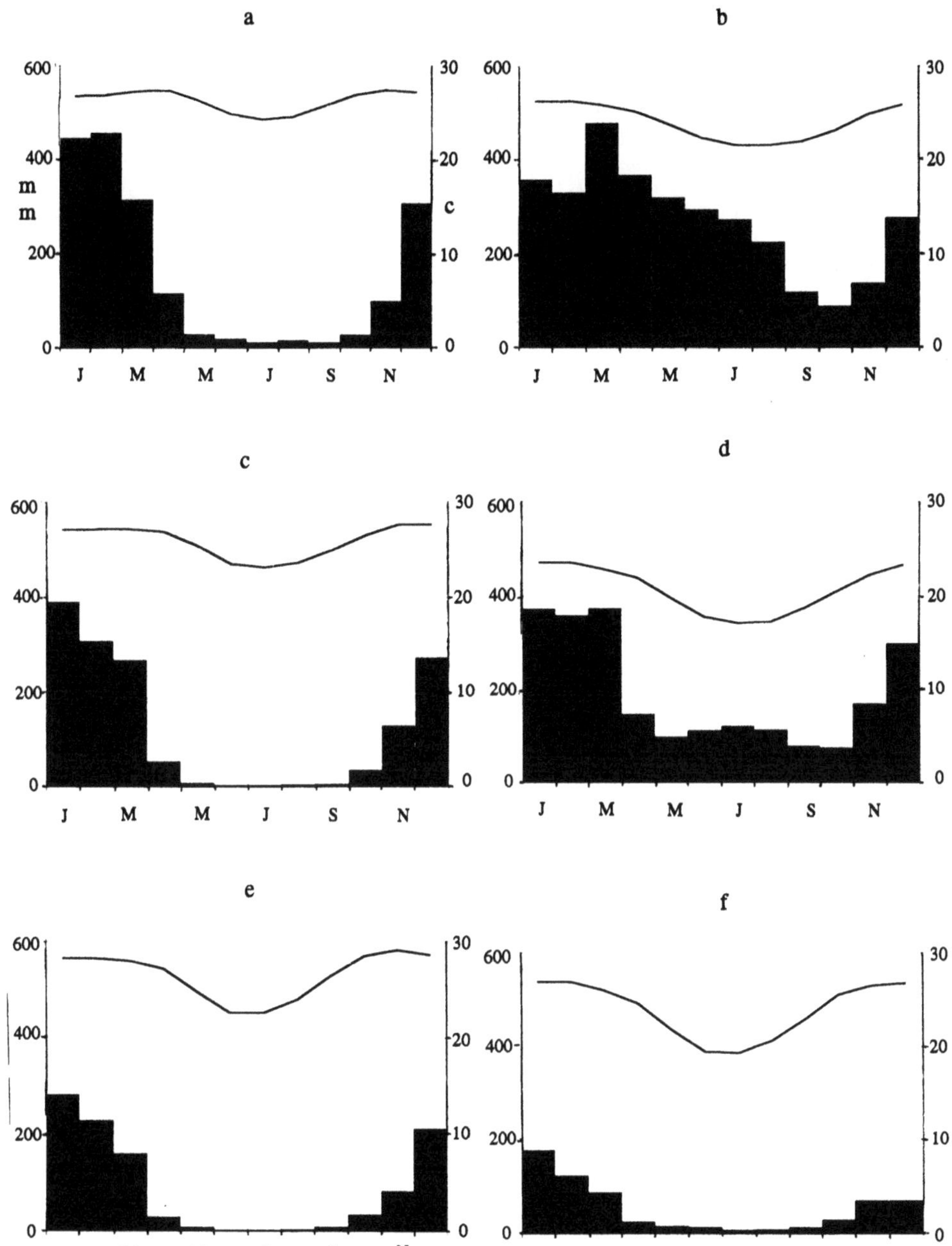


**Figure 1.** Seasonal variation in rainfall and temperature in six regions of Madagascar (Adapted from Richard and Dewar [1991]). Mean monthly rainfall and temperature for six regions were taken from Williams (1990). The bars show the mean regional rainfall (mm) from meteorological stations with records longer than 10 years in each of the six regions. The lines show the mean monthly temperature (°C) from the "best" meteorological stations in each of the six regions for 1951-1960. The regions, with name of a sample station and surface area included in each, are: (a) North (Region #15) - Ambilobe - ?; (b) East Littoral (#3) - Maroantsetra - 10,000km$^2$; (c) Northwest (#11) - Tsaratanana - 120,000km$^2$; (d) East Interior (#1) - Analamazotra, Ranomafana sud - 70,000km$^2$; (e) West (#18) - Isalo - 45,000km$^2$; (f) South (#17) - Ihosy - 120,000km$^2$.

Seasonal behavioral differences will be illustrated with data on activity levels, resting positions, feeding time, and daily travel distance. Instantaneous sampling data were used to determine the percentage of time spent in inactivity, different resting positions, and feeding. The amount of time ruffed lemurs were "inactive" was the mean percentage of 5-min samples in which focal animals were recorded as resting, sleeping, sitting, selfgrooming, or in any other stationary behavior. Four resting positions were used during inactivity: (1) hunched: the animal curled tightly into a ball with the tail wrapped around and covering the limbs, (2) prone: the animal lay with the ventrum stretched full length along a horizontal branch and the tail and limbs hanging down, (3) sunbathe: the animal sat upright or lay on its back with arms extended laterally, orienting and exposing the black skin and fur on the chest and ventrum to the sun, and (4) upright: all other stationary postures (see Pereira et al., 1988). Daily travel distance, or daily path length (DPL), was determined by manually marking and subsequently measuring the total distance traveled by the focal animal during a dawn-to-dusk observation session.

Three seasons were distinguished in this study (see Results). Behavioral data were used from selected months in each season: May-June 1988 were used for cool-wet months, November 1988 for the transitional-dry season, and December 1988-January 1989 for warm-wet months. DPL was measured on 9 days in May-June, 4 days in November, and 2 days in December.

Average monthly rainfall and temperature were determined from records collected at the Maroantsetra meteorological station, located on the mainland about 5km from the study site, by the Government Meteorological Service from 1951 to 1984.

Seasonality can be defined as a predictable, periodic pattern of change in a phenomenon and can be divided into "constancy", in which a state is the same for all seasons in all years, and "contingency", in which a state is different for each season but the pattern is the same for all years (Colwell, 1974).

## RESULTS

### Climatic Seasonality

Mean annual rainfall in the region of the study site is 3709mm with 250 days of rain. Rain falls in all months of the year with a peak usually occurring in the austral mid-summer and a trough occurring during a short period at the end of the austral winter (Fig. 2). The amount and distribution of rainfall varies annually, probably due to the occurrence of cyclones. There is a 6.3°C difference between the lowest and highest average monthly temperatures. Temperature appears to vary in a predictable seasonal cycle.

Williams' (1990) division of the year into four unequal length seasons in Madagascar was modified slightly to adjust for local climatic conditions and for the ruffed lemur reproductive cycle. Three unequal length seasons were distinguished in this study: a cool-wet season from May to August with 35% of the total annual rainfall and the lowest average monthly temperatures (Jul-Aug: 21.0°C), a transitional-dry season from September to November with 11% of total annual rainfall and ascending temperatures (Sep 21.5°C - Nov 24.6°C), and a warm-wet season from December to April with 54% of total annual rainfall and the highest average monthly temperatures (Feb 26.3°C). Ruffed lemurs breed in the cool-wet and give birth in the transitional-dry season (Morland, 1991).

### Activity Levels

Ruffed lemurs spent more of their time being "inactive" in the cool-wet season (mean=64.2±0.07%) than they did in the transitional-dry (mean=50.8±0.05%) or warm-wet (mean=55.7±0.07%) seasons, and were significantly less active in cool-wet than in warm-wet months (Sign test: $N(+)=0$; $N=6$; $P<0.05$).

### Resting Positions

Resting positions changed seasonally (Fig. 3). In the cool-wet season, ruffed lemurs used the hunched sitting and upright positions about equally, and were not observed to rest in the prone position. In contrast, ruffed lemurs usually rested in an upright or prone position in the warm-wet season, and spent significantly less time in hunched sitting than in the cool-wet months (Sign test: $N(+)=1$; $N=8$; $P<0.05$). All resting positions were used in the transitional-dry season. Ruffed lemurs spent 2% of their resting time sunbathing during the cool-wet months, about four times as much as in the transitional-dry or warm-wet season (Sign test: $N(+)=1$; $N=8$; $P<0.05$).

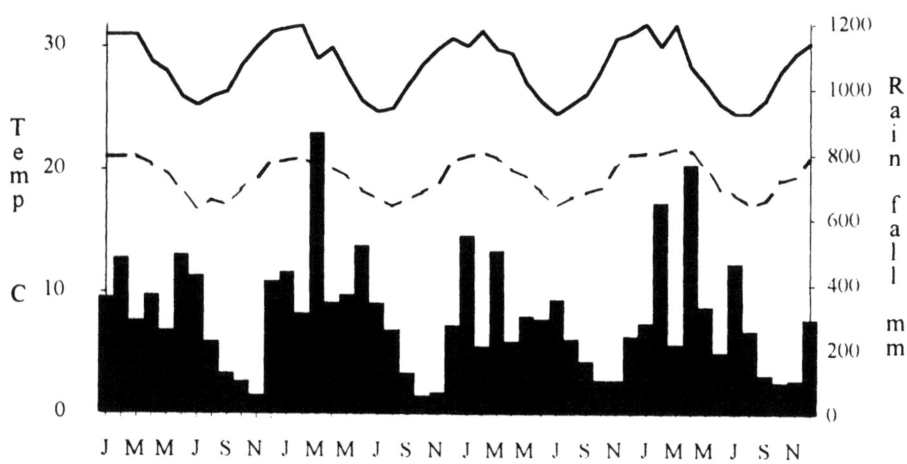


**Figure 2.** Seasonal variation in climate in the region of the study site. Monthly rainfall and mean temperature maxima (solid line) and minima (dotted line) recorded over a four year period (1977 to 1980) in Maroantsetra. Data were obtained from the Centre de Calcul et de Documentation of the Direction de la Meterologie Nationale.

### Feeding

Ruffed lemurs are highly frugivorous: fruit made up the bulk of the diet and there were no seasonal differences in the percentage of fruit in the diet (Fig. 4). However, there were seasonal differences in the total amount of time ruffed lemurs spent feeding: significantly less time was spent feeding in the warm-wet season than in the transitional-dry (Sign test: $N(+)=0$; $N=6$; $P<0.05$) or cool-wet seasons (Sign test: $N(+)=0$; $N=6$; $P<0.05$).

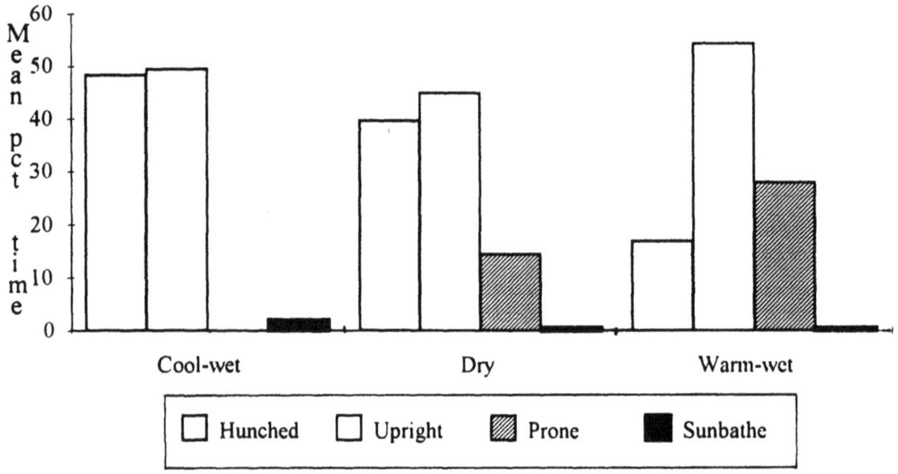


**Figure 3.** Seasonal variation in resting postures: mean percentage of time focal animals spent in each of four resting positions during the cool-wet, transitional-dry, and warm-wet seasons (see text for definitions of positions). The number of male and non-lactating female focal animals observed per season was: cool-wet = 10, dry = 6, and warm-wet = 10.

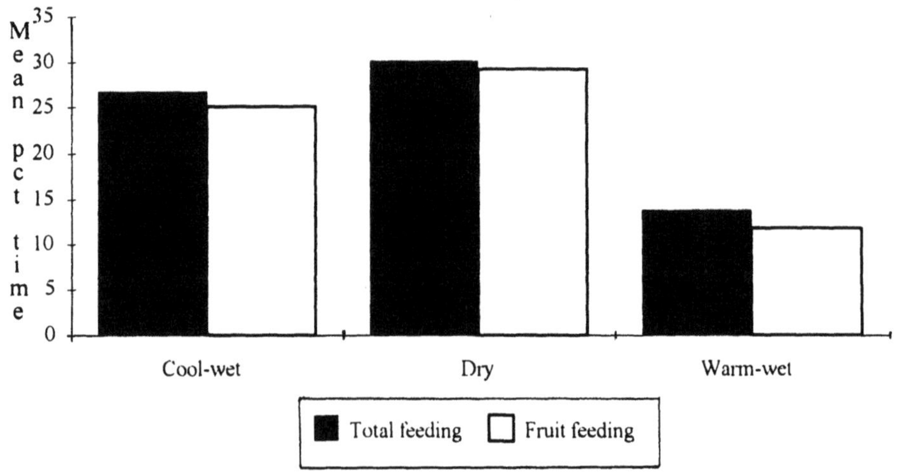


**Figure 4.** Seasonal variation in feeding behavior. The left bars show the mean percentage of time focal animals spent feeding in the cool-wet, transitional-dry, and warm-wet seasons. The right bars show the mean percentage of time focal animals spent feeding on fruit in each season. The number of focal animals observed per season was: cool-wet = 11, dry = 6, and warm-wet = 10.

### Daily Path Lengths

Daily travel distances varied seasonally in the small sample of DPLs measured. Average DPLs were significantly shorter (Sign test: $N(+)=0$; $N=6$; $P<0.01$) in the cool-wet season (mean = 892±492m) than in the transitional-dry (mean = 2450±701m) and warm-wet months (mean = 2931±1286m).

## DISCUSSION

### Climatic Seasonality in Madagascar

The tropical climates in which most primate species live are characterized by consistently high rainfall, temperature, and humidity throughout the year (Richards, 1964). Tropical rain forests receive at least 100mm of rain every month with dry periods occurring in short, unpredictable spells. Differences between average temperatures in the warmest and coolest months do not exceed 5°C (Newman, 1990; Whitmore, 1990). Madagascar's eastern escarpment and coastal lowland rain forests have a subequatorial climate with an average of at least 100mm of rain per month for 9-11 months of the year and annual ranges in mean monthly temperatures of about 5-6.5°C (Jenkins, 1987; Williams, 1990). Annual variations in rainfall and temperature are even more extreme in other regions of Madagascar. Thus, most lemurs may be exposed to more marked seasonal fluctuations in climatic conditions (rainfall and temperature) than primates in other tropical rain forests.

Rainfall patterns vary widely among regions and across years in Madagascar (see Fig. 1 and Fig. 2). Ranges in annual temperatures also vary regionally but, in contrast to rainfall, there appears to be a predictable, contingent pattern (see Colwell, 1974) of temperature variation island-wide: temperatures consistently vary from minima in June-August to maxima in December-February (Williams, 1990).

It is difficult to differentiate the direct effects of climate changes on behavior from indirect effects through the influence of climate on fluctuations in the food supply. However, the lack of periodicity in fruit production during 22 months in a southeast Malagasy rain forest led Overdorff (1991) to conclude that climate does not have a strong effect on phenological patterns. Overdorff (in press) and others (Pollock, 1975; Andrianisa, 1989; Schatz, unpubl. report; Sterling and Carlson, unpubl. data) reported that seasonal peaks in the abundance of foods used by lemurs are variable from year to year in Malagasy rain forests, and many canopy species produce flowers and fruit on prolonged, irregular, asynchronous, or alternate year cycles. At my study site, the abundance and availability of foods used by *Varecia* also varies across years (Andrianisa, 1989; Morland, 1991; Sterling and Carlson, unpubl. data). Longer-term studies will be needed to determine whether there are consistent relationships between variations in patterns of food abundance and behavior in frugivorous rain forest lemurs.

### Ambient Temperature and Behavior

The influence of ambient temperature on behavior has been reported previously in many birds and mammals (e.g., *Macaca arctoides*: Dahl and Smith, 1985; *Atelerix frontalis*: Gillies et al., 1991). However, relatively little attention has been paid to the possible relationship between changes in ambient temperature or other thermal conditions and behavioral variation in lemurs. My field observations indicate that *V. v. variegata* exhibited many behavioral changes during the cool-wet season, probably in response to changes in energy requirements during the annual period of low ambient temperatures.

A variety of qualitative evidence suggests that ambient temperature has a major influence on lemur behavior. Lemurs have an extensive and varied repertoire of behavioral, postural, and social activities which appear to serve as thermoregulatory mechanisms (see Pollock, 1989; Richard and Dewar, 1991). For example, "sunning" is an important aspect of daily activity in many lemur species in some seasons (e.g., Sussman, 1974; Richard, 1978; Tattersall, 1982), and *V. variegata* sunned more often in the cool-wet season than at other times of year. Sunning has been shown to provide radiative heat gain in other dark-skinned, dark-furred mammals (Harri and Korhonen, 1988).

Lemurs also use postural changes to regulate thermal conductance and maintain homeothermy over a wide temperature range. Individuals of several strepsirhine species were observed to adopt different resting positions depending on ambient temperature during measurements of metabolic rate (*Perodicticus potto*: Hildwein and Goffart, 1975; *Nycticebus coucang*: Müller, 1979; *Eulemur fulvus*: Daniels, 1984). Similarly, ruffed lemurs spent more time resting in the hunched sitting position during the cool-wet season, presumably to reduce exposed surface area and minimize heat loss, and more time resting in a prone position which probably aided heat dissipation in the warm-wet season.

Finally, many lemurs engage in "social thermoregulation": diurnal lemurs often huddle in groups to rest and sleep (Jolly, 1966; Richard, 1978; see Tattersall, 1982) while some nocturnal lemurs, especially the small-bodied *Microcebus* spp. (Martin, 1973) and *Cheirogaleus* spp. (Petter, 1988), form daytime sleeping groups of two to ten animals (see Tattersall, 1982). Macaques were reported to derive thermal benefits from huddling by reducing exposed surface area (Dahl and Smith, 1985) and lemurs probably use the same strategy.

**Lemur Thermoregulation**

When ambient temperatures drop, homeotherms must increase their metabolic rates (Schmidt-Nielsen, 1983). However, strepsirhines have lower basal metabolic rates than other primates and many species have relatively low body temperatures (e.g., Müller, 1985; Richard and Nicoll, 1987; Daniels, 1984; McNab and Wright, 1987; Pollock, 1989; but see McCormick, 1981). Moreover, it has been suggested previously that Malagasy primates do not maintain stable body temperatures (see Müller, 1985; Tattersall, 1982; Pollock, 1989). A testable extension of this hypothesis is that lemurs have a broader thermoneutral zone than other primates and allow their body temperatures to drop with declining ambient temperatures in order to conserve energy. However, since lemurs exhibit marked behavioral responses to changes in ambient temperature, I predict that they rely primarily on behavioral (e.g., reducing activity and increasing sunning) rather than strictly physiological mechanisms for temperature regulation. To return to the paradox presented at first, I suggest that island-wide patterns of seasonality in temperature (rather than in rainfall) account for the ubiquity of seasonal behavioral variation in lemurs.

These hypotheses have implications for lemur social organization. Individuals may derive benefits associated with energy conservation from close associations with conspecifics via social thermoregulation. Forming huddles or sleeping groups requires a high degree of tolerance and affiliation among individuals. Social thermoregulation is exhibited by species with gregarious and non-gregarious social systems despite major differences among them in the frequency and type of intraspecific associations. If lemurs have evolved a variety of behavioral adaptations for temperature regulation, I predict that a minimum level of association with conspecifics to permit social thermoregulation is advantageous, especially for small bodied lemurs.

Ruffed lemurs provide an exception: their behavior was influenced by ambient temperature, and yet adults were not observed to exhibit social thermoregulation on Nosy Mangabe. Huddling and sleeping in contact were observed infrequently, and only between mothers and offspring and between infant siblings. Semi-free ranging ruffed lemurs in a temperate forest also were observed huddling very rarely, and did so significantly less frequently than ring-tailed lemurs (Pereira et al., 1988). There may be a relationship between a relatively large body size, the absence of huddling, and the low levels of interindividual affiliation characteristic of *Varecia* at my lowland rain forest site (Morland, 1991). *Varecia* living in cooler climates and/or at higher elevations might exhibit both huddling and higher levels of interindividual affiliation.

### Suggestions for Future Research

It is difficult to distinguish the relative influence of individual environmental variables on behavior since they often are interdependent and co-vary. Moreover, long-term studies are needed to distinguish seasonal patterns of annual, periodic, contingent variation from short-term or unpredictable fluctuations. The relationships among external thermal conditions, thermoregulation, and behavior also are likely to be complex and difficult to measure, but the ideas discussed in this paper suggest it would be very fruitful to undertake an integrated study using experimental and field research to examine thermoregulation in lemurs. Experimental work should be conducted to test the hypothesis that lemurs have a broader thermoneutral zone than haplorhines. Also, the relative importance of behavioral and physiological responses to changes in thermal conditions (e.g., cloud cover, wind chill, ambient temperature, annual temperature range) should be determined among lemurs. Experimental studies also could measure the impact of the presence of conspecifics - and opportunities to huddle - on individual rates of heat loss among lemurs (see Dahl and Smith, 1985). Finally, field studies can be used to document the relationship between behavioral changes and thermal conditions across lemur species, and to compare seasonal changes in the type and frequency of thermoregulatory behavior exhibited by lemurs in different regions of Madagascar. A comprehensive study of thermoregulation and behavior will enhance our understanding of lemur physiology and should provide many insights into the complexity and determinants of strepsirhine social organization.

## ACKNOWLEDGEMENTS

Field work was conducted under the aegis of the Ecole Supérieure des Sciences Agronomiques of the Université d'Antananarivo, and with the permission and support of the Ministère des Affaires Etrangères, Direction des Eaux et Forêts, and former Ministère de l'Enseignement Supérieur and Ministère de la Recherche Scientifique et Technologique pour le Développement. I thank my field assistants, Victor Baba and Fortunat Toto, and the staff of the Service des Eaux et Forêts in Maroantsetra for help in the field. Alison Richard provided guidance and support throughout the study and reviewed several versions of the manuscript. I also thank Godfrey Bourne, Claire Hemingway, John Morland, Eleanor Sterling, and the editors and reviewers for their useful comments. Field work was supported by the National Geographic Society, National Science Foundation (BNS-8612154), Explorer's Club, Sigma Xi, Wenner-Gren Foundation, and World Wildlife Fund-US.

## REFERENCES

Altmann, J., 1974, Observational study of behavior: sampling methods, *Behaviour* 49:227.

Anderson, C.M., 1989, The spread of exclusive mating in a chacma baboon population, *Am. J. Phys. Anthropol.* 78:355.

Andrianisa, J.A., 1989, Observations phénologiques dans la Réserve Spéciale de Nosy Mangabe Maroantsetra. Memoires de fin d'etudes, Université d'Antananarivo, Madagascar.

Cavallini, P. and Lovari, S., 1991, Environmental factors influencing the use of habitat in the red fox *Vulpes vulpes*, *J. Zool., Lond.* 223:323.

Chapman, C., 1988, Patterns of foraging and range use by three species of neotropical primates, *Primates* 29:177.

Colwell, R.K., 1974, Predictability, constancy, and contingency of periodic phenomena, *Ecology* 55:1148.

Conner, D.A., 1983, Seasonal changes in activity patterns and the adaptive value of haying in pikas (*Ochotona princeps*), *Can. J. Zool.* 61:411.

Dahl, J.F., and Smith, E.O., 1985, Assessing variation in the social behavior of stumptail macaques using thermal criteria, *Am. J. Phys. Anthrop.* 68:467.

Daniels, H.L., 1984, Oxygen consumption in *Lemur fulvus*: deviation from the ideal model, *J. Mamm.* 65:584.

Gautier-Hion, A., 1980, Seasonal variations of diet related to species and sex in a community of *Cercopithecus monkeys*, *J. Anim. Ecol.* 49:237.
Gillies, A.C., Ellison, G.T.H. and Skinner, J.D., 1991, The effect of seasonal food restriction on activity, metabolism, and torpor in the South African hedgehog *Atelerix frontalis*, *J. Zool. Lond.* 223:117.
Harri, M. and Korhonen, H., 1988, Thermoregulatory significance of basking behaviour in the racoon dog, (*Nyctereutes procyonoides*), *J. Therm. Biol.* 13:169.
Hildwein, G. and Goffart, M., 1975, Standard metabolism and thermoregulation in a prosimian *Perodicticus potto*, *Comp. Biochem. Physiol.* 50A:201.
Hladik, A., 1980, The dry forest of the west coast of Madagascar: climate, phenology, and food available for prosimians, *in*: "Nocturnal Malagasy Primates", P. Charles-Dominique, H.M. Cooper, A. Hladik, C.M. Hladik, E. Pages, G.F. Pariente, A. Petter-Rousseaux, J.J. Petter and A. Schilling, eds., Academic Press, New York.
Hladik, C.M., Charles-Dominique, P. and Petter, J.J., 1980, Feeding strategies of five nocturnal prosimians in the dry forest of the west coast of Madagascar, *in*: "Nocturnal Malagasy Primates", P. Charles-Dominique, H.M. Cooper, A. Hladik, C.M. Hladik, E. Pages, G.F. Pariente, A. Petter-Rousseaux, J.J. Petter and A. Schilling, eds., Academic Press, New York.
Jenkins, M.D., 1987, "Madagascar, an Environmental Profile", IUCN, Gland.
Jolly, A., 1966, "Lemur Behavior", University of Chicago Press, Chicago.
Jolly, A., 1984, The puzzle of female feeding priority, *in*: "Female Primates: Studies by Women Primatologists," M. Small, ed., Alan R. Liss, Inc., New York.
Korn, H., 1989, The annual cycle in body weight of small mammals from the Transvaal, South Africa, as an adaptation to a subtropical seasonal environment, *J. Zool., Lond.* 218:223.
McCormick, S.A., 1981, Oxygen consumption and torpor in the fat-tailed dwarf lemur, *Cheirogaleus medius*: rethinking prosimian metabolism, *Comp. Biochem. Physiol.* 68A:605.
McNab, B.K., and Wright, P.C., 1987, Temperature regulation and oxygen consumption in the Philippine tarsier *Tarsius syrichta*, *Physiol. Zool.* 60:596.
Martin, R.D., 1973, A review of the behavior and ecology of the lesser mouse lemur, *in*: "Comparative Ecology and Behaviour of Primates", R.P. Michael and J.H. Crook, eds., Academic Press, London.
Meyers, D., 1992, Seasonal and habitat related variation in ranging behavior in the golden crowned sifaka (*Propithecus tattersalli*), *Am. J. Phys. Anthropol.* 14:122.
Morland, H.S., 1991, Social Organization and Ecology of Black and White Ruffed Lemurs (*Varecia variegata variegata*) in Lowland Rain Forest, Nosy Mangabe, Madagascar, Ph.D. thesis, Yale University, New Haven.
Müller, E.F., 1979, Energy metabolism, thermoregulation and water budget in the slow loris (*Nycticebus coucang*, Boddaert 1785), *Comp. Biochem. Physiol.* 64A:109.
Müller, E.F., 1985, Basal metabolic rates in primates - the possible role of phylogenetic and ecological factors, *Comp. Biochem. Physiol.* 81A:707.
Newman, A., 1990, "Tropical Rainforest", Facts on File, New York.
Overdorff, D.J., 1991, Ecological Correlates to Social Structure in two Prosimian Primates: *Eulemur fulvus rufus* and *Eulemur rubriventer*, Ph.D. thesis, Duke University, Durham.
Overdorff, D.J., (in press), Similarities, differences, and seasonal patterns in the diets of *Eulemur rubriventer* and *Eulemur fulvus rufus* in the Ranomafana National Park, Madagascar, *Int. J. Primatol.*
Pereira, M.E., Seeligson, M.L., and Macedonia, J.M., 1988, The behavioral repertoire of the black-and-white ruffed lemur, *Varecia variegata variegata* (Primates: Lemuridae). *Folia Primatol.* 51:1.
Petter, J.J., 1988, Contribution à l'étude du *Cheirogaleus medius* dans la forêt de Morondava, *in*: "L'Equilibre des Ecosystèmes Forestiers à Madagascar: Actes d'un Seminaire International", L. Rakotovao, V. Barre and J. Sayer, eds., IUCN, Gland.
Petter, J.J., Albignac, R. and Rumpler, Y., 1977, "Faune de Madagascar 44: Mammifères Lémuriens (Primates Prosimiens)", ORSTOM/CNRS, Paris.
Petter-Rousseaux, A., 1980, Seasonal activity rhythms, reproduction, and body weight variations in five sympatric nocturnal prosimians, in simulated light and climatic conditions, *in*: "Nocturnal Malagasy Primates", P. Charles-Dominique, H.M. Cooper, A. Hladik, C.M. Hladik, E. Pages, G.F. Pariente, A. Petter-Rousseaux, J.J. Petter and A. Schilling, eds., Academic Press, New York.
Pollock. J.I., 1975, Field observations on *Indri indri*: A preliminary report, *in*: "Lemur Biology", I. Tattersall and R.W. Sussman, eds., Plenum Press, New York.
Pollock, J.I., 1989, Intersexual relationships amongst prosimians, *Hum. Evol.* 4:133.
Raemaekers, J., 1980, Causes of variation between months in the distance traveled daily by gibbons, *Folia Primatol.* 34:46.

Rasmussen, D.T., 1985, A comparative study of breeding seasonality and litter size in eleven taxa of captive lemurs (*Lemur* and *Varecia*), *Int. J. Primatol.* 6:501.
Richard, A.F., 1978, "Behavioral Variation: Case Study of a Malagasy Lemur", Bucknell University Press, Lewisburg.
Richard, A.F. and Dewar, R.E., 1991, Lemur ecology, *Ann. Rev. Ecol. Syst.* 22:145.
Richard, A.F. and Nicoll, M.E., 1987, Female social dominance and basal metabolism in a Malagasy primate, *Propithecus verreauxi*, *Am. J. Primatol.* 12:309.
Richards, P.W., 1964, "The Tropical Rain Forest", Cambridge Univ. Press., Cambridge.
Robitaille, J.F. and Baron, G., 1987, Seasonal changes in the activity budget of captive ermine, *Mustela erminea* L., *Can. J. Zool.* 65:2864.
Russell, R.J., 1975, Body temperatures and behavior of captive cheirogaleids, *in*: "Lemur Biology" I. Tattersall and R.W. Sussman, eds., Plenum Press, New York.
Schatz, G., Final report on the floristic inventory of Nosy Mangabe, Madagascar, unpubl. report.
Schmidt-Nielsen, K., 1983, "Animal Physiology", Cambridge University Press, Cambridge.
Struhsaker, T.T., 1975, "The Red Colobus Monkey", The University of Chicago Press, Chicago.
Sussman, R.W., 1974, Ecological distinctions in sympatric species of Lemur, *in*: "Prosimian Biology," R.D. Martin, G.A. Doyle, and A.C. Walker, eds., Duckworth, London.
Sussman, R.W. and Richard, A., 1974, The role of aggression among diurnal prosimians, *in*: "Primate Aggression: Territoriality and Xenophobia", R.L. Holloway, ed., Academic Press, New York.
Tattersall, I., 1982, "The Primates of Madagascar", Columbia University Press, New York.
Terborgh, J., 1983, "Five New World Primates: A Study in Comparative Ecology," Princeton University Press, Princeton.
Waser, P.M., 1975, Monthly variations in feeding and activity patterns of the mangabey, *Cercocebus albigena* Lyddeker, *East. Afr. Wildl. J.* 13:249.
Whitmore, T.C., 1990, "An Introduction to Tropical Rain Forests," Oxford University Press, Oxford.
Williams, J.B., 1990, "Some Temporal and Regional Variations of Climate in Madagascar," Overseas Development Natural Resources Institute, Chatham.

# SEASONAL ADJUSTMENT OF GROWTH RATE AND ADULT BODY WEIGHT IN RINGTAILED LEMURS

Michael E. Pereira

Duke University Primate Center
3705 Erwin Road
Durham, NC 27705
U.S.A.

## ABSTRACT

Photoperiodic cues provide nonequatorial animals a reliable schedule for adaptive changes in metabolism, developmental rate, reproductive effort, and behavior. This study discovered seasonal adjustments of growth rate and adult body weight in semi-free-ranging ringtailed lemurs (*Lemur catta*) that were correlated with changes in rate of hair growth, thermoregulatory behavior, and social behavior but independent of changes in food availability. Infants increased body weight with 3 to 12g/day, accelerating growth rate annually just after summer solstices. Adult males and females gained weight one to two months later, just before longterm suppressions of growth rate in immatures. Correlations between this set of results and the annual schedule of rainfall in Madagascar led me to propose a new model of life-history strategy for the large-bodied lemurs. Harsh but predictable seasonality of nutritive resources seems to have led these primates to evolve life histories comprehensively geared toward the conservation of energy, including suppression of metabolic rate for much of the Malagasy dry season. This life-history tactic would be analogous to diverse torpid states that enhance survival and reproduction in nonprimate mammals living in harshly seasonal environments. The overall life-history strategy, however, requires rates of infant growth that are uncommonly high among group-living primates relative to maternal size. Like many other animal parents, lemur mothers suffer high costs of investment to help offspring meet a deadline: lemur infants must grow large and perhaps store energy before the dry season, so they can safely reduce metabolic rate and grow little during the dry season. The high cost of lactation is suggested to have promoted wide-spread evolution of female dominance and male-committed infanticide among lemurid and indrid species. Continued research on the environmental regulation of life histories will be required before a deep understanding of primate natural history can be achieved.

## INTRODUCTION

The reproductive physiology of most lemurs is entrained to circannual photoperiodic cycles. Translocation from southern to northern latitudes shifts the annual schedule for reproduction by about six months in almost all extant species (e.g., Rasmussen, 1985). In the 1970s, van Horn (1975) and colleagues (van Horn and Resko, 1977; van Horn and Eaton, 1979) experimentally investigated photoperiodic

*Lemur Social Systems and Their Ecological Basis,* Edited by
P.M. Kappeler and J.U. Ganzhorn, Plenum Press, New York, 1993

effects on lemur estrous cycling and showed that female ringtailed lemurs (*Lemur catta*) maintained on unchanging long-day light:dark cycles become acyclic and that even small reductions in daylength thereafter trigger estrous cycling. Since these pioneering efforts, few advances have been made toward a fuller understanding of photoperiodic regulation of primate life histories (Lindburg, 1987; Chik et al., 1992). This chapter contributes by presenting preliminary analyses of a six-year database on growth rate and adult body weight in semifree-ranging ringtailed lemurs. It is complemented by another initial report from my longitudinal research on development and life history in ringtailed lemurs (Pereira, 1993).

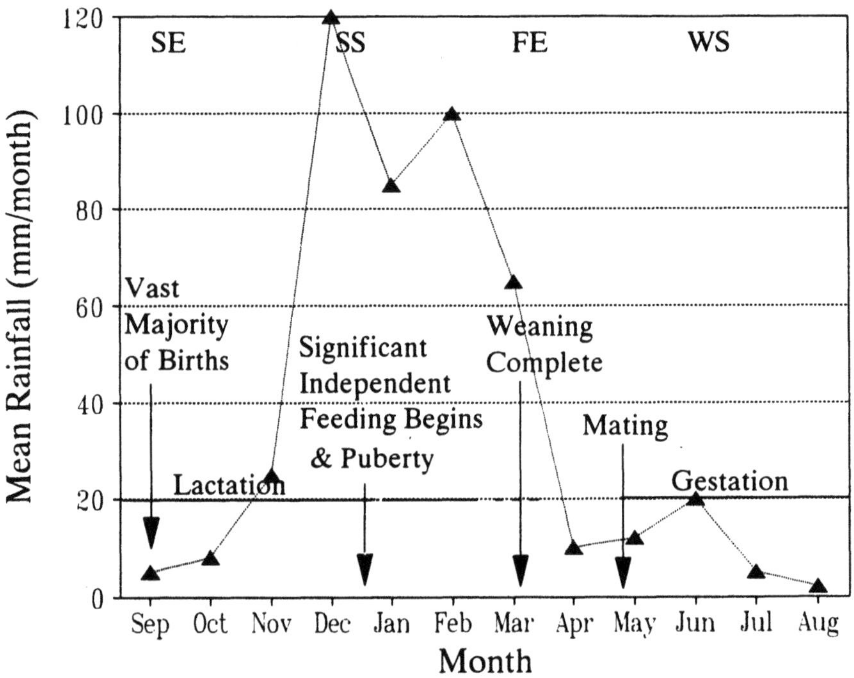


**Figure 1.** Rainfall seasonality in southwestern Madagascar, reproductive schedule, and developmental milestones in ringtailed lemurs (rainfall data from Griffiths and Ranaivoson, 1972). SE = spring equinox, SS = summer solstice, FE = fall equinox, WS = winter solstice.

Seasonal reproduction in the Lemuridae under natural photoperiodic cycles is a canalized trait: several species maintained in diverse captive colonies for forty or more generations have developed little variation in the timing of reproduction. This indicates fundamental adaptedness for observed reproductive schedules in the natural environment, and reasonable adaptive arguments are easy to advance after juxtaposing schedules for Malagasy rainfall with milestones of lemur reproduction and development (e.g., Fig. 1). Annually, for example, births in all of the Lemuridae occur just a few months before the rainy season, a timing that allows immatures to begin contributing significantly to their own nourishment when tender, high-quality foods are most plentiful (i.e., new leaves, flowers, and fruits). Indeed, similarities in the seasonal timing of lactation and weaning across extant lemurs, despite huge variation in developmental rate, led Petter-Rousseaux (1968) to propose the general hypothesis that lemur reproductive schedules function to provide lactating mothers and their developing infants ready access to high-quality foods.

Full consideration of seasonality and lemur life histories indicates that several major effects probably do conjointly determine annual schedules for certain species, but also that cross-taxonomic similarities in the timing of reproduction and development might result from different sets of selection pressures. Ringtailed lemurs, for example, are restricted to southern Madagascar, where four months of limited rain are followed annually by eight dry months (Griffiths and Ranaivoson, 1972). Also, ringtails are the most gregarious prosimian primate, which likely exacerbates for them aspects of foraging competition (van Schaik, 1989). Given their profound birthing synchrony (Pereira, 1991) and high rates of infant growth (Pereira et al., 1987; also results), then, one can indeed reasonably argue that females of this species could not nourish infants effectively on any other schedule, leave alone also be prepared to conceive again just after weaning! Few aspects of this particular argument, however, apply in consideration of, say, nongregarious mouse lemurs in Madagascar's eastern rain forest.

Further consideration of Figure 1 suggests for ringtailed lemurs that high rates of *juvenile* survival might also be achieved on few other schedules. Infants are weaned before the eight-month harsh season. Now, the transition to independence is generally difficult for immature primates (Pereira and Fairbanks, 1993); but, dry season foods can be particularly difficult to find, hotly contested, tough to process, and low in nutritive value (Janson and van Schaik, 1993). Hampered in every domain by inexperience, young juveniles also find most matters further complicated by their status as smallest and weakest in their group. Attempting to survive their first harsh season, immatures clearly stand to benefit from beginning their careers as independent foragers with a few months of wet-season gorging.

Juvenile ringtailed lemurs might further reduce their risks of premature mortality by postponing the energetic expenses of growth, timing most growth to occur during the next season of plentiful food (Table 1). This strategy would be most likely to evolve if the phenologies of lemur foods themselves were entrained to photoperiod, preventing odd, out-of-season rain from significantly influencing food supply. Anticipating that photoperiod might regulate growth, I predicted that the data would show both a decline in growth rate following the photoperiod of the Malagasy rainy season and relative inflexibility of growth rate in response to variation in food supply across the dry season photoperiod. Complementary predictions advanced on growth rate are listed in Table 1.

**Table 1.** Predictions on growth rates and adult body weights of ringtailed lemurs.

(1.) Maximal growth rates are sustained across ringtailed lemurs' first six months of life.
  (1a.) Increased nutrition elevates growth rate during the photoperiod of the Malagasy rainy season.

(2.) Growth rates decline significantly after the autumnal equinox, which heralds the onset of the Malagasy dry season.
  (2a.) Growth rates are relatively inflexible in response to variation in food supply across the photoperiod of the Malagasy dry season.

(3.) Adult male and female ringtailed lemurs gain weight after the summer solstice, in preparation for mating competition, gestation, lactation, and the harsh season.

Finally, before analyzing any of the weight data, I also advanced a simple prediction concerning possible seasonal changes in adult male and female body weight. Adult ringtails of both sexes compete intensely for dominance, access to resources, and group membership (e.g., Jolly, 1966; Koyama, 1988, 1991, 1992; Vick and Pereira, 1989; Pereira and Weiss, 1991; Sauther, 1991; Pereira, 1993). Especially considering the ecological context of these phenomena (Fig. 1), I expected that both males and females gained weight after summer solstices (Table 1). A similar phenomenon was described long ago in male squirrel monkeys (DuMond, 1968), members of a primate species that shows several parallels in natural history to that of ringtailed lemurs (Hrdy, 1981; Jolly, 1984). Again, evidence of photoperiodic control over these phenomena would be obtained if changes in nutritive supply did not obliterate or change the direction of seasonal adjustments in body weight.

## METHODS

### Subjects, Forest Enclosures, and Provisioning Regimes

The first study group, Lc1 Group, was established at the Duke University Primate Center (DUPC) in 1977 and has occupied forest enclosures since 1981. In 1987, Lc2 Group was established in an enclosure adjacent to Lc1 Group. Most of its original members were unrelated to those of Lc1 Group; the others were unfamiliar, distant relatives. Group histories have been detailed further elsewhere (e.g., Taylor and Sussman, 1985 and papers cited earlier).

In total, DUPC enclosures include more than 30 contiguous hectares of forest. Each resident lemur group forages extensively among 20 to 30 species of trees and, once daily, receives a small amount of Purina monkey chow, scattered widely at one of several provisioning sites in its core range.

From 1981 until 1988, the enclosure-living lemurs at the DUPC were maintained under High Provisioning (HP). At the end of 1988, a stepwise reduction of provisions was begun in the two enclosure areas pertinent to this study (Table 2). Throughout this program, provisions were measured only roughly (levels in buckets) and probably also varied as weekday caretakers differed from those on week-ends, during staff vacations, etc. Also, for the three winter months each year of Low Provisioning (LP), we increased provisions slightly beyond the indicated amount, to counteract the relatively low temperatures experienced by lemurs in North Carolina.

**Table 2.** Regimes of daily provisioning.

| Provisioning Regime | Dates | Approximate Amounts of Food Provided | Approximate No. Lemur Recipients |
|---|---|---|---|
| High (HP) | 1986-1988 | 3.6 kg chow<br>4.0 kg fruit | 40 |
| Medium | 1989 | 2.0 kg chow | 40 |
| Low (LP) | 1990-1992 | 1.0 kg chow | 40 |

### Data Collection and Analysis

Throughout the six-year period, each study group was detained at four- to six-week intervals by delivering provisions entirely within large cages in the centers of core ranging areas. Thereafter, each lemur was caught in a net and weighed by looping the net over the hook of a hanging scale (Hanson). Periodic capture disrupted neither social relations nor animal-observer habituation, and this method was also used to obtain cross-sectional weight data from an unprovisioned group of wild ringtailed lemurs (F Group) at the Berenty Reserve in southeast Madagascar (Pereira and Crowley, unpubl. data).

No analysis was conducted until the end of the six-year study, when data from the HP period were balanced by data collected during LP. Only data from apparently healthy subjects experiencing HP or LP were submitted to analysis. Exclusion of data from the intermediary year of Medium Provisioning minimized effects of the early HP on weights during the LP period. Only data from subjects born in the first, large cluster of births each year (cf. Pereira, 1991) were admitted to analyses of the timing of growth. Sexual maturation begins in the DUPC population around 16 months of age; but, growth is completed during the fourth year of life (Pereira, unpubl. data). For analyses of adult weight, therefore, data were taken only from subjects greater than three years of age.

This first set of analyses is presented graphically. Only sign testing was done, after comparing the weights of each subject measured in each of two adjacent months, considering each pair of adjacent months across the study. Pooling data from the two provisioning regimes, I effectively asked whether significant proportions of subjects in a given class (e.g., infants) increased or decreased growth rate or weight across a month-to-month transition *regardless of their level of nutrition*. Because increases or decreases could have occurred, two-tailed sign tests were applied (alpha level = 0.05). Data for a given monthly transition taken from the same adult in different years were considered independent. Each condition (e.g., August-to-September/adult females/HP) was characterized by two years of data from each of two social groups (4 group-years).

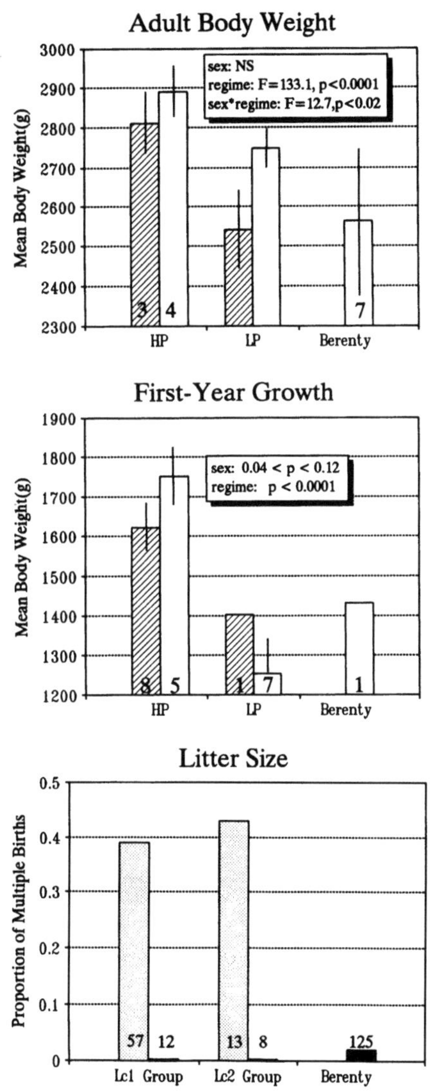


**Figure 2.** Effects of reduction of provisions. Top panel: Mean annual weights of 3 females (hatched bars) and 4 males (open bars) that were fully adult during early HP as well as later LP periods. Also, mean weight of wild adults (6 females, 1 male) at end of dry season at Berenty. Middle panel: Mean weights attained by females (hatched) and males (open) in first year of life under two regimes, and weight of wild one year-old. Bottom panel: Proportions of births producing twins under HP (open), LP (filled), and at Berenty in 1992 (black). Sample sizes and standard deviations of means shown.

## RESULTS

### Planes of Nutrition

Analyses of adult weight, of immature weight at one year of age, and of litter size all indicated that the reduction of provisions (HP to LP) significantly reduced the nutrition that was available to the lemurs (Fig. 2). Also, the plane of nutrition during recent years of Low Provisioning (LP) approximated the nutritive conditions experienced by the wild group whose members were captured and weighed at the Berenty Reserve.

### Seasonal Adjustment of Growth Rate

Infants gained approximately 4 to 8g of body weight per day across their first 7 months of life. Regardless of level of provisioning, growth was accelerated in the fourth month of life, around the time that immatures began contributing significantly to their own nutrition (Fig. 3). With increased food availability (HP), however, semi-independent infants gained weight significantly faster (months 4-7; HP: mean=7.5 g/day (N=5 infants); LP: mean=4.6 g/day (N=10 infants); F=9.19, P<0.01). Finally, in the eighth month of life, regardless of their plane of nutrition, juveniles exhibited dramatic reductions in growth rate. Immatures' weights increased roughly 2g/d thereafter.

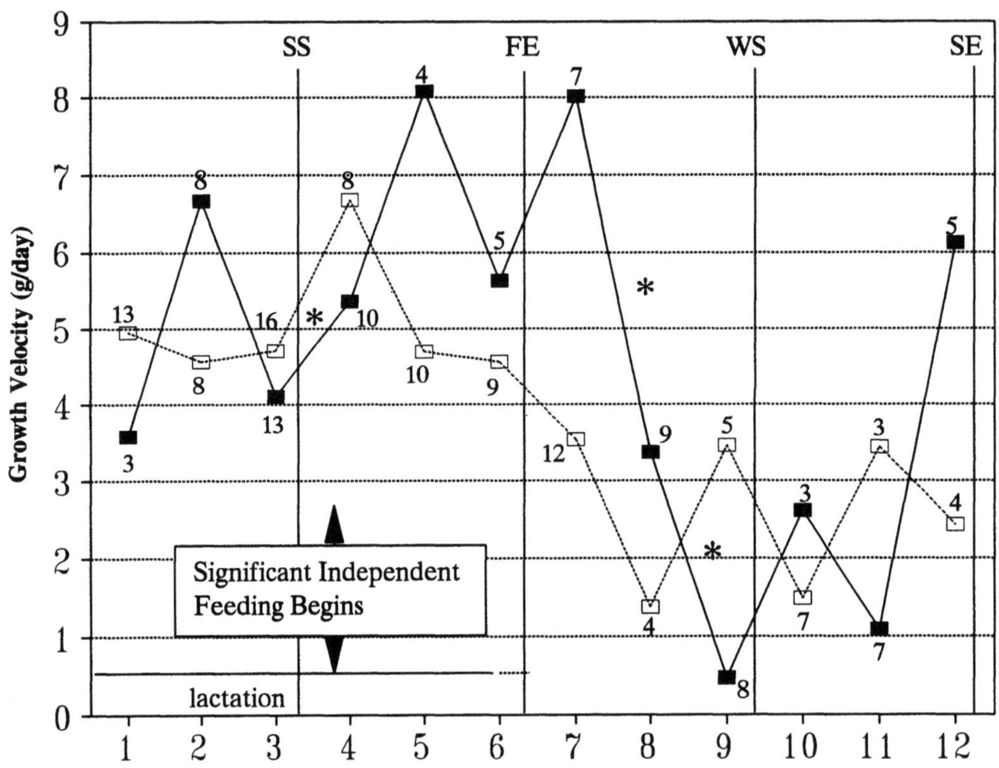


**Figure 3.** Mean growth velocities across first year of life under HP (solid) and LP (dashed), for infants whose first months of life occurred between mid-March and the end of April. Asterisks indicate when significant proportions of subjects from both regimes showed unidirectional change in growth rate from one month to next (see also text). Solstices, equinoxes, and sample sizes indicated.

Whereas reduced growth rates between November and March might have been caused by the unusually low temperatures experienced by these immature lemurs living at 35 degrees latitude, four separate analyses indicate that temperature was unlikely to have been the principal mechanism for the change. First is the juxtaposition of growth rates under HP and LP itself (Fig. 3). Very cold weather, relative to that in southern Madagascar, is uncommon enough in North Carolina so that the great nutrition available under HP should have counteracted temperature effects better than the lower plane of nutrition (LP). Yet, if any difference existed in wintertime growth rates, it would have been that immatures reduced rates of growth *further* under HP than under LP. This trend suggests the existence of compensatory effects, with reductions in growth rate after the equinox being, in part, a function of the amount of growth already achieved.

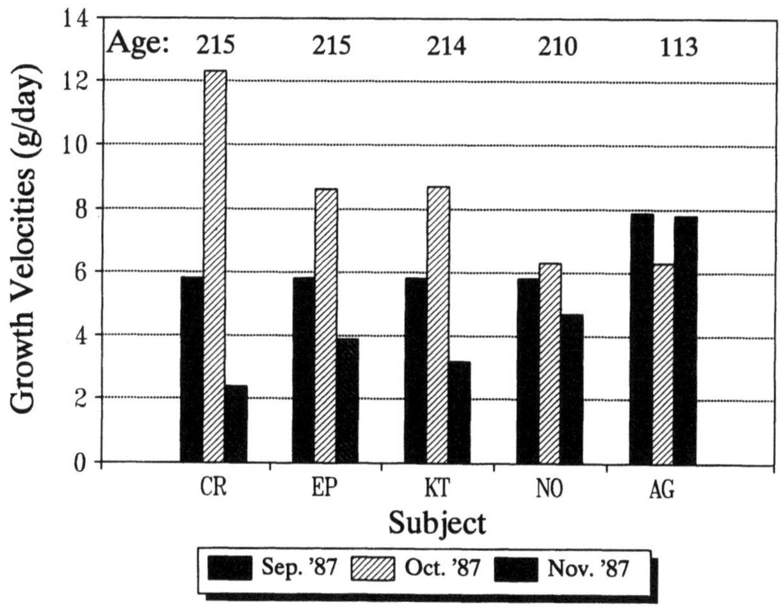


**Figure 4.** Autumnal growth velocities of infant cohort in 1987, the only year in study when mean daily temperature in Nov. (11.6 °C) was not appreciably lower than in Oct. (12.6 °C). Subjects' median ages in days during this period shown.

Second, the weather in 1987 allowed for an elaboration of this analysis, providing one case where mean daily temperature declined only slightly in November (Fig. 4). Even that year, subjects born early, i.e., large juveniles, exhibited the usual substantial reduction in growth rate in November. That the much younger, smaller infant failed to show the reduction suggests that the general decline in growth was not caused by an inadvertent reduction of provisions.

Third, data from all seven subjects born late during birth seasons failed to corroborate an age effect ($F=1.76$, NS). But, 14 infants weighing less than 1,200g by Novembers exhibited significantly smaller growth-rate reductions in Novembers (mean = 1.77 g/day) than did 19 heavier infants (mean = 4.77 g/day; $F=7.45$, $P<0.01$). Cold temperature alone would have reversed these results. As with the first analysis, these latter two suggest compensatory effects, with size-related stimuli modulating a basic schedule prescribed by photoperiod.

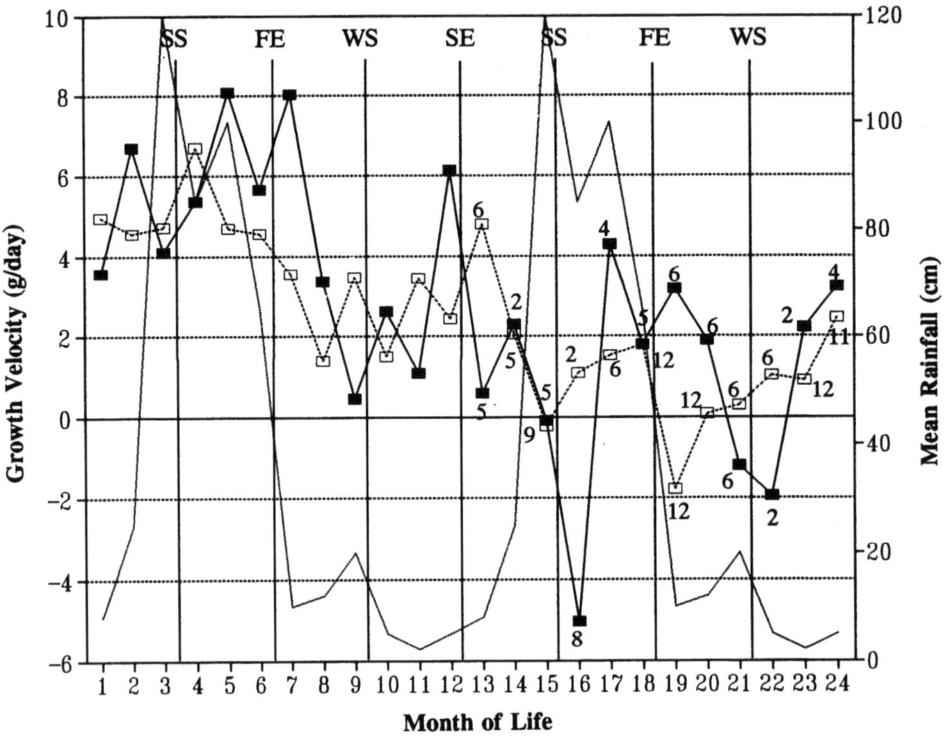


**Figure 5.** Mean growth velocities across first two years of life under HP (solid) and LP (dashed) and mean monthly rainfall values for the same photoperiods in southern Madagascar (dotted). See also Fig. 3.

Finally, rates of growth through the second year of life indicated that time of year *per se* continued to play a major role (Fig. 5). Growth rates remained relatively low after infancy; but, only between months 15 and 20 did they show (a) sustained elevation under both provisioning regimes and (b) greater acceleration in response to HP. Moreover, the timing of the latter effect was precisely 12 months after its first occurrence, i.e., months 5-8, and longitudinal patterns again suggested compensatory effects. In any case, 18 month-olds are not only more experienced but also much larger and higher in dominance status than are weanlings (Pereira, 1993); thus, temperature effects should have been comparatively small for them, especially under HP conditions.

## Seasonal Adjustments in Adult Body Weight

The pattern of monthly change in adult male weight was unaffected by level of provisioning (Fig. 6). As predicted, males showed a significant increase in body weight following the summer solstice, two to three months prior to the mating season. Under the naturalistic level of provisioning (LP), male weights increased about 9% on average, before plummeting during and after mating seasons. Males consistently recovered body weight after winter solstices and showed a significant increase just after spring equinoxes, when the Duke Forest provided flowers and new leaves.

Interpretation of changes in adult female body weight (Fig. 7) requires attention to the fact that each year nearly all females gave birth to fullterm infants (70-100g; Pereira, unpubl. data). Consequently, most female weights included the growing weights of fetuses from November to March, inclusive. Unlike for males, however, the change in provisioning level seemingly revealed a switch in metabolism, foraging strategy, or both that occurred early in lactation, when females gained weight dramatically under HP. Just after that, female weights declined significantly, during the month following infants' significant growth acceleration. Finally, like males, females showed a significant increase in weight between Augusts and Septembers, just before complete weaning of infants.

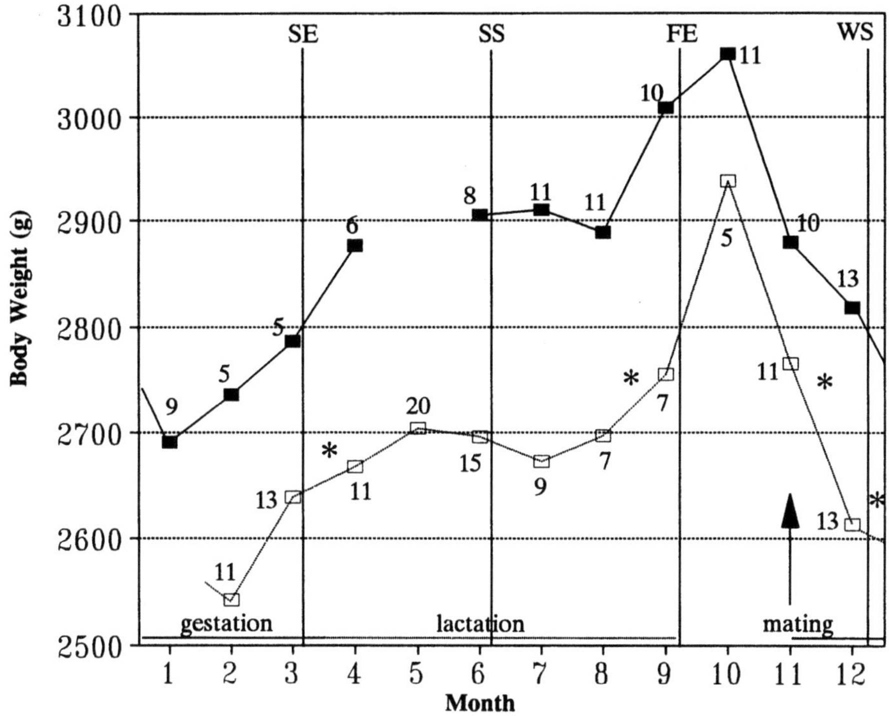


**Figure 6.** Mean weights of adult males under HP (solid) and LP (dashed). Asterisks indicate when significant proportions of subjects from both regimes showed unidirectional weight change from one month to next. Note: By chance, males captured for May weighings during HP were a different subset from those captured in Aprils and Junes, leading to spurious appearance of transient decline in weight at that time.

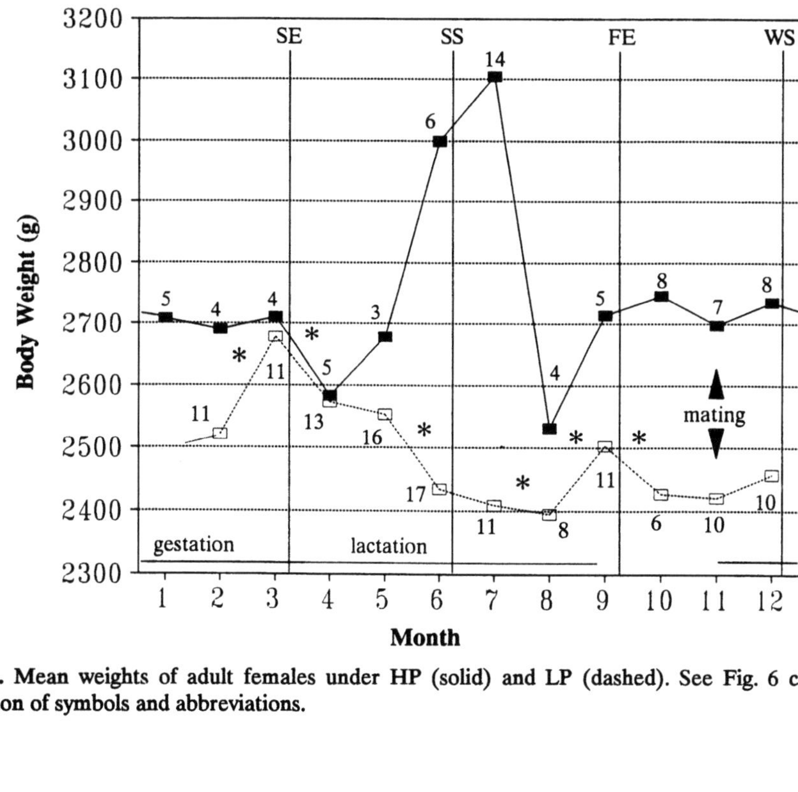


**Figure 7.** Mean weights of adult females under HP (solid) and LP (dashed). See Fig. 6 caption for explanation of symbols and abbreviations.

## DISCUSSION

Systematic variation of provisioning allowed for an experimental analysis of seasonal changes in growth rate and adult body weight in forest-living ringtailed lemurs. Nutrient availability under LP approximated conditions experienced by a wild group of ringtailed lemurs, excepting probable attentuation of seasonal variation. Prior years of HP, by contrast, provided a much higher plane of nutrition that must have significantly attenuated seasonal variation in the availability of nutrients.

Seasonal adjustments of growth rate and adult body weight were revealed that were consistent across individuals, social groups, years, and provisioning regimes. These findings almost certainly reflect direct or indirect photoperiodic regulation of the physiological systems underlying these characters. Having virtually ruled out major temperature effects, the only other possible causal agent would have been phenology in the Duke Forest. One version of this -- that the forest provided so little natural food in certain seasons that lemur weights were consistently suppressed at that time of year -- was ruled out by daily provisioning, including augmentation of winter provisions during the LP period. The converse possibility -- that seasonal bonanzas in natural foraging provided extra nutrition leading to consistent weight enhancements -- might help to explain male weight gains as spring seasons approached but was unlikely to have influenced other results significantly. Greatest opportunity in natural foraging seemed to occur for our lemurs during April-May transitions, when most trees had young leaves and flowers, and Septembers, when acorns (*Quercus* spp.) and other fall fruits became available. These brief food flushes occurred well before or just after most significant weight enhancements, however, and provisioning, especially during HP, should have attentuated their effects. Our lemurs always preferred provisions over natural foods, and one could consume and metabolize only so much on a given day.

### Photoperiodic Regulation of Lemur Life-History Cycles

The summer solstice emerges from these and other analyses as an important milestone in the annual life-history cycle of ringtailed lemurs. A separate work showed that, in addition to photoperiodic control of breeding seasonality, group-living females tightly synchronize their estrous cycles annually (Pereira, 1991). Three to four cycles are generally needed for such biological coupling (Winfree, 1980; McClintock, 1983), and precisely that amount of time exists between the summer solstice and functional estrus in ringtailed lemurs. Also, immature ringtails at the DUPC invariably begin genital marking in July (Pereira, unpubl. data) and adults seem to increase rates of investigation of female scent-marks at that time (Kaestle and Pereira, unpubl. data). In the present study, immatures consistently exhibited growth spurts just after summer solstices and adults of both sexes began gaining weight one to two months later. Given that female reproductive physiology is entrained to photoperiod, it seems particularly likely that the observed female weight gains resulted in part from direct hormonal and metabolic responses to changing photoperiod. Male weight gains would have been indirect effects of photoperiod, if male physiology responded to cues from females (cf. van Horn, 1980).

Six-month shifts in reproductive schedule following south-to-north translocation indicate that life histories in all of the Lemuridae and in many other lemurs are regulated to some degree by photoperiod. The extent to which these species share with ringtails photoperiodic control of growth, adult body weight, and perhaps other phenomena (see below) remains to be discovered. Some important parallels are already evident, however. Mouse and dwarf lemurs (Cheirogaleidae), for example, respond to photoperiodic stimuli heralding the dry season by fatting, decreasing metabolic rate, and in some cases, becoming torpid (Petter-Rousseaux, 1974, 1980; Russell, 1975; McCormick, 1981; Foerg, 1982; Foerg and Hoffman, 1982). As in ringtailed lemurs, immatures in these taxa grow rapidly before the autumnal equinox and hardly at all just after it (*Cheirogaleus medius*: Foerg, 1982; *Mirza coquereli*: K. Stanger, unpubl. data).

Photoperiodic control of metabolic rate, developmental rate, fatness, activity level, foraging strategy, and other behavior are well-known among non-primate animal taxa (Hoffman, 1981; Lee and Zucker, 1988; Bronson, 1989). By contrast, such forms of life history regulation have hardly begun to receive attention in primate research. The remainder of this chapter is an attempt to show that they will require illumination if we are to achieve a deep and comprehensive understanding of primate behavioral ecology and sociobiology.

**The Functioning of Seasonal Adjustments**

Why do adult ringtailed lemurs gain weight each summer? Consider first the relatively familiar phenomenon of seasonal weight gain by males. DuMond (1968) described the annual fatting of adult male squirrel monkeys prior to mating seasons and noted its coincidence with dominance contests among males. Hrdy (1981) suggested that such seasonal restriction of male weight gain and dominance behavior has evolved in small primates whose food supply is seasonally variable. Natural history parallels between squirrel monkeys and the Lemuridae (see also Jolly, 1984) led me to predict the premating weight gains observed in male ringtailed lemurs. These weight gains were tightly correlated with testicular enlargement (Pereira, unpubl. data), which correlates in turn with increased levels of circulating testosterone (Bogart et al., 1977). This set of phenomena corresponds closely with the premating changes seen in males of the seasonally-breeding macaques (e.g., Glick, 1979; Matsubayashi and Enmoto, 1983) and the hypothesis that seasonal male weight gain evolved in these primates because it enhances male mating success remains compelling. Important species differences may exist in the metabolic and anatomical nature of these weight increases, however, and these remain to be investigated (see below).

Why do female ringtailed lemurs gain weight each summer? Fatting in the Cheirogaleidae reminds us that lemurid females might also store energy in the form of adipose tissue before dry seasons (*n.b.*: lack of subcutaneous depots does not indicate that fat storage is unimportant; Pond, 1978). In their cases, fatting would occur just prior to conception, and Figure 1 shows how this might function as a reproductive tactic. Each year, successful summer and fall foraging must enhance females' chances for reproductive success. But if energy storage is important, highly successful foraging before dry seasons might also maximize rate of reproduction by enhancing the probability of weaning survivable juveniles in *consecutive* years. Elsewhere, we presented a corollary unfamiliar for seasonally-breeding taxa: lemuroid males are inclined to kill infants to which they are certainly unrelated (Pereira and Weiss, 1991; see also van Schaik and Kappeler, 1993). The existence of this tactic suggests that full reproductive success in consecutive years is difficult for female lemurs to achieve.

The rates and scheduling of growth observed in lemurs well support these arguments. Across their first three months of life, while completely dependent on mothers' milk, ringtailed infants gained 4 to 8g of body weight per day on average. Nine different infants grew between 9 and 12g per day during their third and/or fourth months, no doubt contributing substantially to the weight losses sustained by females beginning midway through lactation (Fig. 7). Other lemurid infants grow this fast or faster (Benirschke and Miller, 1981; Pereira et al., 1987; Kappeler, 1992).

Analyzing data from across the prosimian sub-order, Kappeler (1992) showed that infant growth rates distinguish neither lemuriform from lorisiform primates, seasonal-breeders from non-seasonal breeders, nor female-dominant species from those lacking adult female dominance over males. Heterogeneity of data within and between classes reduced the import of the analysis, however. "Seasonal breeder," for example, batched lemurs that can reproduce twice in one rainy season with others that must mate, gestate, and lactate before the rains begin and also galagos whose seasonality is much broader than that of lemurs, bespeaking perhaps its relationship to different environmental factors. The analysis also failed to address that all large-bodied lemurs (Lemuridae and Indridae) are gregarious, eat little or no animal matter, and show nearly constant maternal carriage of dependent infants, whereas all lorisiform primates are nongregarious, depend heavily on insect prey, and show little maternal carriage. With such diversity of life history unaccounted for, the failure of statistical tests to attain significance remains uninterpretable (cf. Kappeler, 1992).

Different allometric analyses lie beyond the scope of this chapter, but some would likely indicate important roles for seasonality and other variables upon which lemur biologists have focussed attention. Specifically, contrasts emerge in comparing the Lemuridae and Indridae with group-living anthropoid primates whose mothers carry their infants. Lemurid growth rates do not differ from those of infant Cercopithecinae, for example (Glassman et al., 1984; Rutenberg and Coelho, 1988; Strum, 1991; Sackett and Ruppenthal, 1992). Because these monkeys' mothers are an order of magnitude larger than are lemur mothers, however, the cost of lactation would be roughly an order of magnitude greater for the lemurs, all else being equal. This estimate may err to the high side of reality; but, rates of infant growth in a comparably-sized, group-living monkey (*Cebus albifrons*: 2-3 g/day; Wilen and Naftolin, 1978) suggest that lactation is at least three times more costly to lemur mothers than to monkey mothers. Both of these estimates actually correspond well with results of an allometric analysis of postnatal growth rate across the major primate taxa (Kirkwood, 1985).

The growth data presented in this chapter and related observations at the DUPC and in the wild lead me to propose a new model of basic life-history strategy for the Lemuridae and Indridae. Harsh but predictable seasonality of nutritive resources seems to have led ringtails and other large-bodied lemurs to evolve life histories that are comprehensively geared toward the conservation of energy. All prosimians yet investigated show resting rates of metabolism that are low for mammals of their sizes (Daniels, 1984; Müller, 1985; Young et al., 1990). But also, seasonal suppression of growth rate in ringtailed lemurs is invariably associated with reductions in hair growth, general activity, and aggressivity (Pereira, unpubl. data), which suggests that further reduction of metabolic rate across much of the Malagasy dry season helps immature and adult Lemurids and Indrids survive and reproduce (cf. Dark and Zucker, 1983).

Lemurs probably achieve significant reduction of metabolic rate by reducing not only costly behavioral activities but also body temperature. In response to cold ambient temperature, lemurs spend much time in a distinctive hunched posture (McNab, 1988) and in group huddles, two behaviors that appreciably reduce body-surface exposure. Lemurs reduce heat loss further by reducing blood circulation in peripheral body parts, creating a cool "shell" and allowing overall body temperature to be lowered while deep body temperature is maintained (Müller, 1985; McNab, 1988). Despite fairly mild winters in North Carolina, this phenomenon is reflected in DUPC lemurs by progressive loss of tail-length to frostbite with age, especially in chronically subordinate animals.

Finally, diel heterothermy probably also helps lemurs conserve energy. The behavioral and physiological capacities outlined above could allow significant energy savings daily, as in *Tupaia glis*, where reduction of body temperature each night substantially reduces energy expenditure (Bradley and Hudson, 1974). I expect that the wellknown early-morning sunning, midday sleeping, and general lounging behavior of the diurnal lemurs (e.g., Jolly, 1966) will one day be shown to relate to daily adjustments of metabolic rate (cf. Morland, 1993). Given lemurs' physiological adjustments to cool temperatures, the cathemeral schedule of activity in some species might also prove to be primarily a tactic to conserve energy (cf. Engqvist and Richard, 1991). Many nocturnal mammals show unusually low metabolic rates, including most prosimians and *Aotus* (Müller, 1985).

Traits correlated with seasonal adjustment of growth rate and adult body weight probably include significant fatting as the dry season begins and, demonstrably, rates of infant growth that are uncommonly high relative to maternal size among group-living primates (Brizzee and Dunlap, 1986; Kappeler 1992). Like many other animal parents, lemur mothers essentially suffer high costs of investment to help offspring meet a deadline: lemur infants must grow large, and perhaps also store energy, before dry season, so they can safely reduce metabolic rate and grow little during dry season. Patterns of growth in infant ringtails support this view: (a) significant growth accelerations occurred just after solstices, when infants began supplementing mothers' milk with some independent feeding, (b) the accelerations were accentuated when extra nutrition was available, and (c) growth rates plummeted after fall equinoxes -- earlier in larger infants -- regardless of the amount of nutrition made available artificially. Recent research on nonprimate mammals forewarns against glib predictions regarding when energy stores are used (Loudon and Racey, 1987). But, the huge weight gains exhibited by female ringtails early in lactation under HP seem to reveal metabolic and/or behavioral switches that should guard wild mothers (see LP, Fig. 7) against complete exhaustion of their fat stores. I expect that female ringtailed lemurs use stored energy earlier, to help effect low rates of activity and metabolism throughout much of gestation and the harsh season.

Large lemurs' putative seasonal suppression of metabolic rate would be analogous to the diverse torpid states that enhance survival and reproduction in large and small non-primate mammals coping with marked seasonality of resources (Eisenberg, 1981; Loudon and Racey, 1987; Boyce, 1988). The hypothesis that large-bodied lemurs have evolved this strategy recalls the idea that female dominance evolved in some lemurs because females incur costs in elevating metabolic rate to reproduce (Richard and Nicoll, 1987). The present view casts Richard and Nicoll's (1987) idea in a new light. If lemurs suppress metabolic rates across dry seasons (see also Morland, 1993), we might expect to see elevated rates of metabolism in pregnant females. Prior to arousal, all physiological systems in hibernating rodents operate at barely detectable levels except those dedicated to reproduction (Bronson, 1989). Pregnant lemurs may similarly fuel reproduction selectively, thereby elevating overall rate of metabolism. If semi-torpidity helps adult lemurs to *survive* their long dry season, then elevated rates of metabolism in gestating females would represent a cost of reproduction, by definition. An important issue raised by the present interpretation, however, is that lemur metabolic rates may not be quite as low overall as formerly suspected: most available data were gathered during the photoperiod of Malagasy dry seasons (B. McNab, pers. comm.).

**Seasonality and Social Behavior**

This is the first extensive longitudinal analysis of growth and adult body weight in a large-bodied lemur, and it carries important implications for our developing understanding of lemur social organization. The growth data support the hypotheses that costs of reproduction for female lemurs exceed those experienced by most other female primates (Jolly, 1984; Young et al., 1990) and that these costs have promoted the evolution of female dominance among Lemuriformes (Jolly, 1984; Richard, 1987). Additional laboratory research will be needed to corroborate that evidence, however. Specifically, we must determine how infant lemurs achieve their relatively high rates of growth. Is it accomplished via richer milk composition or greater rate of milk delivery relative to maternal size (see Buss et al., 1976)? Or, rather, is it due chiefly to greater growth efficiencies in immature lemurs (cf. Sackett and Ruppenthal, 1992)? Simultaneously, future field work should investigate whether the challenge to juveniles of surviving their first dry seasons likely generates the need for such rapid growth, as I have argued here.

Another aspect of lemur social organization illuminated by the present results is a mode of aggression between group mates termed *targeted aggression* (Vick and Pereira, 1989). Each year in DUPC groups of ringtailed lemurs, mature females target individual peers for chronic harassment and sometimes social eviction (Pereira, 1993; Vick and Pereira, 1989 discuss species differences). The phenomenon also occurs regularly at Berenty (Koyama, 1991, 1992; Pereira, unpubl. data) but less saliently at the Beza-Mahafaly Reserve (Sauther, pers. comm.), where development is slower and groups are smaller and possibly nonterritorial (Sussman, 1991). Because all known episodes of targeted aggression have begun in either premating or birth seasons (N=97; Pereira, 1993), it seems likely that photoperiod also regulates the expression of agonistic relations between females in this primate. Avoidance of targeting throughout gestation could function as follows: while it is important for reproducing females to limit their numbers in a given range or territory, it is crucial for them *not* to implement the required tactic of hyperaggressivity early in the dry season, when energy is conserved assiduously.

Finally, the present work suggests operational and functional affinities between premating and lactation season episodes of targeted aggression. Foraging efficiency is most likely reduced significantly for targeted females, diminishing their chances of reproductive success by interfering with the gaining of weight before conception or its sustenance through lactation. The targeted aggression of each season also threatens infant survivability. Birth season episodes have caused many infant deaths directly (Vick and Pereira, 1989; Koyama, 1992; Pereira, 1993 and unpubl. data) and constant stress before mating can delay estrus (Pereira, unpubl. data), leading targets' infants to be smallest and therefore subordinate to their peers (Pereira, 1993) and hardest pressed to gain weight before dry season. In either case, targeted females' options are to sustain their agonistic relationships and suffer their consequences, to try and reverse negative relations, or to depart their groups and try to take up residence elsewhere. The difficulty of relocation in a territorial population (Koyama, 1991; Pereira, unpubl. data) contributes pivotally to the development of targeted aggression, favoring both eviction and its resistance.

## CONCLUSIONS

With this chapter, I have tried to show how the study of proximate factors and environmental regulation of individual life histories is an essential part of our efforts to achieve meaningful depth of understanding of primate natural history. Photoperiodic cues provide a reliable schedule not only for reproduction, but also for adaptive changes in growth rate, metabolism, and behavior in ringtailed lemurs, just as they do for many nonprimate mammals, the Cheirogaleidae, and other lemuroid and anthropoid primates. Responsiveness along the basic schedule should be modulated by information specific to individuals, including age, nutritive status, and physical condition. Over the years, for example, light-weight infant ringtails at the DUPC have exhibited relatively small reductions of growth rate in November (e.g., Fig. 4), reflecting an interaction between photoperiod and size-related stimuli in determining individual growth rate.

Correlated seasonal adjustments of growth rate, adult body weight, and behavior in ringtailed lemurs led me to propose a new model of life-history strategy in the large-bodied lemurs, Lemuridae and Indridae. The harshness and predictability of the Malagasy dry season are suggested to have favored the evolution of a coadapted set of photoperiodically-regulated traits in these primates, including seasonal energy storage, reproduction, and modulation of metabolic rate, growth rate, temperature regulation, fat metabolism, activity level, foraging, and aggression. The general life-history strategy requires rates of infant growth that are uncommonly high relative to maternal size among group-living primates, and the consequent high cost of lactation has helped to promote the wide-spread evolution of female dominance and male-committed infanticide among lemurid and indrid species.

Looking to the future, only research in controlled environments will be able to disclose in detail how life histories in various primates are influenced by photoperiod and how other experiential factors interact in each case to determine characters like growth rate, body weight, metabolism, reproductive effort, aggressivity, and other social behavior. Meanwhile, field studies should be designed to characterize behavioral strategies as unequivocally as possible. What are the principal environmental challenges engaged by adult males, by adult females, and by immatures? When in the annual cycle does each challenge and its respective solutions occur? And, what is it, precisely, over which primates in each class actually compete?

## ACKNOWLEDGEMENTS

I extend deep appreciation and gratitude to the DUPC technicians whose assistance made it possible to weigh and measure scores of forest-living lemurs monthly over the past seven years: especially L. Martin, K. Alford-Madden, D. Brewer, and F. Stewart; also B. Coffman and B. Hess. I also thank the other technicians and myriad students who pitched in on occasion and acknowledge the entire DUPC staff for its caring and expert maintenance of the world's most diverse and unusual colony of captive primates. I thank past and current Directors, Drs. E. Simons and K. Glander, for the opportunities they have afforded me to research at the Duke facility. And, special thanks go to Terri-Jean Pyer, whose indefatigable partnership promotes my work immeasurably. This research was sponsored by the NICHD (R29-HD23243), the Chicago Zoological Society, and the Duke University Primate Center. This is DUPC Publication 548.

## REFERENCES

Benirschke, K. and Miller, C.J., 1981, Weights and neonatal growth of ringtailed lemurs (*Lemur catta*) and ruffed lemurs (*Lemur variegatus*), *J. Zoo Anim. Med.* 12:107.

Bogart, M.H., Kumamoto, A.T. and Lasley, B.L., 1977, A comparison of the reproductive cycle of three species of *Lemur*, *Folia Primatol.* 28:134.

Boyce, M.S., 1988, Evolution of life histories: theory and patterns from mammals, *in*: " Evolution of Life Histories of Mammals," M.S. Boyce, ed., Yale University Press, New Haven.

Bradley, S.R. and Hudson, J.W., 1974, Temperature regulation in the tree shrew *Tupaia glis*, Comp. *Biochem. Physiol.* 48A:55.

Brizzee, K.R. and Dunlap, W.P., 1986, Growth, *in*: " Comparative Primate Biology, Vol. 3: Reproduction and Development," J. Erwin and G. Mitchell, eds., Alan R. Liss, New York.

Bronson, F.H., 1989, "Mammalian Reproductive Biology," University of Chicago Press, Chicago.

Buss, D.H., Cooper, R.W. and Wallen, K., 1976, Composition of lemur milk, *Folia Primatol.* 26:301.

Chik, C.L., Almeida, O.F.X., Libre, E.A., Booth, J.D., Renquist, D. and Merriam, G.R., 1992, Photoperiod-driven changes in reproductive function in male rhesus monkeys, *J. Clin. Endocrinol. Metabol.* 74:1068.

Daniels, H., 1984, Oxygen consumption in *Lemur fulvus*: deviation from the ideal model, *J. Mammal.* 65:584.

Dark, J. and Zucker, I., 1983, Short photoperiods reduce winter energy requirements of the meadow vole, *Microtus pennsylvanicus*, *Physiol. Behav.* 31:699.

DuMond, F.V., 1968, The squirrel monkey in a seminatural environment, *in*: "The Squirrel Monkey," L.A. Rosenblum and R.W. Cooper, Academic Press, New York.

Eisenberg, J.F., 1981, "The Mammalian Radiations," University of Chicago Press, Chicago.

Engqvist, A. and Richard, A., 1991, Diet as a possible determinant of cathemeral activity patterns in primates, *Folia Primatol.* 57:169.

Foerg, R., 1982, Reproduction in *Cheirogaleus medius*, *Folia Primatol.* 39:49.

Foerg, R. and Hoffman, R., 1982, Seasonal and daily activity changes in captive *Cheirogaleus medius*, *Folia Primatol.* 38:259.

Glassman, D., Coelho, A., Carey, K. and Bramblett, C., 1984, Weight growth in savannah baboons: A longitudinal study from birth to adulthood, *Growth* 48:425.

Glick, B.B., 1979, Testicular size, testosterone level, and body weight in male *Macaca radiata*, Maturational and seasonal effects, *Folia Primatol.* 32:268.

Griffiths, J.J. and Ranaivoson, R., 1972, Madagascar *in*: "Climates of Africa," J.F. Griffiths, ed., Elsevier, New York.

Hoffmann, K., 1981, Photoperiodism in vertebrates, *in*: "Biological Rhythms, Handbook of Behavioral Neurobiology," Vol 4, J. Aschoff, ed., Plenum Press, New York.

Hrdy, S.B., 1981, "The Woman That Never Evolved," Harvard University Press, Cambridge.

Janson, C.H. and van Schaik, C.P., 1993, Slow and steady wins the race: ecological risk aversion in juvenile primates, *in*: "Juvenile Primates: Life History, Development and Behavior," M.E. Pereira and L.A. Fairbanks, eds., Oxford University Press, New York.

Jolly, A., 1966, "Lemur Behavior: A Madagascar Field Study," University of Chicago Press, Chicago.

Jolly, A., 1984, The puzzle of female feeding priority, *in*: "Female Primates: Studies by Women Primatologists," M. Small, ed., Alan R. Liss, Inc., New York.

Kappeler, P.M., 1992, "Female dominance in Malagasy primates,", Ph.D. thesis, Duke University, Durham.

Kirkwood, J.K., 1985, Patterns of growth in primates, *J. Zool., Lond.* 205:123.

Koyama, N., 1988, "Mating behavior of ring-tailed lemurs (*Lemur catta*) at Berenty, Madagascar, *Primates*, 29:163.

Koyama, N., 1991, Troop division and inter-troop relationships of ring-tailed lemurs (*Lemur catta*) at Berenty, Madagascar, *in*: "Primatol. Today," A. Ehara, T. Kimura, O. Takenaka, and M. Iwamoto, eds., Elsevier, Amsterdam.

Koyama, N., 1992, Some demographic data of ring-tailed lemurs (*Lemur catta*) at Berenty, Madagascar, *in*: "Social Structure of Madagascar Higher Vertebrates in Relation to Their Adaptive Radiation," S. Yamagishi, ed., Osaka City University, Osaka.

Lee, T.M. and Zucker, I., 1988, Vole infant development is influenced perinatally by maternal photoperiodic history, *Am. J. Physiol.*, 255, RICP 24:831.

Lindburg, D.G., 1987, Seasonality of reproduction in primates, *in*: "Comparative Primate Biology, Vol. 2B: Behavior, Cognition, and Motivation," J. Erwin and G.D. Mitchell, eds., Alan R. Liss., New York.

Loudon, A.S.I. and Racey, P.A., (eds.), 1987., "Reproductive Energetics in Mammals", Clarendon Press, Oxford.

Matsubayashi, K. and Enomoto, T., 1983, Longitudinal studies on annual changes in plasma testosterone, body weight and spermatogenesis in adult Japanese monkeys (*Macaca fuscata fuscata*) under laboratory conditions, *Primates*, 24: 521.

McClintock, M.K., 1983, Pheromonal regulation of the ovarian cycle: enhancement, suppression, and synchrony, *in*: "Pheromones and Reproduction in Mammals," J. Vandenberg, ed., Academic Press, New York.

McCormick, S.A., 1981, Oxygen consumption and torpor in the fat-tailed dwarf lemur (*Cheirogaleus medius*): rethinking prosimian metabolism, *Comp. Biochem. Physiol. A* 68:605.

McNab, B.K., 1988, Energy conservation in a tree-kangaroo (*Dendrolagus matschiei*) and the red panda (*Ailurus fulgens*), *Physiol. Zool.* 61:280.

Morland, H.S., 1992, Seasonal behavioral variation and its relationship to thermoregulation in ruffed lemurs (*Varecia variegata variegata*), *in*: "Lemur Social Systems and Their Ecological Basis," P.M. Kappeler and J.U. Ganzhorn, eds., Plenum Press, New York.

Müller, E.F., 1985, Basal metabolic rates in primates - the possible role of phylogenetic and ecological factors, *Comp. Biochem. Physiol. A* 81:707.

Pereira, M.E., 1991, Asynchrony within estrous synchrony among ringtailed lemurs (Primates: Lemuridae), *Physiol. Behav.* 49:47.

Pereira, M.E., 1993, Agonistic interaction, dominance relation, and ontogenetic trajectories in ringtailed lemurs, *in*: "Juvenile Primates: Life History, Development, and Behavior," M. E. Pereira and L. A. Fairbanks, eds., Oxford University Press, New York.

Pereira, M.E. and Fairbanks, L.A.(eds.), 1993, "Juvenile Primates: Life History, Development, and Behavior," Oxford University Press, New York.

Pereira, M.E. and Weiss, M.L., 1991, Female mate choice, male migration, and the threat of infanticide in ringtailed lemurs, *Behav. Ecol. Sociobiol.* 28:141.

Pereira, M.E., Klepper, A. and Simons, E.L., 1987, Tactics of care for young infants by forest-living ruffed lemurs (*Varecia variegata variegata*): Ground nests, parking, and biparental guarding, *Am. J. Primatol.* 13:129.

Petter-Rousseaux, A., 1968, Cycles genitaux saisonniers des lémuriens malgaches, *in*: "Cycles Genitaux Saisonniers du Mammifères Sauvages," R. Canivenc, ed., Masson, Paris.

Petter-Rousseaux, A., 1974, Photoperiod, sexual activity and body weight variation of *Microcebus murinus* (Miller 1777), *in*: "Prosimian Biology," R.D. Martin, G.A. Doyle and A.C. Walker, eds., Duckworth, London.

Petter-Rousseaux, A., 1980, Seasonal activity rhythms, reproduction, and body weight variations in five sympatric nocturnal prosimians, in simulated light and climatic conditions, *in*: "Nocturnal Malagasy Primates," P. Charles-Dominique, H.M. Cooper, A. Hladik, C.M. Hladik, E. Pages, G.F. Pariente, A. Petter-Rousseaux, J.J. Petter and A. Schilling, eds., Academic Press, New York.

Pond, C.M., 1978, Morphological aspects and the ecological and mechanical consequences of fat deposition in wild vertebrates, *Ann. Rev. Ecol. Syst.* 9:519.

Rasmussen, D.T., 1985, A comparative study of breeding seasonality and litter size in eleven taxa of captive lemurs (*Lemur* and *Varecia*), *Int. J. Primatol.* 6:501.

Richard, A.F., 1987, Malagasy prosimians: female dominance, *in*: "Primate Societies," B.B. Smuts, D.L. Cheney, R.M. Seyfarth, R.W. Wrangham and T.T. Struhsaker, eds., University of Chicago Press, Chicago.

Richard, A.F. and Nicoll, M.E., 1987, Female social dominance and basal metabolism in a Malagasy primate, *Propithecus verreauxi*, *Am. J. Primatol.* 12:309.

Russell, R.J., 1975, Body temperatures and behavior of captive Cheirogaleids, *in*: "Lemur Biology," I. Tattersall and R. Sussman, eds., Plenum Press, London.

Rutenberg, G. and Coelho, A., 1988, Neonatal nutrition and longitudinal growth from birth to adolescence in baboons, *Am. J. Phys. Anthropol.* 75:529.

Sackett, G.P. and Ruppenthal, G.C., 1992, Growth of nursery-raised *Macaca nemestrina* infants: Effects of feeding schedules, sex, and birth weight, *Am. J. Primatol.* 27:177.

Sauther, M.L., 1991, Reproductive behavior of free-ranging *Lemur catta* at Beza Mahafaly Special Reserve, Madagascar, *Am. J. Phys. Anthropol.* 84:463.

Strum, S.C., 1991, Weight and age in wild olive baboons, *Am. J. Primatol.* 25:219.

Sussman, R.W., 1991, Demography and social organization of free-ranging *Lemur catta* in the Beza Mahafaly Reserve, Madagascar, *Am. J. Phys. Anthropol.* 84:43.

Taylor, L.L. and Sussman, R.W., 1985, A preliminary study of kinship and social organization in a semifree-ranging group of *Lemur catta*, *Int. J. Primatol.* 6:601.

van Horn, R.N., 1975, Primate breeding season: photoperiodic regulation in captive *Lemur catta*, *Folia Primatol.* 24:203.

van Horn, R.N., 1980. Seasonal reproductive patterns in primates, *in*: "Progress in Reproductive Biology," P.O. Hubinont, ed., Karger, Basel.

van Horn, R.N. and Eaton, G.G., 1979, Reproductive physiology and behavior in prosimians *in*: "The Study of Prosimian Behavior," G.A. Doyle and R.D. Martin, eds., Academic Press, New York.

van Horn, R.N. and Resko, J.A., 1977, The reproductive cycle of the ring-tailed lemur *Lemur catta*: sex steroid levels and sexual receptivity under controlled photoperiods, *Endocrinol.* 101:1579.

van Schaik, C.P., 1989, The ecology of social relationships amongst female primates, *in*: "Comparative Socioecology: The Behavioural Ecology of Humans and Other Mammals," V. Standen and R.A. Foley, eds., Blackwell Scientific Publ., Boston.

van Schaik, C.P. and P.M. Kappeler, 1993, Life history, activity period and lemur social systems, *in*: "Lemur Social Systems and Their Ecological Basis," P.M. Kappeler and J.U. Ganzhorn, eds., Plenum Press, New York.

Vick, L.G. and Pereira, M.E., 1989, Episodic targeting aggression and the histories of *Lemur* social groups, *Behav. Ecol. Sociobiol.* 25:3.

Wilen, R. and Naftolin, F., 1978, Pubertal age, weight, and weight gain in the individual female New World monkey (*Cebus albifrons*), *Primates* 19:769.

Winfree, A.T., 1980, "The Geometry of Biological Time," Springer, New York.

Young, A.L., Richard, A.F. and Aiello, L.C., 1990, Female dominance and maternal investment in strepsirhine primates, *Am. Nat.* 135:473.

# SEXUAL SELECTION AND LEMUR SOCIAL SYSTEMS

Peter M. Kappeler

Department of Zoology
Duke University
Durham, NC 27706
U.S.A.

## ABSTRACT

The lack of sexual size dimorphism among the lemurs of Madagascar affects agonistic relations between males and females, and therefore contributes to the most salient feature of lemur social systems: female dominance over males. The purpose of this chapter is to examine the nature and mechanisms of intrasexual selection in polygynous lemurs to illuminate the relationship between sexual selection and sex differences in dominance. First, I test the hypothesis that the lack of sexual size dimorphism in polygynous lemurs is a result of high viability costs. This hypothesis predicts that males in polygynous lemur species should have significantly larger canines than females, if male combat contributes importantly to variance in male reproductive success. Male-biased sexual canine dimorphism occurs in only 4 polygynous species, however. The reverse is true in *Propithecus diadema* and significant sex differences in canine size are absent in 8 other species. Moreover, there is no heterogeneity in the average degree of sexual canine dimorphism among lemurs with different mating systems. The general lack of sexual size dimorphism in polygynous lemurs is therefore not the result of high viability costs. More likely, it reflects weak selection on characters associated with male combat. The widespread absence of sexual dimorphism in body and canine size may still be reconcilable with sexual selection theory if male-male competition in polygynous lemurs is primarily post-copulatory, i.e., if it takes the form of sperm competition. In contrast to the prediction of this hypothesis, however, I found that males of solitary, pair- and group-living lemurs do not differ significantly in average relative testes size during the breeding season. Together, these comparative analyses suggest that intrasexual selection is of similar intensity in both monogamous and polygynous lemurs. Possible reasons for reduced variance in male reproductive success in non-monogamous species include scramble polygyny competition in some solitary species, as well as the existence of male-female pair-bonds in some group-living species. Female dominance occurs in species where males have superior weapons, but it does not characterize all polygynous lemurs. Thus, reduced intrasexual selection may have facilitated female dominance, but sexual selection theory alone cannot provide a sufficient explanation for the evolution of female dominance.

*Lemur Social Systems and Their Ecological Basis*, Edited by
P.M. Kappeler and J.U. Ganzhorn, Plenum Press, New York, 1993 

## INTRODUCTION

Female dominance over males is a salient and unique feature of lemur social systems (Jolly, 1984; Richard, 1987; Kappeler, 1993). The ability of all adult females to consistently evoke submissive behavior from all adult males in dyadic agonistic interactions has been demonstrated in *Lemur catta, Indri indri, Varecia variegata* and *Eulemur coronatus* (Jolly, 1966; Budnitz and Dainis, 1975; Pollock, 1979, 1989; Kappeler, 1990a; Pereira et al., 1990; Kaufman, 1991), and there is suggestive circumstantial evidence for female dominance in *Propithecus verreauxi, P. diadema, Microcebus murinus* and *Phaner furcifer* (Richard and Heimbuch, 1975; Charles-Dominique and Petter, 1980; Richard and Nicoll, 1987; Pages-Feuillade, 1988; Wright, 1988). Although female dominance is not characteristic of all lemurs (e.g., *Eulemur fulvus:* Pereira et al., 1990), male dominance, which is typical for polygynous mammals, has not been reported for any Malagasy primate.

Recent attempts to explain the adaptiveness of female dominance in lemurs have focussed exclusively on female biology (Jolly, 1984; Richard and Nicoll, 1987; Pollock, 1989; Young et al., 1990). In contrast, the role of lemur males in the evolution of female dominance is still poorly understood (but see Pollock, 1979; Hrdy, 1981). In particular, the absence of male dominance, i.e., the unability of males in many lemur species to elicit submissive behavior from females, has not been reconciled with sexual selection theory. In this chapter, I will try to illuminate the mechanisms underlying this deviation of lemurs from the typical mammalian pattern by analyzing interspecific variation in three sexually selected traits across prosimian primates.

Sexual selection theory, originally developed by Darwin (1871), explains the adaptiveness of secondary sexual characters. They are thought to have evolved because they contribute to an individual's reproductive success, either by increasing its attractiveness to members of the opposite sex, or by improving its ability to compete with members of the same sex. Because males of most mammals make a smaller parental investment (Trivers, 1972) and/or have higher potential reproductive rates (Clutton-Brock, 1991a) than females, their reproductive success is limited by the number of females they can inseminate (Bateman, 1948). Thus, males will typically compete among themselves for access to fertile females, i.e., they are subject to intrasexual selection.

The generally accepted explanation of intersexual dominance relations is based on this aspect of sexual selection theory. It posits that male dominance in polygynous mammals is a byproduct of intrasexual selection, because competition among males favors the development of morphological and behavioral traits that increase agonistic power (Darwin, 1871; Hrdy, 1981; Fedigan, 1982; Smuts, 1987a). Social dominance typically rests on superior force (Hand, 1986), which in turn is based on these intrinsic traits. Thus, if these traits contribute to variance in male reproductive success, there will be selection for larger male size, strength, weaponry, stamina and aggressiveness, which contribute to male agonistic superiority in intersexual conflict.

The distribution of behavioral and morphological sex differences in a wide range of mammals is consistent with this explanation. First, males are the larger sex in the majority of mammals, and sexual size dimorphism increases with the extent of male polygyny (Crook, 1972; Clutton-Brock et al., 1977; Alexander et al., 1979; Clutton-Brock, 1985; Barrette and Vandal, 1990). Second, selection for male competitive abilities has also resulted in the development of larger weapons, such as horns, antlers or canine teeth in males (Geist, 1966; Harvey et al., 1978; Clutton-Brock et al., 1980; Packer, 1983). Third, males are more likely to escalate intrasexual conflicts because they compete for higher stakes than females. Since aggressive males are more likely to be successful in contests with other males, sexual selection favors behavioral features associated with aggressive competition, and, as a result, males are believed to be the more aggressive sex in intrasexual conflict (Collias, 1953; Bouissou, 1983; Reinhardt et al., 1987). Finally, male dominance has been demonstrated in many mammals (Archer, 1988), including a number of anthropoid primates (Kummer, 1968; Hausfater, 1975; Harcourt, 1979; Jones, 1980; Noe et al., 1980; Fedigan and Baxter, 1984).

A few exceptions to this general mammalian pattern exist, however. For example, females are larger than males in several polygynous species of bats, cetaceans, artiodactyls, pinnipeds, lagomorphs, and rodents (Ralls, 1976). However, although females have been reported to dominate males in some of these species, none of them are characterized by invariable female dominance (Kappeler, 1993). Competition among female mammals for access to males is rare because they cannot increase their reproductive success by mating with more than one male, except for species in which success in sperm competition is heritable. Thus, intrasexual selection on females cannot provide an analogous explanation for larger female size and female dominance (Ralls, 1976). Instead, it has been suggested that larger female size confers a reproductive advantage by reducing the energetic costs of relative maternal investment (Ralls, 1976), or a competitive advantage in intersexual competition for food (Jolly, 1984; Tilson and Hamilton, 1984). These hypotheses do not account for the occurrence of male-male competition, however.

The unusual constellation of morphological and or behavioral sex differences in these "sex-reversed" species raises several questions about the opportunity for sexual selection. For example, do these males not compete for access to females, or are size and weaponry not important determinants of male reproductive success? Or, if pre-copulatory competition exists, which other evolutionary forces override its effects, and why? The lemurs of Madagascar provide a unique opportunity for a comparative study of these questions because they form a monophyletic group of 30 species that represent several types of mammalian social and mating systems (Tattersall, 1982; Richard, 1987; Richard and Dewar, 1991). Furthermore, many of their closest living relatives, the African and Asian lorises, are characterized by male dominance and larger male size (Bearder, 1987; Nash et al., 1989; Kappeler, 1991), and therefore provide a convenient outgroup for comparison.

I will begin a comparative analysis of evolutionary responses to sexual selection in males of these two groups of prosimian primates with a brief summary of interspecific variation in sexual size dimorphism. A discussion of possible causes for the lack of larger male size will lead over to an examination of interspecific variation in two other sexually selected traits: canine size and testes size. I will examine the predictions of sexual selection theory for interspecific variation in these characters across prosimians before discussing possible reasons for the deviation of lemur males from the expected mammalian pattern. I will conclude this chapter with a general discussion of the adaptiveness of sex differences in dominance.

## SEXUAL DIMORPHISM IN BODY SIZE

Patterns of sexual size dimorphism among prosimian primates have only recently been described in detail (Kappeler, 1990b, 1991; Jenkins and Albrecht, 1991). I will therefore briefly summarize sex differences in body mass and supplement them with a few new analyses. I collected data on adult male and female body mass for 33 prosimian species from the records of the Duke University Primate Center (DUPC) and the literature to determine the degree of sexual size dimorphism. I determined the mean mass for both sexes of each species, calculated the degree of sexual size dimorphism as the ratio of mean male and female mass, and tested each species for significant sex differences in body mass (Table 1).

**Methodological Considerations.** Sexual selection theory makes specific predictions about interspecific variation in sexually selected traits as a function of species differences in variance in male lifetime reproductive success. Because such information on reproductive success is not available for primates, a categorical distinction among species with different mating systems has been employed in several previous studies (Clutton-Brock et al., 1977; Leutenegger, 1978; Alexander et al., 1979). The crudest distinction is based on the fact that variance in male reproductive success is typically lower in monogamous than in polygynous species (Clutton-Brock, 1985; 1988). If more detailed information is available, it is possible to make a further distinction among species with different types of polygynous mating systems (Clutton-Brock, 1991b). Mating systems are frequently inferred from information on social organization, especially if mating behavior is cryptic and/or genetic data are not available (Clutton-Brock, 1989). Because there is not necessarily a close correspondence between social organization and mating system (Rowell and Chism, 1986), analyses based on these classification criteria have to be considered preliminary, however.

**Table 1.** Sexual size dimorphism, sexual canine dimorphism and relative testes size in prosimian primates[1].

| SPECIES | SSD | N | SCD | N | RTS | N | MS |
|---|---|---|---|---|---|---|---|
| Allocebus trichotis | 1.08 | 4 | | | | | S |
| Cheirogaleus medius | 1.00 | 48 | 0.94 | 12 | -0.232 | 15 | S |
| Microcebus murinus | 0.83 | 60 | 0.99 | 100 | 0.091 | 10 | S |
| Microcebus rufus | 1.02 | 28 | 0.94 | 21 | | | S |
| Mirza coquereli | 1.02 | 19 | | | 0.497 | 11 | S |
| Daubentonia madagascariensis | 1.07 | 4 | S | | | | |
| Hapalemur griseus | 1.05 | 19 | 1.00 | 23 | -0.332 | 5 | P |
| Eulemur coronatus | 1.02 | 19 | 1.11 | 22 | 0.060 | 8 | G |
| Eulemur fulvus | 1.00 | 107 | 1.18 | 37 | 0.274 | 8 | G |
| Eulemur macaco | 0.97 | 43 | 1.02 | 14 | 0.326 | 8 | G |
| Eulemur mongoz | 1.01 | 27 | 1.20 | 13 | -0.315 | 9 | P |
| Eulemur rubriventer | 1.06 | 8 | 1.05 | 21 | -0.611 | 4 | P |
| Lemur catta | 1.01 | 40 | 1.19 | 19 | 0.350 | 12 | G |
| Varecia variegata | 0.99 | 81 | 1.04 | 10 | 0.112 | 11 | P |
| Lepilemur leucopus | 1.12 | 24 | | | | | S |
| Lepilemur mustelinus | 1.04 | 7 | 1.33 | 13 | | | S |
| Lepilemur ruficaudatus | 0.90 | 6 | 1.18 | 11 | | | S |
| Avahi laniger | 0.79 | 8 | 0.95 | 11 | | | P |
| Indri indri | 1.00 | 18 | | | | | P |
| Propithecus diadema | 0.96 | 14 | 0.88 | 12 | | | G |
| Propithecus tattersalli | 0.96 | 9 | | | | | G |
| Propithecus verreauxi | 0.98 | 22 | 1.06 | 12 | | | G |
| Galago elegantulus | 1.07 | 24 | | | | | S |
| Galago moholi | 1.17 | 45 | 1.06 | 86 | -0.414 | 14 | S |
| Galago senegalensis | 1.11 | 101 | | | | | S |
| Galagoides alleni | 1.11 | 22 | | | | | S |
| Galagoides demidovii | 1.17 | 12 | 1.28 | 56 | | | S |
| Galagoides zanzibaricus | 1.17 | 26 | 1.11 | 18 | | | S |
| Otolemur crassicaudatus | 1.20 | 84 | 1.23 | 40 | -0.193 | 9 | S |
| Otolemur garnettii | 1.18 | 38 | 1.08 | 59 | 0.072 | 10 | S |
| Arctocebus calabarensis | 1.07 | 11 | 1.00 | 10 | | | S |
| Loris tardigradus | 1.00 | 18 | 0.98 | 15 | 0.154 | 4 | S |
| Nycticebus coucang | 1.01 | 29 | 1.04 | 28 | -0.203 | 8 | S |
| Nycticebus pygmaeus | 1.23 | 12 | 1.11 | 10 | 0.175 | 4 | S |
| Perodicticus potto | 0.96 | 11 | 1.01 | 53 | 0.190 | 3 | S |

[1] The ratio of mean male and female body mass (SSD), canine height (SCD) and relative testes size (RTS), along with the respective sample sizes (N). The mating system classification (MS) distinguishes between solitary (S), pair-living (P) and group-living (G) species. Significant sex differences in body mass were demonstrated in *M. murinus, G. moholi, G. demidovii, G. zanzibaricus, O. crassicaudatus, O. garnettii* and *N. pygmaeus*. Significant sex differences in canine height were demonstrated in *P. diadema, E. fulvus, E. mongoz, L. catta, L. leucopus, L. mustelinus, G. moholi, G. senegalensis, G. alleni, G. demidovii, G. zanzibaricus, O. crassicaudatus* and *O. garnettii*.

As information on prosimian mating behavior is sparse (Tattersall, 1982; Richard, 1987), I based the present classification of prosimian mating systems primarily on reports of group size and composition. This classification assumes that species in which only one pair of adults is permanently associated have a monogamous mating system (Clutton-Brock, 1989). I distinguished among the remaining polygynous species between solitary and group-living species. The main reason for this distinction is that the mating system of solitary prosimians, which has been termed spatial polygyny (Bearder, 1987), may effectively resemble that of anthropoids with harem polygyny (Clutton-Brock, 1989). In both of these groups of species, a single male typically monopolizes access to several females. The mating system of group-living species is fundamentally different because several adult males may compete for and achieve matings with individual females (Clutton-Brock, 1989; 1991b).

At this point, a brief digression on comparative analyses is also indicated. Statistical dependence between related species is a common problem in interspecific comparisons because species with recent common ancestors are more likely to have similar character states than more distantly related species (Clutton-Brock and Harvey, 1984; Felsenstein, 1985; Harvey and Pagel, 1991). It is therefore not justified to use species values as the units for statistical analysis, unless one is willing to defend unrealistic models of speciation. Several statistical methods have been developed to deal with the problems of non-independence among species. As a first step, nested analysis of variance can be employed to describe how the total interspecific variation in a continuous character is distributed among taxonomic levels. Several phylogenetic contrast methods, which find mutually independent comparisons within a true bifurcating phylogeny, are then available to control for phylogenetic bias in analyses of discrete and continuous characters (Harvey and Pagel, 1991).

The purpose of phylogenetic contrast methods is to examine whether pairs of discrete or continuous characters have co-evolved or not. Appropriate methods for a mixture of discrete and continuous data, such as mating system and testes size, for example, are presently not available, however. I will therefore consider mating system as a treatment, using species values as the units of analysis. This decision is supported by the results of nested ANOVAs, which revealed that most of the interspecific variance in sexual size dimorphism, sexual canine dimorphism and relative testes size is located within genera (Kappeler, 1990b, 1991, 1992 and unpubl. data). Thus, closely related species display considerable variation in these characters, suggesting that these three traits evolved largely independent of phylogenetic constraints. Nevertheless, I will implement some phylogenetic control by also performing analyses based on generic means of species with the same mating system, so that e.g., all polygynous *Eulemur* species contribute one value.

For the present analyses, I employed non-parametric tests (Siegel and Castellan, 1988) to compare the degree of sexual size dimorphism among subsets of species with different mating systems, among members of different mating systems in the phylogenetically controlled sample, and between lemurs and lorises that share the same mating system.

**Patterns of Sexual Size Dimorphism**

With the exception of *Microcebus murinus*, who have significantly larger females, lemurs generally lack significant sexual size dimorphism, whereas males of most bushbabies (Galagidae) and some lorises (Lorisidae) are significantly larger than females (Table 1). There are complex interactions between taxonomic affiliation and mating system in their effect on sexual size dimorphism in prosimians. In contrast to the predictions of sexual selection theory, there is no significant difference in sexual size dimorphism among solitary, pair- and group-living lemurs (Kruskal-Wallis $H = 1.82$, NS). However, there is significant heterogeneity in sexual size dimorphism if solitary lorises are included in the comparison ($H = 8.77$, $P = 0.032$, Fig. 1).

Comparisons between polygynous lemurs and lorises, and between solitary lemurs and lorises revealed that lorises in both samples exhibit on average a significantly higher degree of sexual size dimorphism than lemurs (polygynous: $z=2.61$, $P=0.0089$; solitary: $z=1.96$, $P=0.049$). The mating system had no effect on sexual size dimorphism in the phylogenetically-controlled sample, however.

**Why Do Lemurs Lack Sexual Size Dimorphism?**

The lack of male-biased size dimorphism in polygynous lemurs is puzzling because it is not in agreement with the predictions of any of the major hypotheses traditionally proposed to explain the evolution of sexual dimorphism. First, the lack of sexual size dimorphism in recent lemurs is most likely not a result of their relatively small size and associated allometric constraints. Among extant prosimians, the degree of sexual size dimorphism varies independently of body size among species (Kappeler, 1990b). Moreover, simulations by Godfrey et al. (1993) demonstrated that the giant subfossil lemurs were also not sexually dimorphic, indicating that the lack of sexual size dimorphism is a characteristic trait of all Malagasy primates, ranging in size from mouse lemurs to gorillas.

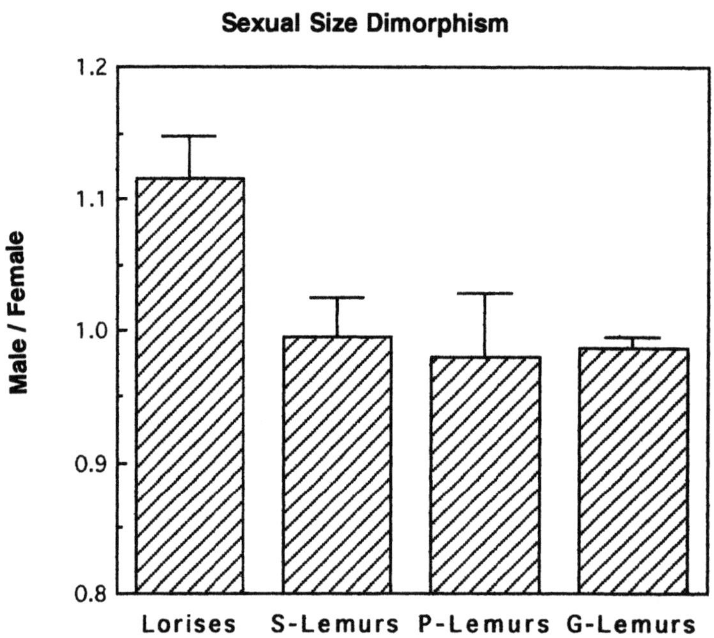


**Figure 1.** Average degree of sexual size dimorphism (± SE) for prosimians with different mating systems. See Tab. 1 for classification and sample size.

Second, although all lemurs lack male-biased sexual size dimorphism, this pattern is not simply a non-adaptive consequence of their common ancestry. One could argue that the ancestor of the monophyletic lemurs was monomorphic, and that this state was maintained during their subsequent adaptive radiation (e.g., Cheverud et al., 1985). In this case, a comparison with the closely related Lorisiformes should reveal that most of the interspecific variation in sexual size dimorphism is located between these two superfamilies. However, most variance in dimorphism among prosimians occurs within genera, and not at higher taxonomic levels (Kappeler, 1990b), suggesting that even closely related species are not phylogenetically constrained in their evolutionary response to selective forces acting on male and female body size.

This conclusion must be qualified, however, because of the observed variation in the direction and magnitude of sexual size dimorphism between lemurs and lorises. After controlling for mating system, lorises exhibit on average a significantly larger degree of male-biased sexual size dimorphism than lemurs, despite similar variation within genera. Thus, there are consistent differences between higher taxonomic groups, but to attribute them to phylogenetic inertia provides only a description and not an explanation of an interesting phenomenon (Kappeler, 1991).

Finally, the absence of sexual size dimorphism in polygynous lemurs does not agree with the predictions of sexual selection theory. Several explanations for this discrepancy are possible. First, it may be that size and strength contribute little to male competitive ability in lemurs (Clutton-Brock, 1985). Speed and agility, for example, may be more important qualities for chases among relatively small mammals in a three-dimensional environment. However, many other mammals faced with similar problems, including the closely related bushbabies, exhibit moderate sexual size dimorphism, suggesting that superior size can provide benefits in direct competitive interactions among small arboreal male primates.

Second, the development and maintenance of a sexually selected trait such as large body size, may be associated with high viability costs. Across mammals, the degree of sexual size dimorphism is positively correlated with the degree of male-bias in adult mortality (Promislow, 1992), suggesting that male size is constrained by natural selection. Energetic demands during prolonged periods of food scarcity, which are characteristic of many lemur habitats (Hladik, 1980), may ultimately be one factor contributing to these potential costs in lemurs (Hladik et al., 1980; Jolly, 1984; see also Hrdy, 1981).

One prediction of this hypothesis is that sexually selected traits with smaller viability costs should be developed in males of polygynous lemurs. Canine teeth are an obvious candidate for such a trait because they are the most effective weapons of primates, contribute importantly to physical superiority, and have presumably low maintenance costs (Kay et al., 1988; Plavcan and van Schaik, 1992). Patterns of injuries and direct observations during the mating season indicate that lemurs actually use their canines as weapons (Jolly, 1967; Richard, 1974; Foerg, 1982a, b; Colquhoun, 1987). Thus, polygynous lemurs should exhibit male-biased sexual canine dimorphism, if intrasexual selection acting on determinants of competitive ability is not eliminated altogether. The existence of sexual canine dimorphism would also be consistent with the hypothesis that large male body size is costly.

## SEXUAL DIMORPHISM IN CANINE SIZE

Because published data on prosimian canine size (Swindler, 1976; Gingerich and Ryan, 1979; Kieser and Groeneveld, 1990) did not allow comprehensive interspecific comparisons, I measured several craniodental dimensions in a total of 1309 adult specimens from 42 prosimian species at the collections of the United States National Museum in Washington, D. C., the American Museum of Natural History in New York, N.Y. and the British Museum of Natural History in London, U. K. For the following analyses, I included only species for which at least 10 specimens were measured, and used the subspecies with the largest sample to represent species with several subspecies. These selection criteria, which control for sampling error and geographical variation (Albrecht et al., 1990), reduced the final sample to 915 specimens from 31 species (Table 1).

I used canine height, measured with digital calipers as the distance between the tip of the canine to the cemento-enamel junction on the buccal side, to determine a functionally meaningful index of sexual canine dimorphism. This is a contrast to previous studies (Harvey et al., 1978; Kay et al., 1988), which used the occlusal dimensions to estimate canine size. I chose canine height because it is presumably the primary determinant of the tooth's suitability as a weapon. I expressed sexual canine dimorphism as the ratio of mean male and female canine height and determined sex differences with a t-test. Interspecific comparisons among subsets of species parallel those outlined above for sexual size dimorphism.

## Patterns in Sexual Canine Dimorphism

Prosimians exhibit great interspecific variation in sexual canine dimorphism. Among lemurs, females of *Propithecus diadema* have significantly larger canines than males, whereas male *Lepilemur leucopus, L. mustelinus, Lemur catta, Eulemur fulvus* and *E. mongoz* are endowed with significantly larger canines than females. Male and female canine height are not significantly different in the remaining 12 lemur species in this sample. None of the true lorises (Lorisidae) exhibit sexual canine dimorphism, whereas seven out of eight bushbabies (Galagidae) are characterized by significant male-biased canine dimorphism.

In contrast to sexual size dimorphism, the average degree of sexual canine dimorphism does not vary significantly among Malagasy and non-Malagasy prosimians with different mating systems ($H=1.44$, NS, Fig. 2). The mating system has also no effect on differences in sexual canine dimorphism among lemurs ($H=0.29$, NS). There is also no difference in average sexual canine dimorphism between polygynous ($z=0.19$, NS) or solitary ($z=0.26$, NS) lemurs and lorises. However, there is highly significant variation in sexual dimorphism in canine size among all prosimians at the taxonomic level of the family ($H=18.57$, $P=0.002$), even in the phylogenetically-controlled sample ($H=13.14$, $P=0.022$).

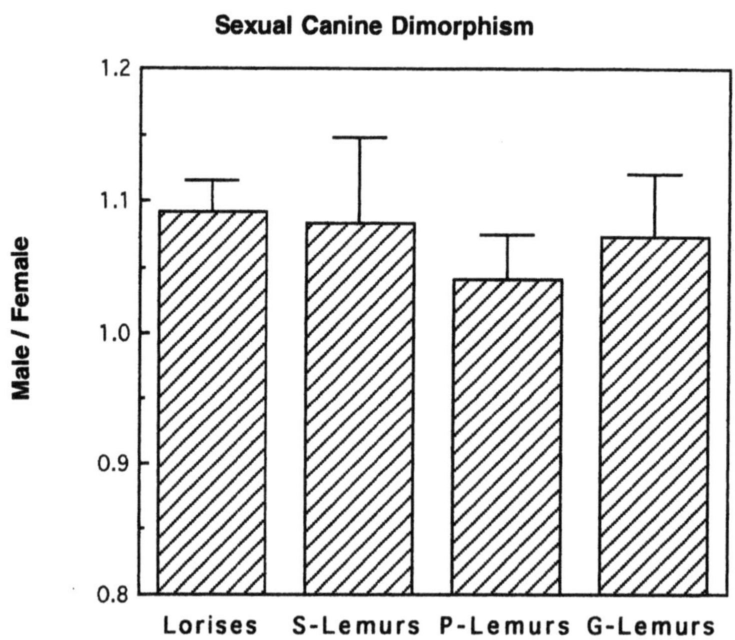


**Figure 2.** Average degree of sexual canine dimorphism (± SE) for prosimians with different mating systems. See Tab. I for classification and sample size.

## Determinants of Sexual Canine Dimorphism

The observed interspecific variation in sexual canine dimorphism among prosimians is difficult to interpret because the prediction, that males of polygynous species have larger canines than females, was not generally confirmed. This result may indicate (1) that canines are not a target of intrasexual selection, (2) that males and females are subject to selection that results in similar canine size, (3) that male canine size is also constrained by viability costs, or (4) that sex differences in canine size are strongly influenced by non-adaptive evolutionary forces.

There is indeed evidence that canine size and sex differences in canine size are influenced by phylogenetic effects. There is highly significant variation in sexual canine dimorphism, as well as in relative canine size of both males and females among prosimian families. The Indridae and Cheirogaleidae, in particular, are charcterized by small relative canine size in both sexes, as well as by a low degree of sexual canine dimorphism (Kappeler, 1992).

Sexual canine dimorphism of prosimians is unlikely to be affected by allometric effects. In contrast to anthropoids, the degree of sexual canine dimorphism among extant prosimians is not positively correlated with body size (Kappeler, 1992). Moreover, the giant subfossil lemurs lacked sexual canine dimorphism that is sufficient to sex individual specimens (L. Godfrey, pers. comm.), also indicating that lack in interspecific variation in sexual canine dimorphism among extant lemurs is not the result of allometric constraints.

The lack of sexual canine dimorphism in many polygynous lemurs is also unlikely to be a result of high viability costs of male canines. The presence of similar degrees of sexual canine dimorphism in some lemurs and virtually all galagos indicates that prosimian males are capable of developing and maintaining moderately larger canines than females. Furthermore, males of many anthropoid species possess canines that are about twice as large as those of females (Plavcan and van Schaik, 1992), obviously without interference with food processing or with other viability costs.

It may also be possible that many lemurs lack sexual canine dimorphism because the nature of aggressive competition among females selects for large canine size in females as well. An analysis of the effects of frequency and intensity of agonistic interactions between male and female anthropoids revealed that sex differences in the intensity of conflicts parallel sex differences in canine size. The relative canine size of lemur females is not unusual, however, compared to that of anthropoid females with similar competitive regimes (Plavcan and van Schaik, 1992; Plavcan, van Schaik and Kappeler, unpubl. data).

Moreover, relative female canine size and sexual canine dimorphism are also not related to female dominance. Females have smaller canines than males in species with invariable female dominance, such as *Lemur catta*, as well as in species where sex has no consistent effect on the outcome of male-female agonistic interactions, such as *Eulemur fulvus*. In other species with female dominance, such as *Varecia variegata*, males and females do not differ significantly in canine size. Together, these interspecific differences suggest that relative female canine size in polygynous lemurs corresponds to that of anthropoid females, and that the lack of sexual canine dimorphism in many species is due to a deviation of male canine size. This hypothesis could be tested by comparing male and female phenotypic variances in different canine dimensions between lemurs and other primates (e.g., Plavcan and Kay, 1988).

The absence of sexual dimorphism in body and canine size in most polygynous lemurs may therefore indicate that variance in male reproductive success is little affected by physical, pre-copulatory competition (Clutton-Brock, 1991b; Harvey, 1991). In other words, post-copulatory competition, which does not necessarily entail physical combat, may be a relatively important mode of reproductive competition in polygynous lemurs. A reduction in the relative importance of pre-copulatory competition could for example arise if the identity of copulating males is primarily determined by female choice.

## RELATIVE TESTES SIZE AND SPERM COMPETITION

With the possible exception of orangutans (Rodman and Mitani, 1987), female non-human primates cannot be forced into copulation. They can terminate copulatory attempts simply by sitting down, or by refusing to cooperate in other necessary ways (Smuts, 1987b). Lemur females, in addition, have the agonistic power to repel male advances actively. If females choose to copulate with several males, as happens in most lemur societies, male-male competition will occur primarily among their ejaculates (Curtsinger, 1991). The existence of this type of post-copulatory competition should be reflected by predictable species differences in relative testes size (Harcourt et al., 1981; Kenagy and Trombulak, 1986).

Males with the ability to produce more and/or larger ejaculates have a greater probability of prevailing successfully in sperm competition (Møller, 1988, 1991; Birkhead and Hunter, 1990). Because these abilities are positively correlated with testes size, sexual selection will favor the evolution of larger testes whenever sperm competition is a significant mechanism of male-male competition (Harvey and Harcourt, 1984; Ginsberg and Huck, 1989). Thus, lemur males of group-living species should have relatively larger testes than males of other species. Moreover, males of group-living lemurs should have relatively larger testes than other multi-male primates, if male-male competition in lemurs is primarily or exclusively post-copulatory.

**Interspecific Variation in Prosimian Testes Size**

I measured the greatest length and width of both testes of 153 prosimian males at the DUPC with hand-held sliding calipers to obtain estimates of testes size in a total of 18 species. The measurements used in this analysis were recorded during the breeding season, as determined by observations of mating behavior and/or vaginal smears from females housed with the male subjects. I averaged the four measurements from each individual to calculate the volume of a spherical ellipsoid to obtain an estimate of its average testes volume. These values were then averaged across all individuals of a species.

The body mass of each male was recorded immediately after the testes measurements and averaged for each species, as well. I used these data to calculate relative testes size (Table 1). I determined relative testes size as the vertical deviation (residual) from the best-fit line based on a least-square regression of (log) mean testes volume on (log) mean male body mass, and used it as the dependent variable in the same set of interspecific comparisons outlined above for sexual dimorphism.

Prosimians also display great interspecific variability in testes size. A large proportion of this variation is explained by interspecific variation in body size ($r=0.68$, $P<0.0001$, $N=18$), but a significant proportion of the variance remains unexplained by allometric effects (ANCOVA: difference among slopes: $F_{17,117}=1.12$, NS; difference among adjusted means: $F_{17,134}=46.75$, $P<0.0001$). Contrary to the predictions of sexual selection theory, solitary, pair- and group-living prosimians do not exhibit significant differences in relative testes size ($H=6.31$, NS; Fig. 3). Similarly, there is no significant variation in relative testes size among lemurs with these three types of mating system ($H=4.47$, NS). The comparison between polygynous lemurs and lorises revealed a trend towards larger relative testes size in lemurs, but it was statistically not significant ($z=1.47$, NS). The same result was obtained in a comparison between solitary lemurs and lorises ($z=0.34$, NS).

**Prosimian Testes Size and Mating Systems**

These analyses of relative testes size among prosimians revealed that interspecific variation in this sexually selected trait does also not completely agree with the predictions of sexual selection theory. If males of solitary prosimians defend and monopolize access to several females, sperm competition should be absent. Thus, relative testes size of these males should be similar to that of males in monogamous species, as is the case with single-male anthropoids (Harvey and Harcourt, 1984). This prediction cannot be tested with lorises because all of them are solitary. Solitary lemurs and lorises do not differ significantly in relative testes size, suggesting that sperm competition is of similar intensity in these two groups of species.

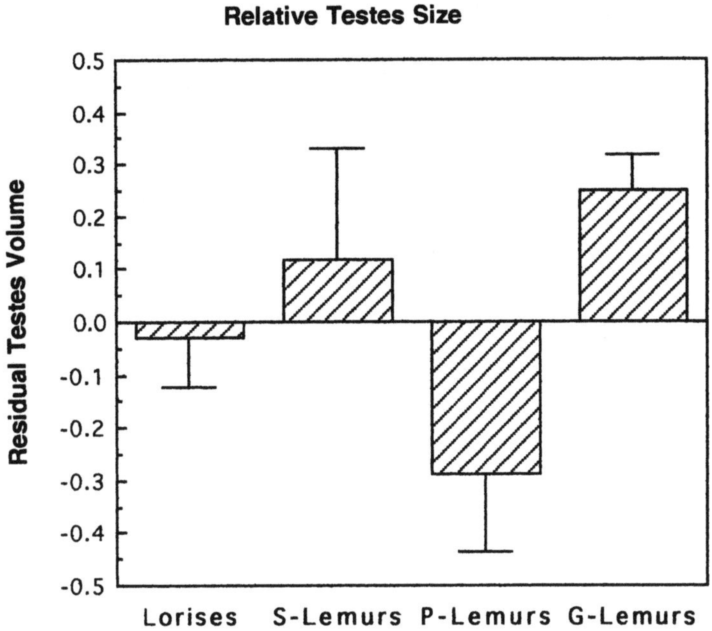


**Figure 3.** Relative testes size in prosimians with different mating systems. See Tab. 1 for classification and sample size.

The lack of a significant difference in relative testes size between solitary and pair-living lemurs is in accordance with the predictions of sexual selection (Harvey and Harcourt, 1984). However, it is not expected that both of these groups do not differ significantly from group-living species. This result is not affected by an alternative classification of *Eulemur mongoz* and *Varecia variegata* as group-living (see below for classification problems). It is not clear whether solitary lemurs have unusually large testes, whether group-living lemurs have unusually small testes, or both, because direct comparisons with anthropoids are presently not possible. Existing data for prosimians express testes size as the average volume, whereas existing analyses for anthropoids employ testes mass as an indicator of testes size (Harvey and Harcourt, 1984; Møller, 1988).

This problem precludes a direct test of the hypothesis that group-living lemurs have relatively larger testes than multi-male anthropoids. However, the observation that average relative testes size of group-living lemurs is not significantly different from that of monogamous lemurs suggests that it is not unusually large. Thus, this comparison of relative testes size among lemurs with different mating systems also demonstrated that males in polygynous lemurs do not respond to intrasexual selection in the predicted fashion.

The large interspecific variation in relative testes size among solitary prosimians supports the notion that mating systems cannot always be reliably inferred from the observed social organization. The existence of sperm competition is likely in some solitary lemurs because their average relative testes volume is more similar to group-living than to pair-living species. This conclusion in turn suggests that males in some solitary species are unable to monopolize access to females. Scramble competition polygyny, where solitary males search for receptive females and possibly guard them temporarily (Schwagmeyer, 1988; Clutton-Brock, 1989; see also Sterling, 1993), is therefore indicated as a likely mating system of some species, such as *Mirza coquereli,* for example, who have relatively large testes.

## SYNTHESIS: INTRASEXUAL SELECTION IN LEMURS

The main conclusion from these comparative analyses is that sexually selected morphological traits in polygynous lemurs are statistically indistinguishable from those of monogamous species. This is in sharp contrast to anthropoids, where multi-male species display a considerable degree of sexual size and canine dimorphism and have on average much larger testes than both monogamous and single-male species. Anthropoid species in which a single male typically defends a group of females exhibit on average the largest degree of sexual size and canine dimorphism, but their relative testes size does not exceed that of most monogamous species. Pair-living anthropoids are characterized by a low degree of sexual size dimorphism, a low degree of sexual canine dimorphism and relatively small testes size (Harvey and Harcourt, 1984). Thus, the relationship among sexual size dimorphism, sexual canine dimorphism and relative testes size varies between polygynous lemurs and anthropoids.

One possible explanation for the apparently unusual response of polygynous lemurs to intrasexual selection is that variance in male reproductive success is reduced through social mechanisms. More specifically, it is possible that groups of lemurs actually consist of multiple male-female pairs, rather than multiple males and females (van Schaik and Kappeler, 1993). This hypothesis is supported by the unskewed sex ratio of lemur groups, by the existence of strong social bonds between pairs of individual males and females, and by the observations of intra-populational switches from family groups to multi-male, multi-female groups and analogous intraspecific flexibility in several species.

This interpretation of lemur social systems suggests that primary male partners of individual females have a high probability of siring their offspring. In this case, variation in reproductive success among males would be much reduced because all males with such a pair-bond are likely to reproduce. Competition among males would be largely devaluated, and, as a consequence, the phenotypic expression of sexually selected traits may resemble that of monogamous species. This hypothesis makes testable predictions about association patterns and the distribution of paternity in lemur groups. If confirmed, it would provide an explanation for the unusual patterns of sexually selected traits in group-living lemurs. It would not account for the lack of sexual dimorphism and the relatively large testes of most solitary lemurs, however.

Two possible explanations for the interspecific variation in sexually selected traits among solitary lemurs focus on alternative strategies of male-male competition. First, it is possible that males of some species do not defend female ranges for various reasons. Because females do not occur in groups, the best male mating strategy is to search widely for receptive females, i.e., their mating system should be classified as scramble competition polygyny, which has different consequences for sexually selected traits than other types of polygyny (Clutton-Brock, 1991b).

Second, if so-called central males defend a range that includes ranges of several females, sperm competition may nevertheless occur because females are not clustered and may be difficult to defend against peripheral males, especially if female reproductive activity is synchronized to some extent, as in most Malagasy primates (Richard, 1987). In addition, physiological mechanisms of male-male competition may be widespread among solitary lemurs. For example, controlled experiments with captive *Microcebus murinus* males demonstrated that the urine of dominant males contains volatile substances that induce a reduction of body size, testes size and testosterone titers in subordinant and in naive males (Schilling et al., 1984; Perret and Schilling, 1987; Schilling and Perret, 1987). Indirect male-male competition through such chemical signals may also devaluate pre-copulatory competition and explain the lack of sexual size and canine dimorphism in some solitary lemurs. It is not clear, however, why these mechanisms of male-male competition are apparently not prevalent among lorises.

A crucial assumption in the previous analyses is that the assignment of individual species to different mating systems is correct. Very little is known about the social organization of the solitary prosimians, and, as these analyses demonstrated, lumping them into a single category may not be justified. Some non-solitary lemurs present a classification problem because they have been observed in both family units and larger associations. This flexibility, which is not known from anthropoids, was observed within single populations (*E. mongoz:* Sussman and Tattersall, 1976; Tattersall, 1976; Harrington, 1978; *V.variegata rubra:* Rigamonti, 1993) and among different populations in different habitats (*V. variegata variegata:* Morland, 1991; White, 1991). Detailed long-term field studies of different populations are required to obtain information about the "typical" social and mating system of these species. Moreover, genetic studies of virtually all prosimians are needed to obtain information about mating patterns and variance in reproductive success.

The cost of sexual selection-hypothesis has received mixed evidence as an explanation for the lack of sexual size dimorphism. Some polygynous species, such as *Lemur catta* and *Lepilemur* spp., have indeed male-biased sexual canine dimorphism, but many other species do not. Moreover, the presence of sexually dimorphic primates in other seasonally unfavorable habitats indicates that ecological factors alone are not responsible for a lack of sexual dimorphism. On the other hand, the degree of sexual size dimorphism in baboons is positively correlated with the average annual rainfall (Popp, 1983; see also Dunbar, 1990). High viability costs of large body size may therefore be a necessary, but not sufficient, condition for a reduction in sexual dimorphism.

A complete absence of sexual size dimorphism is difficult to explain, however, because two types of sexual selection may favor large male size, especially if increased size is associated with large costs. First, if size provides males with a competitive advantage, the potential reproductive payoffs for the largest males in the population will be especially high. Second, the ability to develop and maintain large body size under unfavorable conditions may be an honest indicator of "good genes", and may therefore be selected by female choice (Kirkpatrick, 1987; Kirkpatrick and Ryan, 1991).

Other Malagasy mammals could provide important comparative information to distinguish between the cost of sexual selection hypothesis and the social explanations for the lack of sexual size dimorphism suggested above. A lack of sexual size dimorphism in non-primate Malagasy mammals would be consistent with the notion that ecological factors are important determinants of male and female body size, and would support the hypothesis that environmental forces in Madagascar impose large costs on male body size, if it deviates from the ecological optimum. The presence of sexual size dimorphism in other polygynous Malagasy mammals, on the other hand, would point to lemur-specific social processes as the primarily cause of the absence of sexual size dimorphism among polygynous lemurs.

## SEXUAL SELECTION AND DOMINANCE

Finally, what can these studies of intrasexual selection contribute to an illumination of the problem of female dominance in lemurs? The observation that solitary, pair- and group-living lemurs do not differ in average sexual size dimorphism, sexual canine dimorphism and relative testes size indicates that the intensity of intrasexual selection is of similar low intensity in all lemurs. Because the absence of a male size advantage is probably an important condition for female dominance, the unusually low intensity of intrasexual selection in polygynous lemurs may have facilitated the evolution of female dominance. In sexually monomorphic gibbons, for example, females are also able to displace males from resources (Leighton, 1987). However, female dominance is not characteristic of all lemurs and it occurs in species where males have superior weapons. Little opportunity for intrasexual selection does therefore not provide a sufficient explanation for female dominance. Clearly, additional determinants of sex differences in agonistic asymmetry are operating.

The observation that female dominance is not proximately based on physical superiority raises questions about the general determinants of dominance. The commonly accepted explanation of male dominance as a coincidental byproduct of intrasexual selection is not very compelling. It also suggests that other selective forces may be important, and that the exclusive focus on the dominant sex may not be sufficient to provide a complete explanation. Because sex differences in dominance are primarily defined by submissive behavior, it may be useful to examine the adaptiveness of submissive behavior instead. For this purpose, it may be sufficient to distinguish between a survival and reproduction component of fitness and to relate them to sex differences in submissive behavior.

Such a thought experiment suggests that female anthropoids and lorises that behave submissively in conflicts with larger and stronger males may be at a lower risk of being injured, which may ultimately increase the survival component of female fitness. Immediate consequences of submissive behavior for the reproductive component of female fitness, on the other hand, are not obvious. While this explanation of male dominance is also based on the assumption that sex differences in agonistic power arise for other reasons, unlike the sexual selection explanation, it is adaptive from the females' point of view and it makes testable predictions, e.g., females that consistently fail to display submissive behavior towards adult males should exhibit increased mortality rates.

Because females are physically not superior to males, there are probably no survival benefits to male subordinance in lemurs. Thus, lemur males may be subordinate because to do so may increase the reproduction component of their fitness. However, there is currently no information about the determinants of lifetime reproductive success in both sexes that would suggest a more specific factor. The onset of female dominance during puberty (Pereira, 1993) also suggests that female dominance/male subordinance is functionally related to reproduction. Sex differences in leverage advantage (Hand, 1986), payoff asymmetries (Noe et al., 1991) or ecological needs (Young et al., 1990) may explain why female lemurs are consistently able to evoke submissive behavior from all males without general physical superiority. Detailed studies of female choice could generate specific hypotheses about these ideas.

## ACKNOWLEDGMENTS

I thank Peter Klopfer, Carel van Schaik, Paul Harvey, Alison Jolly, Jörg Ganzhorn, Joe Macedonia, Michael Pereira and Patricia Wright for helpful discussions and comments. This research was supported by a Harry Frank Guggenheim Foundation Dissertation Award and the Duke University Graduate School. This is DUPC publication No. 544.

## REFERENCES

Albrecht, G. H., Jenkins, P. D. and Godfrey, L. R., 1990, Ecogeographic size variation among the living and subfossil prosimians of Madagascar, *Am. J. Primatol.*, 22:1.

Alexander, R. D., Hoogland, J. L., Howard, R. D., Noonan, K. M. and Sherman, P. W., 1979, Sexual dimorphism and breeding systems in pinnipeds, ungulates, primates, and humans, *in*: "Evolutionary Biology and Human Social Behavior," N. A. Chagnon and W. Irons, eds., Duxbury, North Scituate.

Archer, J., 1988., "The Behavioural Biology of Aggression," Cambridge University Press, Cambridge.

Barrette, C. and Vandal, D., 1990, Sparring, relative antler size, and assessment in male caribou, *Behav. Ecol. Sociobiol.* 26:383.

Bateman, A. J., 1948, Intrasexual selection in *Drosophila.*, *Heredity* 2:349.

Bearder, S. K., 1987, Lorises, bushbabies, and tarsiers: diverse societies in solitary foragers, *in*: "Primate Societies," B. B. Smuts, D. L. Cheney, R. M. Seyfarth, R. W. Wrangham and T. T. Struhsaker, eds., University of Chicago Press, Chicago.

Birkhead, T. R. and Hunter, F. M., 1990, Mechanisms of sperm competition, *TREE* 5:48.

Bouissou, M. F., 1983, Hormonal influences on aggressive behavior in ungulates, *in*: "Hormones and Aggressive Behavior," B. B. Svare, ed., Plenum Press, New York.

Budnitz, N. and Dainis, K., 1975, *Lemur catta*: ecology and behavior, *in*: "Lemur Biology," I. Tattersall and R. W. Sussman, eds., Plenum Press, New York.

Charles-Dominique, P. and Petter, J. J., 1980, Ecology and social life of *Phaner furcifer*, *in*: "Nocturnal Malagasy Primates," P. Charles-Dominique, H. M. Cooper, A. Hladik, C. M. Hladik, E. Pages, G. F. Pariente, A. Petter-Rousseaux, J.-J. Petter and A. Schilling, eds., Academic Press, New York.
Cheverud, J. M., Dow, M. M. and Leutenegger, W., 1985, The quantitative assessment of phylogenetic constraints in comparative analyses: sexual dimorphism in body weight among primates, *Evolution* 39:1335.
Clutton-Brock, T. H., 1985, Size, sexual dimorphism, and polygyny in primates, *in*: "Size and Scaling in Primate Biology," W. L. Jungers, ed., Plenum Press, New York.
Clutton-Brock, T. H., 1988, "Reproductive Success: Studies of Individual Variation in Contrasting Breeding Systems," University of Chicago Press, Chicago.
Clutton-Brock, T. H., 1989, Mammalian mating systems, *Proc. Roy. Soc. Lond. B* 236:339.
Clutton-Brock, T. H., 1991a, "The Evolution of Parental Care," Princeton University Press, Princeton.
Clutton-Brock, T. H., 1991b, The evolution of sex differences and the consequences of polygyny in mammals, *in*: "The Development and Integration of Behaviour," P. Bateson, ed., University of Cambridge Press, Cambridge.
Clutton-Brock, T. H., Albon, S. D. and Harvey, P. H., 1980, Antlers, body size and breeding group size in the Cervidae, *Nature* 285:565.
Clutton-Brock, T. H. and Harvey, P. H., 1984, Comparative approaches to investigating adaptation, *in*: "Behavioural Ecology,"J. R. Krebs and N. B. Davies, eds., Blackwell, Oxford.
Clutton-Brock, T. H., Harvey, P. H. and Rudder, B., 1977, Sexual dimorphism, socionomic sex ratio and body weight in primates, *Nature* 269:797.
Collias, N. E., 1953, Social behavior in animals, *Ecology* 34:810.
Colquhoun, I. C., 1987, Dominance and "fall fever": The reproductive behavior of male brown lemurs (*Lemur fulvus*), *Can. Rev. Phys. Anthropol.* 6:10.
Crook, J. H., 1972, Sexual selection, dimorphism, and social organization in the primates, *in*: "Sexual Selection and the Descent of Man," B. G. Campbell, ed., Aldine, Chicago.
Curtsinger, J. W., 1991, Sperm competition and the evolution of multiple mating, *Am. Nat.* 138:93.
Darwin, C., 1871, "The Descent of Man and Selection in Relation to Sex," Murray, London.
Dunbar, R. I. M., 1990, Environmental determinants of intraspecific variation in body weight in baboons (*Papio* ssp.), *J. Zool., Lond.* 220:157.
Fedigan, L., 1982, "Primate Paradigms: Sex Roles and Social Bonds," Eden Press, Montreal.
Fedigan, L. M. and Baxter, M. J., 1984, Sex differences and social organization in free-ranging spider monkeys (*Ateles geoffroyi*), *Primates* 25:279.
Felsenstein, J., 1985, Phylogenies and the comparative method, *Am. Nat.* 125:1.
Foerg, R., 1982a, Reproductive behavior in *Varecia variegata*, *Fol. Primatol.* 38:108.
Foerg, R., 1982b, Reproduction in *Cheirogaleus medius*, *Fol. Primatol.* 39:49.
Geist, V., 1966, The evolution of horn-like organs, *Behaviour* 27:175.
Gingerich, P. D. and Ryan, A. S., 1979, Dental and cranial variation in living Indriidae, *Primates* 20:141.
Ginsberg, J. R. and Huck, U. W., 1989, Sperm competition in mammals, *TREE* 4:74.
Godfrey, L. R., Lyon, S. K. and Sutherland, M.R., 1993, Sexual dimorphism in large-bodied primates: the case of the subfossil lemurs, *Am. J. Phys. Anthropol.* 10:315.
Hand, J. L., 1986, Resolution of social conflicts: dominance, egalitarianism, spheres of dominance and game theory, *Q. Rev. Biol.* 61:201.
Harcourt, A. H., 1979, Social relationships between adult male and female mountain gorillas, *Anim. Behav.* 27:325.
Harcourt, A. H., Harvey, P. H., Larson, S. G. and Short, R. V., 1981, Testis weight, body weight, and breeding system in primates, *Nature* 293:55.
Harrington, J. E., 1978, Diurnal behavior of *Lemur mongoz* at Ampijoroa, *Fol. Primatol.* 29:291.
Harvey, P. H., 1991, Sexual selection, *in*: "Behavioural Ecology," J. Krebs and N. Davies, eds., Blackwell, Oxford.
Harvey, P. H. and Harcourt, A. H., 1984, Sperm competition, testes size, and breeding system in primates, *in*: "Sperm Competition and the Evolution of Animal Mating Systems," R.L. Smith, ed., Academic Press, New York.
Harvey, P. H., Kavanaugh, M. and Clutton-Brock, T. H., 1978, Sexual dimorphism in primate teeth, *J. Zool., Lond* 186:475.
Harvey, P. H. and Pagel, M., 1991, "The Comparative Method in Evolutionary Biology," Oxford University Press, Oxford.
Hausfater, G., 1975, Dominance and reproduction in baboons: a quantitative analysis, *Contrib. Primatol.* 7:1.

Hladik, A., 1980, The dry forest of the west coast of Madagascar: climate, phenology, and food available for prosimians, *in*: "Nocturnal Malagasy Primates: Ecology, Physiology, and Behavior," P. Charles-Dominique, H. M. Cooper, A. Hladik, C. M. Hladik, E. Pages, G. F. Pariente, A. Petter-Rousseaux, J.-J. Petter and A. Schilling, eds., Academic Press, New York.

Hladik, C. M., Charles-Dominique, P. and Petter, J.-J., 1980, Feeding strategies of five nocturnal prosimians in the dry forest of the West coast of Madagascar, *in*: "Nocturnal Malagasy Primates: Ecology, Physiology and Behavior," P. Charles-Dominique, H. M. Cooper, A. Hladik, C. M. Hladik, E. Pages, G. F. Pariente, A. Petter-Rousseaux, J.-J. Petter and A. Schilling, eds., Academic Press, New York.

Hrdy, S. B., 1981, "The Woman that Never Evolved," Harvard University Press, Cambridge.

Jenkins, P. and Albrecht, G., 1991, Sexual dimorphism and sex ratios in Madagascan prosimians, *Am. J. Primatol.* 24:1.

Jolly, A., 1966, "Lemur Behavior," University of Chicago Press, Chicago.

Jolly, A., 1967, Breeding synchrony in wild *Lemur catta*, *in*: "Social Communication among Primates," S. A. Altmann, ed., University of Chicago Press, Chicago.

Jolly, A., 1984, The puzzle of female feeding priority, *in*: "Female Primates: Studies by Woman Primatologists," M. F. Small, ed., A.R. Liss, New York.

Jones, C. B., 1980, The functions of status in the mantled howler monkey, *Alouatta palliata* Gray: intraspecific competition for group membership in a folivorous neotropical primate, *Primates* 21:389.

Kappeler, P. M., 1990a, Female dominance in *Lemur catta*: more than just female feeding priority?, *Fol. Primatol.* 55:92.

Kappeler, P. M., 1990b, The evolution of sexual size dimorphism in prosimian primates, *Am. J. Primatol.* 21:201.

Kappeler, P. M., 1991, Patterns of sexual dimorphism in body weight among prosimian primates. *Fol. Primatol.* 57:132.

Kappeler, P. M., 1992, Female dominance in Malagasy primates, Ph.D. thesis, Duke University, Durham.

Kappeler, P. M., 1993, Female dominance in primates and other mammals, *in*: "Perspectives in Ethology, Vol. X," P. P. G. Bateson, P. H. Klopfer and N. S. Thompson, eds., Plenum Press, New York.

Kaufman, R., 1991, Female dominance in semifree-ranging black and white ruffed lemurs, *Varecia variegata*, *Fol. Primatol.* 57:39.

Kay, R. F., Plavcan, J. M., Glander, K. E. and Wright, P. C., 1988, Sexual selection and canine dimorphism in New World monkeys, *Am. J. Phys. Anthropol.* 77:385.

Kenagy, G. J. and Trombulak, S. C., 1986, Size and function of mammalian testes in relation to body size, *J.Mammal.* 67:1.

Kieser, J. A. and Groeneveld, H. T., 1990, Patterns of sexual dimorphism and of variability in the dentition of *Otolemur crassicaudatus*, *Int. J. Primatol.* 10:137.

Kirkpatrick, M., 1987, Sexual selection by female choice in polygynous animals, *Ann. Rev. Ecol. Syst.* 18:43.

Kirkpatrick, M. and Ryan, M. J., 1991, The evolution of mating preferences and the paradox of the lek, *Nature* 350:33.

Kummer, H., 1968, "Social Organization of Hamadryas Baboons," University of Chicago Press, Chicago.

Leighton, D.R. 1987, Gibbons: territoriality and monogamy, *in*: "Primate Societies," B.B. Smuts, D. L. Cheney, R. M. Seyfarth, R. W. Wrangham and T. T. Struhsaker, eds., University of Chicago Press, Chicago.

Leutenegger, W., 1978, Scaling of sexual dimorphism in body size and breeding system in primates, *Nature* 272:610.

Møller, A. P., 1988, Ejaculate quality, testes size and sperm competition in primates, *J. Hum. Evol.* 17:479.

Møller, A. P., 1991, Concordance of mammalian ejaculate features, *Proc. R. Soc. Lond. B* 246:237.

Morland, H. S., 1991, Preliminary report on the social organization of ruffed lemurs (*Varecia variegata variegata*) in a northeast Madagascar rain forest, *Fol. Primatol.* 56:157.

Nash, L. T., Bearder, S. K. and Olson, T. R., 1989, Synopsis of *Galago* species characteristics, *Int. J. Primatol.* 10:57.

Noe, R., van Schaik, C. and van Hooff, J., 1991, The market effect: an explanation for pay-off asymmetries among collaborating animals, *Ethology* 87:97.

Noe, R., de Waal, F. B. M. and van Hooff, J., 1980, Types of dominance in a chimpanzee colony, *Fol. Primatol.* 34:90.

Packer, C., 1983, Sexual dimorphism: the horns of African antelope, *Science* 221:1191.

Pagès-Feuillade, E., 1988, Modalités de l'occupation de l'espace et relations interindividuelles chez un prosimien nocturne malgache (*Microcebus murinus*), *Fol. Primatol.* 50:204.
Pereira, M. E., 1993, Agonistic interaction, dominance relations, and ontogenetic trajectories in ringtailed lemurs. *in*: "Juvenile Primates: Life History, Development, and Behavior," M. E. Pereira and L. A. Fairbanks, eds., Oxford University Press, New York.
Pereira, M. E., Kaufman, R., Kappeler, P. M. and Overdorff, D. J., 1990, Female dominance does not characterize all of the Lemuridae, *Fol. Primatol.* 55:96.
Perret, M. and Schilling, A., 1987, Intermale sexual effect elicited by volatile urinary ether extract in *Microcebus murinus* (Prosimian, Primates), *J. Chem. Ecol.* 13:495.
Plavcan, J.M. and Kay, R.F., 1988, Sexual dimorphism and dental variability in platyrrhine primates, *Int. J. Primatol.*9:169.
Plavcan, J. M. and van Schaik, C. P., 1992, Intrasexual competition and canine dimorphism in primates, *Am. J. Phys. Anthropol.* 87:461.
Pollock, J. I., 1979, Female dominance in *Indri indri*, *Fol. Primatol.* 31:143.
Pollock, J. I., 1989, Intersexual relationships amongst prosimians, *Hum. Evol.* 4:133.
Popp, J. L., 1983, Ecological determinism in the life histories of baboons, *Primates* 24:198.
Promislow, D. E. L., 1992, Costs of sexual selection in natural populations of mammals, *Proc. Roy. Soc. Lond. B* 247:203.
Ralls, K., 1976, Mammals in which females are larger than males, *Q. Rev. Biol.* 51:245.
Reinhardt, V., Reinhardt, A. and Reinhardt, C., 1987, Evaluating sex differences in aggressiveness in cattle, bison and rhesus monkeys, *Behaviour* 104:58.
Richard, A. F., 1974, Patterns of mating in *Propithecus verreauxi verreauxi*, *in*: "Prosimian Biology," R. D. Martin, G. A. Doyle and A. C. Walker, eds., Duckworth, London.
Richard, A. F., 1987, Malagasy prosimians: female dominance, *in*: "Primate Societies," B.B. Smuts, D. L. Cheney, R. M. Seyfarth, R. W. Wrangham and T. T. Struhsaker, eds., University of Chicago Press, Chicago.
Richard, A. F. and Dewar, R. E., 1991, Lemur ecology, *Ann. Rev. Ecol. Syst.* 22:145.
Richard, A. F. and Heimbuch, R., 1975, An analysis of the social behavior of three groups of *Propithecus verreauxi*, *in*: "Lemur Biology," I. Tattersall and R. W. Sussman, eds., Plenum Press, New York.
Richard, A. F. and Nicoll, M. E., 1987, Female social dominance and basal metabolism in a Malagasy primate, *Propithecus verreauxi*, *Am. J. Primatol.* 12:309.
Rigamonti, M. M., 1993, Home range and diet in red ruffed lemurs (*Varecia variegata rubra*) on the Masoala peninsula, Madagascar, *in*: "Lemur Social Systems and their Ecological Basis," P.M. Kappeler and J.U. Ganzhorn, eds., Plenum Press, New York.
Rodman, P. S. and Mitani, J. C., 1987, Orangutans: sexual dimorphism in a solitary species, *in*: "Primate Societies," B. B. Smuts, D. L. Cheney, R. M. Seyfarth, R. W. Wrangham and T. T. Struhsaker, eds., Chicago University Press, Chicago.
Rowell, T. E. and Chism, J., 1986, Sexual dimorphism and mating systems: jumping to conclusions, *Hum. Evol.* 1:215.
Schilling, A. and Perret, M., 1987, Chemical signals and reproductive capacity in a male prosimian primate (*Microcebus murinus*), *Chem. Sens.* 12:143.
Schilling, A., Perret, M. and Predine, J., 1984, Sexual inhibition in a prosimian primate: a pheromone-like effect, *J. Endocrinol.* 102:143.
Schwagmeyer, P. L., 1988, Scramble-competition polygyny in an asocial mammal: male mobility and mating success, *Am. Nat.* 131:885.
Siegel, S. and Castellan, N. J., 1988, "Nonparametric Statistics for the Behavioral Sciences," McGraw-Hill, New York.
Smuts, B. B., 1987a, Gender, aggression, and influence, *in*: "Primate Societies," B. B. Smuts, D. L. Cheney, R. M. Seyfarth, R. W. Wrangham and T. T. Struhsaker, eds., University of Chicago Press, Chicago.
Smuts, B. B., 1987b, Sexual competition and mate choice, *in*: "Primate Societies," B. B. Smuts, D. L. Cheney, R. M. Seyfarth, R. W. Wrangham and T. T. Struhsaker, eds., University of Chicago Press, Chicago.
Sterling, E. J., 1993, Patterns of range use and social organization in aye-ayes (*Daubentonia madagascariensis*) on Nosy Mangabe, *in*: "Lemur Social Systems and their Ecological Basis," P.M. Kappeler and J.U. Ganzhorn, eds., Plenum Press, New York.
Sussman, R. W. and Tattersall, I., 1976, Cycles of activity, group composition, and diet of *Lemur mongoz* in Madagascar, *Fol. Primatol.* 26:270.
Swindler, D. R., 1976, "Dentition of Living Primates," Academic Press, New York.

Tattersall, I., 1976, Group structure and activity rhythm in *Lemur mongoz* (Primates, Lemuriformes), *Anthropol. Pap. Am. Mus. Nat. Hist.* 53:369.
Tattersall, I., 1982, "The Primates of Madagascar," Columbia University Press, New York.
Tilson, R. L. and Hamilton, W. J. I., 1984, Social dominance and feeding patterns of spotted hyaenas, *Anim. Behav.* 32:715.
Trivers, R. L., 1972, Parental investment and sexual selection, *in*: "Sexual Selection and the Descent of Man," B. Campbell, ed., Aldine, Chicago.
van Schaik, C. P. and Kappeler, P. M., 1993, Life history, activity period and lemur social systems, *in*: "Lemur Social Systems and their Ecological Basis," P.M. Kappeler and J.U. Ganzhorn, eds., Plenum Press, New York.
White, F. J., 1991, Social organization, feeding ecology, and reproductive strategy of ruffed lemurs, *Varecia variegata*, *in*: "Proceedings of the XIII Congress of the International Primatological Society," A. Ehara, T. Kimura, O. Takenaka and M. Iwamoto, eds., Elsevier, Amsterdam.
Wright, P. C., 1988, Social behavior of *Propithecus diadema edwardsi* in Madagascar, *Am. J. Phys. Anthropol.* 75:289.
Young, A. L., Richard, A. F. and Aiello, L. C., 1990, Female dominance and maternal investment in strepsirhine primates, *Am. Nat.* 135:473.

# LIFE HISTORY, ACTIVITY PERIOD AND LEMUR SOCIAL SYSTEMS

Carel P. van Schaik[1] and Peter M. Kappeler[2]

[1]Department of Biological Anthropology and Anatomy
[2]Department of Zoology
Duke University, Durham, NC 27705
U.S.A.

## ABSTRACT

Extant lemuriform and anthropoid primates differ dramatically in social behavior, grouping patterns, social structure, and sexual dimorphism. In this paper, we analyze and contrast grouping and bonding patterns among lemurs, and suggest determinants of social organization that explain many of these differences. Lemurs exhibit three main grouping patterns: solitary individuals, bonded pairs, and larger groups. Transitions between pairs and larger groups occur within several species and even within populations in some of *Eulemur* species and in *Varecia*. Larger groups of these species have an equal adult sex ratio, and spatial associations and social behavior suggest the existence of male-female pair bonds. Larger groups of the consistently diurnal *Lemur* and *Propithecus* also show equal adult sex ratios, but no evidence for pair-bonding. These differences among lemurs are likely due to variation in the activity periods of the cathemeral species, whose groups are hypothesized to break up into pairs during periods of nocturnal activity. The pair bond is therefore the fundamental unit of many lemur groups. This is a special case of the permanent male-female associations found among the great majority of primate species. We suggest that the evolution of permanent male-female associations is favored by selection for infanticide prevention, because the male can effectively protect the females against infanticide attempts by other males. Where females live alone permanently or part of their time (the cathemeral species), this association takes the form of a pair bond. Where they live in groups, it takes the form of permanently bisexual groups. This hypothesis leads us to predict no male-female association in species with absentee maternal care. This prediction is confirmed: the interspecific variation in infant care (carry vs hide) covaries perfectly with grouping pattern among prosimians, as well as among other primate radiations. We conclude that activity period and type of infant care determine grouping and bonding patterns in both prosimian and anthropoid societies. Both cathemeral activity periods and infant carrying among primates were probably made possible by increased precociality.

## INTRODUCTION

The lemurs of Madagascar (Lemuriformes) show a combination of behavioral, morphological and physiological traits that is unique among primates. For example, female dominance is widespread among lemurs (Jolly, 1966; 1984; Richard, 1987; Kappeler, 1990a; Pereira et al., 1990; Kaufman, 1991), but has not been documented in other primates or mammals (Kappeler, 1993a). This salient feature of lemur social structure is paralleled by a general lack of sexual size dimorphism, even in the giant subfossil lemurs (Godfrey et al. 1993), and a trend towards larger female size in some taxa, such as *Microcebus* and *Propithecus* (Kappeler, 1990b; 1991; Jenkins and Albrecht, 1991). This set of behavioral and morphological traits cannot easily be reconciled with the predictions of sexual selection theory.

Lemur social systems also differ from those of other prosimians and anthropoids. First, many lemur species live in pairs. Several of these species facultatively switch to living in larger groups that tend to have equal adult sex ratios. Second, although female philopatry is apparently widespread among lemurs (Richard et al., 1991; Sussman, 1991), the better-known species, such as *Lemur catta* and *Eulemur fulvus*, do not behave like female-bonded species. Agonistic support and post-conflict reconciliation, for example, are rare in lemur groups, even among closely-related females (Pereira, 1992; Kappeler, 1993b,c; Pereira and Kappeler, unpubl. data). With the exception of mothers and daughters, most pairs of related females have weakly developed bonds, i.e., they groom and associate less with each other than with other group members and they have relatively high rates of conflicts (Kappeler, 1993b). In some species, distant relatives are even targeted for intense aggression, frequently resulting in injury and/or eviction from the group (Vick and Pereira, 1989). Thus, several aspects of grouping and bonding patterns in lemurs are also not in agreement with existing theories of social evolution (van Schaik, 1989; Wrangham, 1980; 1987).

What explains these striking differences in social behavior between lemurs and other primates? Can this variation be explained within a single framework, or were the predominant selective forces that shaped the social systems of the isolated lemur radiation different from those operating in anthropoids? The aim of this chapter is to formulate the outlines of an answer. We will start by reviewing lemur grouping and bonding patterns, in order to identify the fundamental features of lemur sociality. This analysis indicates that activity period is a major correlate of grouping pattern and that the male-female bond is the basic unit of the societies of most gregarious species. From this analysis we derive the hypothesis that (i) pair bonds evolved to protect the female from infanticide by strange males, and (ii) the variability in social organization and the presence of groups with approximately equal sex ratio are the product of the unusual activity period. The available data on interspecific differences in patterns of infant care, activity period and estrous synchrony support several predictions of this hypothesis.

## SOCIAL ORGANIZATION AND STRUCTURE IN LEMURS

### Social Organization: Group Composition

There are three major kinds of grouping patterns among lemurs. First, there are larger groups with multiple adult females and their offspring and multiple adult males; some of these groups are of the fission-fusion variety. Second, there are adult male-female pairs with their young. Third, there are societies in which adults are foraging (though not necessarily sleeping) basically alone. Within this type, the size and overlap of male and female ranges can vary widely (Bearder, 1987), but there is not enough information to further subdivide these species (they comprise Bearder's types 1,3, and 4).

As shown by Table 1, which summarizes the results of numerous published field studies, the grouping patterns of extant lemurs display some striking features. First, grouping pattern is a major correlate of activity period. All nocturnal taxa live in societies in which females are not accompanied by other females during the active period, foraging either alone (6 genera) or with a male associate (2 genera). In contrast, both the cathemeral species, which are potentially active at all hours of the day (Tattersall, 1987), and the diurnal species live as pairs or in larger groups (6 genera), although sometimes in a fission-fusion fashion [e.g., some populations of *Eulemur coronatus* (Wilson et al., 1989), *E. macaco* (Colquhoun, 1993) and *Varecia variegata* (Morland 1991a; Rigamonti, 1993)]. In this respect, lemurs resemble mammals in general, in that nocturnal forms are not found in multi-female groups (Clutton-Brock and Harvey, 1977). However, the existence of nocturnal species living as pairs (two genera in the Lemuridae and Indridae) is unusual among nocturnal mammals, which are typically solitary rather than pair-bonded.

**Table 1.** An overview of lemuriform activity periods, social systems, infant care patterns, and reproduction. Based on compilations by Petter et al. (1977), Klopfer and Boskoff (1979), Charles-Dominique et al. (1980) and Tattersall (1982), supplemented with more recent references[1].

| Species | Activity Period | Social System | Infant Care | Birth Period | Litter Size |
|---|---|---|---|---|---|
| *Microcebus murinus* | N | S | P | 4 | 2+ |
| *M. rufus* | N | S | P | | 2+ |
| *Mirza coquereli* | N | S | P | 6 | 2 |
| *Cheirogaleus medius* | N | S | P | 3 | 2+ |
| *C. major* | N | S | P | | 2+ |
| *Allocebus trichotis* | N | S | P | | 2+ |
| *Phaner furcifer* | N | P | P->C | | 1 |
| *Lepilemur dorsalis* | N | S | P | | 1 |
| *L. edwardsi* | N | S | P | | 1 |
| *L. leucopus* | N | S | P | | 1 |
| *L. microdon* | N | S | P | | 1 |
| *L. mustelinus* | N | S | P | | 1 |
| *L. ruficaudatus* | N | S | P | | 1 |
| *L. septentrionalis* | N | S | P | | 1 |
| *Lemur catta* | D | G | C | 3 | 1(2) |
| *Eulemur coronatus* | C | G | C | 2 | 1-2 |
| *E. fulvus* | C | G | C | 3 | 1 |
| *E. macaco* | C | G | C | 3 | 1-2 |
| *E. mongoz* | C | P->G | C | 3 | 1-2 |
| *E. rubriventer* | C | P->G | C | 2 | 1-2 |
| *Varecia variegata* | D (C) | P->G | P | 2 | 2+ |
| *Hapalemur aureus* | C | P | C | | 1 |
| *H. griseus* | C | P->G | (P->)C | 3 | 1 |
| *H. simus* | C | G | (P->)C | | 1 |
| *Avahi laniger* | N | P | C | | 1 |
| *Indri indri* | D | P | C | | 1 |
| *Propithecus diadema* | D | G | C | 3 | 1 |
| *P. tattersalli* | D | G | C | 2 | 1 |
| *P. verreauxi* | D | (P?->)G | C | 2 | 1 |
| *Daubentonia madagascariensis* | N | S | P | (>4) | 1 |

[1]Activity period is N=nocturnal, C=cathemeral, or D=diurnal; Social system is S=solitary, P=pair-living, G=in larger groups, sometimes of unknown adult composition, but usually in the form of multiple pairs (see Figure 1); Infant care pattern during the active period is P=parking (leaving infants hidden in nest, hole, or vegetation) or C=carrying; Birth seasonality is approximated by the period (in months) during which at least 80% of the infants are born (based on DUPC records, Kappeler, unpubl. data); Litter size gives the modal number of offspring born per reproductive event. Additional sources: *Microcebus murinus*: Pagès-Feuillade, 1988; *Allocebus trichotis*: Meier and Albignac, 1991; *Hapalemur spp*: Meier et al., 1987; Wright, 1989a; Wright et al., 1986; *Avahi laniger*: Ganzhorn et al., 1985; Harcourt, 1991; *Daubentonia madagascariensis*: Sterling, 1992; DUPC records; *Propithecus tattersalli* and *P. diadema*: Meyers and Wright, 1993.

A second salient feature of lemur societies revealed by Table 1 is that grouping patterns can vary within species: some populations live in pairs, while others live in larger groups. Such variability is extremely unusual among anthropoids, where bonded monogamy is never accompanied by polygyny in the same species, except occasionally where groups contain only a single female in species that typically live in larger (harem) groups (facultative monogamy; cf. van Schaik and van Hooff, 1983).

Most of the variability within species is found in the cathemeral species. Because diurnal species tend to live in larger groups than nocturnal ones, one would expect that the variation in grouping pattern is related to activity period. Data from *Eulemur mongoz* permit a preliminary evaluation of this suggestion. Mongoose lemurs lived in pairs in the two studies reporting the animals to be nocturnal, whereas they lived in larger groups in the two studies reporting them to be diurnal (Table 2). In one case, this variation was observed in the same population. Thus, the same animals exhibited different grouping patterns in different seasons, depending on whether they were most active at night or during the day, which in turn may depend on the kinds of food they exploit (cf. Enqvist and Richard, 1991).

**Table 2.** Grouping and activity period in *Eulemur mongoz*.

| | Nocturnal | Diurnal |
|---|---|---|
| Pairs | Ampijoroa[1]<br>Anjouan[3] | |
| Larger groups | | Ampijoroa[2]<br>Moheli[3] |

[1]Sussman and Tattersall, 1976;
[2]Harrington, 1978;
[3]Tattersall, 1978.

A third striking feature of lemur societies is the near-equal adult sex ratios in larger groups. Among mammals in general and anthropoids alike, male-male competition usually leads to elimination of all but one or a few adult males (Clutton-Brock, 1989). Not so among the lemurs, where the adult sex ratio of lemur groups approximates unity, regardless of the number of females in the group (Fig. 1). This is true for both the cathemeral *Eulemur* species and the diurnal genera. This remarkable pattern suggests that male contest competition for group membership is somehow eliminated among group-living lemurs. Instead, the even adult sex ratio in combination with the variability in grouping patterns within species and populations suggests that these groups consist of "multiple pairs".

### Bonding Patterns

A test of the idea that lemur groups consist primarily of multiple pairs requires information on bonding patterns. Animals living together in social units tend to develop social relationships that show a predictable frequency of the different possible kinds of interactions. They are said to have a bond if they support each other agonistically (e.g., Wrangham, 1980) or if they exchange other non-agonistic services (e.g., van Hooff and van Schaik, 1992). Spatial proximity between two animals is often a good predictor of the strength of their social bond (Aureli et al., 1989; Smuts, 1985), simply because animals that exchange behaviors usually have to be close together to do this. In this particular case, where spatial proximity is the basic condition for the presumed male role (see below), proximity is the crucial variable, even if it would not predict other aspects of social behavior.

In anthropoids, at least two kinds of bonding patterns are found. In female-bonded societies (van Schaik, 1989; Wrangham, 1980), strong but differentiated bonds are common among adult females, as well as between the females and one or a few adult males. In non-female-bonded societies, males form strong social bonds, female bonds are weak, and bonds between the females and the group's male tend to be strong (Harcourt, 1979; Watts, 1990). In both kinds of societies, exclusive male-female bonds are absent.

The composition of lemur groups suggests that their bonding structure is different from that of anthropoids. In several species, researchers noted strong associations between the sexes: *Eulemur coronatus* (Arbelot-Tracqui, 1983), *E. fulvus* (Harrington, 1975; Tattersall, 1977; Vick, 1977), *E. macaco* (F. Bayart, pers. comm.) and *E. mongoz* (Harrington, 1975;1978). Several more quantitative studies of affiliation or grooming patterns are now available (Fig. 2). The results of several studies on *E. fulvus* (Kappeler, 1993b; McGlynn, 1992; Overdorff, 1991) confirm the qualitative assessment of early field workers. While bonds within the sexes are clearly present, strong male-female bonds, more or less between pairs, are also apparent (Fig. 2a and 2b). Results on *E. coronatus* (Fig. 2c; Kappeler, 1993b) and *V. variegata* (Fig. 2d; Morland, 1991b) show a similar pattern.

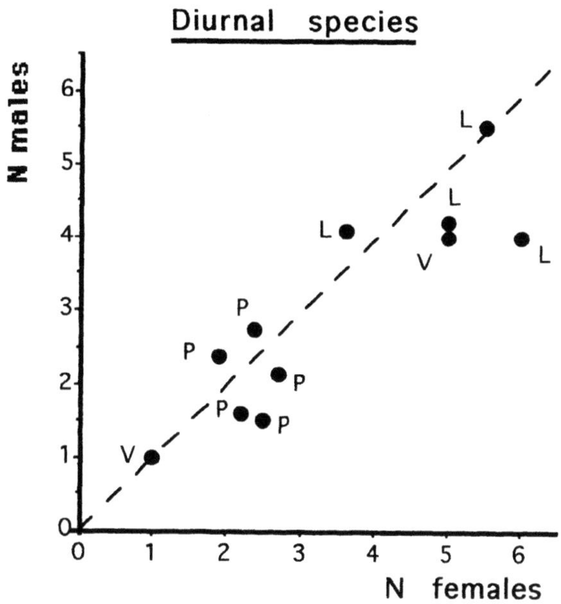


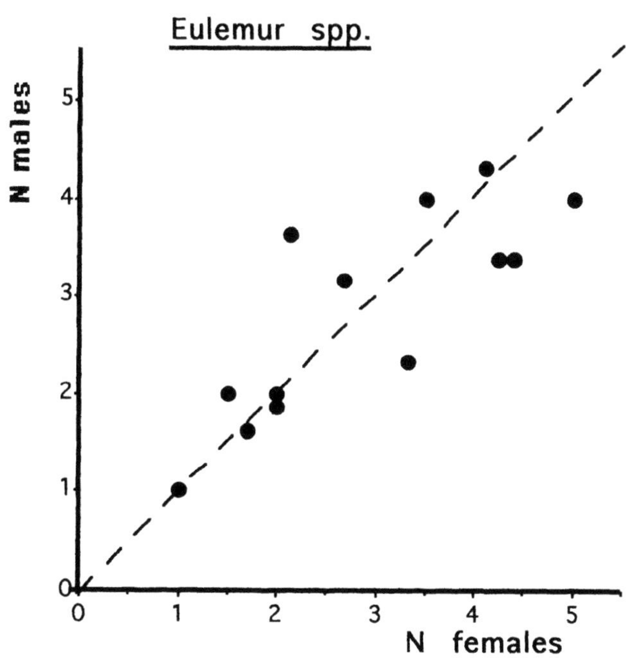


**Figure 1.** Adult sex ratios in wild lemur populations, based on mean compositions of groups determined in reliable censuses. Sources: *Varecia*: (Morland, 1991a; b; White, 1991); *Lemur*: (Budnitz, 1978; Jones, 1983; Sussman, 1977;1991); *Propithecus*: (Jolly et al., 1982; Richard, 1974; Richard et al., 1991; Wright, 1988); *Eulemur*: (Arbelot-Tracqui, 1983; Dague, 1986; Harrington, 1975; Overdorff, 1991; Sussman, 1977; Sussman and Tattersall, 1976; Tattersall, 1977; 1978; Wilson et al., 1989).

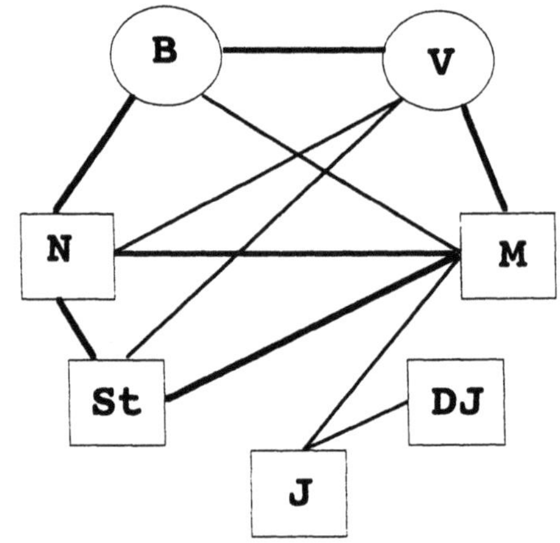


a) *Eulemur fulvus*

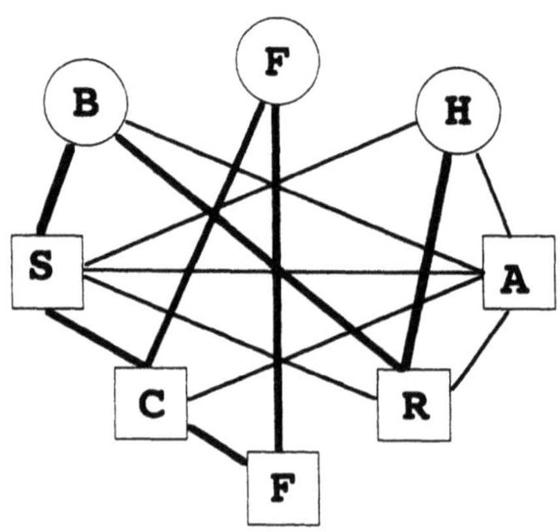


b) *Eulemur fulvus* in captivity

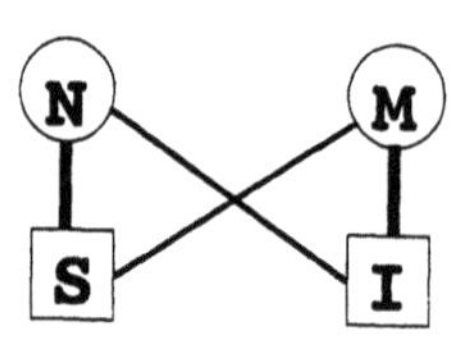


c) *Eulemur coronatus* in captivity

**Figure 2.** Spatial proximity patterns among adults of various wild or free-ranging lemurs. In a, b and d, e, and f, fat lines refer to proximity values in the highest 25 percentile of observed dyads, and thin lines to values in the next highest 25 percentile. In c, the respective 33 percentiles are used. Adult females are represented by circles; adult males by squares. Exact spatial criteria vary among studies.

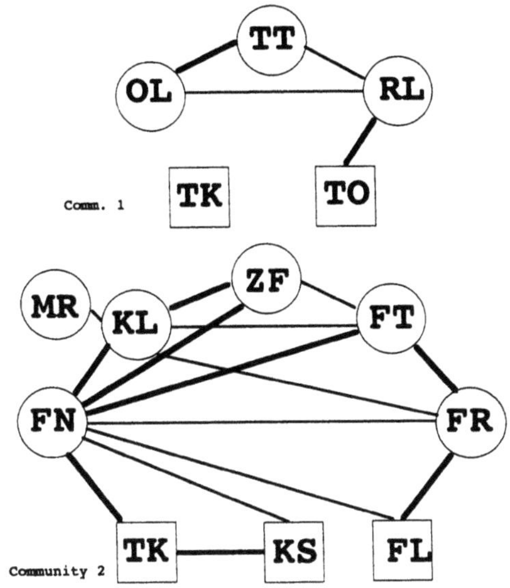


d) *Varecia variegata*

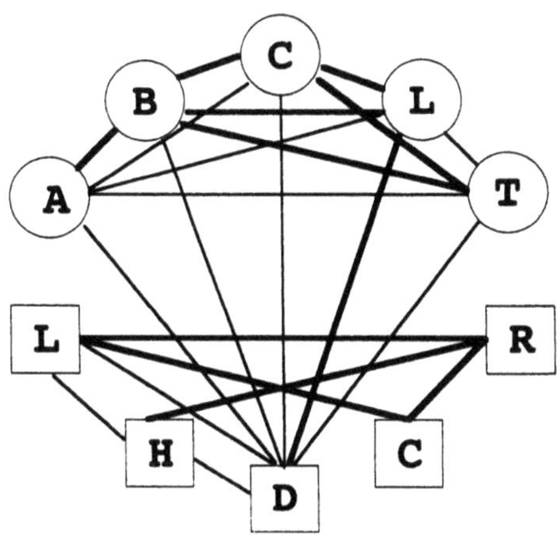


e) *Lemur catta* in captivity

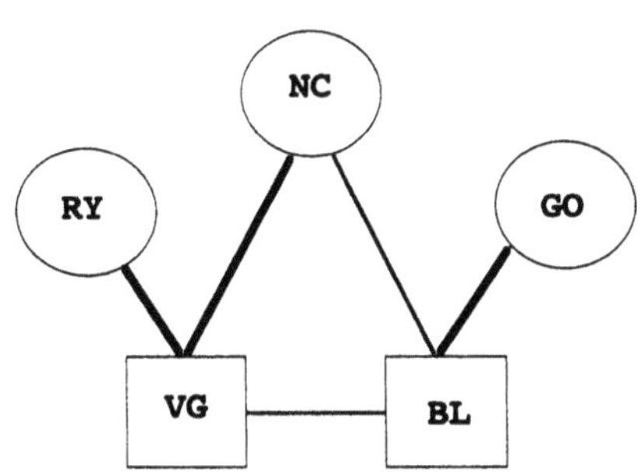


f) *Propithecus tattersalli*

In *Eulemur* groups, males allo-mark females by rubbing their anogenital glands on the back of females (Schilling 1979). The distribution of this marking is non-random. Each male tends to mark only a single female, namely the one he is also closely affiliated with (Fig. 3, Kappeler, unpubl. data).

*Varecia* on Nosy Mangabe exhibited a fission-fusion structure, with so-called core-groups as the basic units (Morland 1991a,b). These core groups basically consist of adult male-female pairs (at least 6 instances; three other pairs consisted of 2 adults of the same sex; Fig. 2d). In Ranomafana National Park, they were observed to form stable pairs (White, 1991), while on the Masoala Peninsula, both pairs and larger groups were observed within a single population (Rigamonti, 1993). Taken together, these studies indicate that *Varecia* and *Eulemur* are quite similar in social organization.

The other diurnal species, however, show a different pattern. Affiliation patterns in a captive group of *L. catta* resemble those for anthropoids with multiple males (Fig. 2e; Kappeler, 1993b): strong female-female associations and male-male associations (which are unusual) and females converging on a single male with whom they associate and have friendly relations (see also Sauther and Sussman, 1993). Interestingly, allo-marking of females by males is absent in *L. catta* (Schilling, 1979). Finally, the nearest neighbor pattern in *Propithecus tattersalli* (Meyers, unpubl. data; Fig. 2f) also indicates a strong bond between the two females and a single male.

These results are clearly preliminary, based on crude analyses of data collected with other questions in mind, and therefore not always entirely appropriate for this purpose. Yet, they indicate that the cathemeral species and *Varecia* have clear pair bonds, although they are not necessarily exclusive, whereas the results for the diurnal *Lemur* and *Propithecus* show an anthropoid-like bonding pattern. It should be emphasized that social pair bonds do not imply exclusive mating relations in a group, although one expects some sort of relationship between the two.

It should be pointed out here that strong bonds between the sexes are not incompatible with strong bonds among members of the same sex. In several species with larger groups, females are clearly tolerant, often being in spatial or grooming association (examples in Fig. 2). In these cases, collective defense of territorial boundaries, especially by females, is observed (Petter et al., 1977 on lemurs in general) in *L. catta* (Jolly et al., 1993; Mertl-Milhollen et al., 1979) and in *Varecia variegata* (Morland, 1991a). Likewise, male bonds can be present (e.g., *E. fulvus*). The existence of these within-sex bonds implies a truly two-tiered social organization, at least in some species. Above the level of the pair, there is the level of the group. However, these within-sex bonds still tend to be different from those observed in anthropoids (Pereira and Kappeler, unpubl. data).

This review of grouping and bonding patterns allows us to develop a new classification of lemuriform societies that uses theoretically meangingful criteria, and to list the ecological factors with which they are associated:
(i) Solitary: foraging alone, sleeping can be with others (only nocturnal species do this);
(ii) Bonded pairs: one male and one female form a pair as a social unit, and possibly also a mating unit (both nocturnal, cathemeral, and diurnal species do this);
(iii) Multiple pairs: a group consists of multiple adults of both sexes that show a tendency toward pairwise social bonds in addition to various intrasexual bonds; these pairs may also be preferential mates (cathemeral species and *Varecia*);
(iv) Groups with multiple adults of both sexes but without the subdivision into bonded pairs (only the diurnal *L. catta* and *Propithecus ssp.*).

## THE ADAPTIVE SIGNIFICANCE OF LEMUR SOCIAL STRUCTURE

This review of lemur grouping (social organization) and bonding (social structure) patterns yields two major organizing principles of lemur societies. First, the pair bond is the fundamental unit in many species (types ii and iii above). Second, the variable social organization, so characteristic for lemurs, is related to their activity period.

In order to explain these features of the typical lemur society we have developed the hypothesis outlined in Fig. 4. The cathemeral lifestyle leads to alternation of diurnal and nocturnal activity periods. This effect of activity period relies on a very basic, though incompletely understood rule: while among diurnal species females can either be solitary or gregarious, females of nocturnal species are never consistently associated with other females during the period of activity. Like among anthropoids (Wright, 1989b), these "solitary" females are accompanied by males. Thus, a female must form a fairly exclusive social bond with one male, if she is to have a male associate during the nocturnal phase. As a result, the gregarious phase represents a multiple-pair situation.

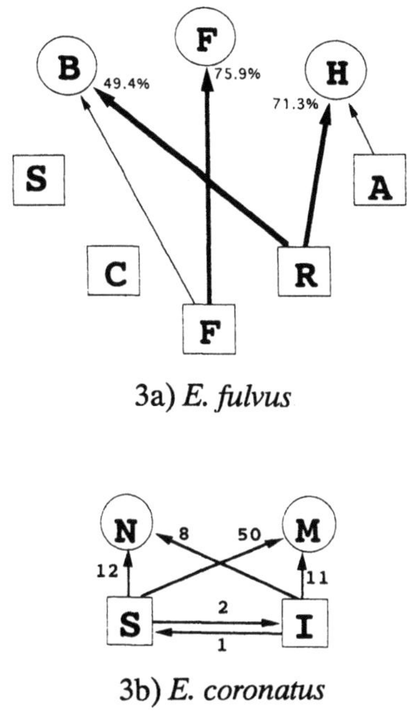


**Figure 3.** Patterns of allo-marking in *Eulemur fulvus* and *E. coronatus*. Adult females are represented by circles; adult males by squares. Each individual served as a focal animal for at least one year. All focal animals were observed for equal amounts of time (for details see Kappeler, 1993b). Arrows in a) indicate which males were responsible for more than 75% of all scent-marks deposited on each female (for B, N=87; for F, N=58, and for H, N=108). In b), the number of all observed incidents of allo-marking are depicted.

This social system has profound consequences for physiology, morphology and social behavior. In order to maintain the pair bond, sexual contest competition between males must be minimized. If it were not, one or a few males would be able to monopolize matings and make it unprofitable for a male to remain the special associate of a female. Lemurs have developed a unique solution to this unique problem, namely estrous synchrony. The significant weakening of sexual contest competition potential has led to sexual monomorphism, which in turn sets up the conditions for female dominance.

Thus, this hypothesis directly addresses all the features that set the typical lemur apart from the typical anthropoid. The rest of this chapter is devoted to examining the validity of this idea, and to examine variability among lemurs.

### The Pair Bond and Infanticide Avoidance

As we noted above, many lemur species with permanently or temporarily gregarious females exhibit strong bonds between individual males and females. What selective forces are responsible for the evolution of this bond? Are there immediate ecological or social advantages to the female of being in spatial association with a male? In a discussion of single-female societies among diurnal anthropoids, van Schaik and Dunbar (1990) juxtaposed several possible male services that would be so valuable to the female

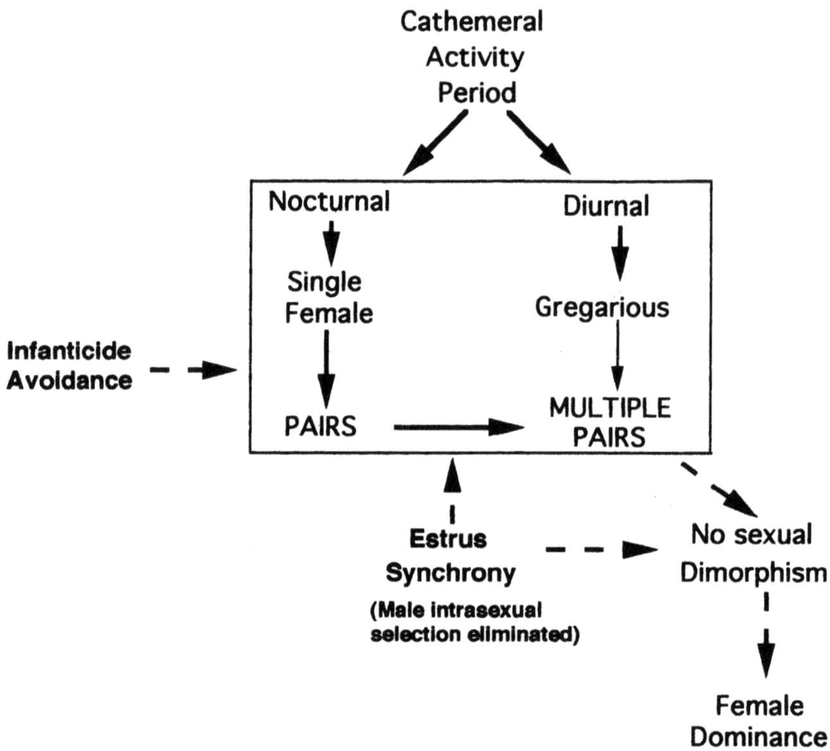


**Figure 4.** Schematic representation of hypothesized determinants of lemur social organization. See text for detailed discussion.

that they may select for continuous spatial association between individual males and females. These services were vigilance and protection against predators, defense of an exclusive territory with its food resources, and prevention of infanticide by invading males. Their preliminary tests supported only the infanticide-prevention hypothesis (van Schaik and Dunbar, 1990). This idea can be extended to cover all situations, regardless of whether females are gregarious or solitary (van Schaik and Kappeler, unpubl. data). Here we will discuss this hypothesis specifically as it applies to lemurs.

Infanticide by incoming males is remarkably widespread in primates (Hiraiwa-Hasegawa, 1988). In most cases, the male who kills is a recent immigrant to the group. The best supported hypothesis posits that infanticide represents a male reproductive strategy. This explanation assumes that females return to receptivity earlier after infanticide than if their infants remain unharmed. This assumption has been tested in langurs and found correct (Sommer, 1987). It also assumes that the male who kills the infants has some probability of subsequently mating with the female.

Alternative hypotheses have been put forward to explain this unusual phenomenon. It has been suggested that male infanticide is a result of high population density or a consequence of disturbance leading to pathological behavior (Hausfater and Hrdy, 1984). However, these alternatives are not supported in their predictions. First, population density is not related to infanticide (in langurs: Newton, 1986). Second, the pathology hypthesis, unlike the male reproductive strategy hypothesis would not predict that infants are subject to infanticidal attacks only during the period in which killing them will speed up the return to estrus of the mother. In langurs, this is in fact the case (Sommer, 1987). Moreover, the infanticidal male will become very tolerant of infants once he has fathered offspring in his group (Pereira and Weiss, 1991). Thus, male infanticide probably represents an adaptive male reproductive strategy (for further discussion, see Hausfater and Hrdy, 1984).

Males who have fathered infants would incur appreciable loss of fitness if the infants they sired are lost due to infanticide. Thus, they are expected to protect their (likely) progeny if they can. Likely fathers are the most suitable protectors of infants, because they are always more closely related to the infants than other females, who may be vulnerable to infanticide themselves and thus less inclined to take risks in defending other infants. Females may also be less of a match to larger and more aggressive males, although this is not an important argument for lemurs. Accordingly, female coalitions have not been found to be effective in preventing infanticide among langurs (Hrdy, 1977). Female counterstrategies are subtle, and mainly limited to early return to estrus following the takeover of the group by a new male or to active mating with multiple males (Hrdy and Whitten, 1987).

Reproduction in most lemurs is highly seasonal (Petter-Rousseaux, 1962; Rasmussen, 1985). Thus, the elimination of infants will not result in immediate reproductive opportunities. However, females in many of the larger species do not reproduce every year (Petter and Peyrieras, 1970; Pollock, 1979; Richard et al.,1991; Wright, 1988; Meyers and Wright, 1993), so infanticide could increase the probability of receptivity in the subsequent year. In other species with high birth rates, females lose many of their infants (Sussman, 1991). Thus, males may benefit from infanticide by shortening the female's interbirth interval or by improving the survival prospects of the next infant. Curiously, *post partum* estrus has been observed in captive *Propithecus* (E. Simons, pers. comm.). This would make it possible to conceive the next infant immediately after the loss of the previous one.

Successful infanticide by likely non-fathers has been observed in several lemur species: *P. diadema* (Wright, 1988), *E. fulvus* and *L. catta* (Pereira and Weiss, 1991; Kappeler and Pereira, unpubl. data), *Varecia variegata* and *E. macaco* (A. Katz, pers. comm.). Moreover, various other behavior patterns suggest that infanticide has been a significant force in lemur social evolution: copulations with multiple males (*L. catta, E. fulvus, E. coronatus, V. variegata*; Colquhoun, 1987; Koyama, 1988; Morland, 1991b; Sauther, 1991; Kappeler, unpubl. data), even with members of neighboring groups (*L. catta*, Sauther, 1991; *P. verreauxi*, Richard, 1985), aggressive female resistance to immigrating males during the birth season (*L. catta*; Pereira and Weiss, 1991), and highly increased aggressiveness of mothers against males (*Hapalemur griseus*; Petter et al., 1977). Nest-guarding (*V. variegata*; Morland, 1990; Pereira et al., 1987) and frequent relocation of infants during the period of maternal activity (*Microcebus murinus, Lepilemur leucopus*; Russell, 1977) is compatible with either an anti-infanticide or an anti-predation function.

## A Test of the Infanticide Avoidance Hypothesis

The opportunity for infant protection depends on the mode of infant care. Where females are spatially associated with their infants, for instance because they carry them around, males can associate spatially with the females to protect them against potentially infanticidal males. By contrast, where females leave their young in nests or hide them somewhere as they forage, males will not provide greater protection to their putative offspring by associating with the female. Active adults are easier to locate than inactive infants, but infant parking is not necessarily safer since, once detected by predators or infanticidal males, there is no one to protect the infant. However, the important point is that males cannot protect hidden infants by associating with the female. Thus, one would predict that the mode of infant care determines whether spatial association of males and their female mates, and thus pair bonds, are found.

This prediction is confirmed by the distribution of pair bonding among the nocturnal and cathemeral lemurs (Table 1). In all four genera in which females carry infants, they are always accompanied by at least one male, whereas in six of the seven genera in which females do not carry their infants, they are solitary during the active period. This is not just a byproduct of the reproductive biology of nesting species. In *Phaner* and *Hapalemur* the infants are born in nests, and stay there during the first few days (Petter et al., 1977). However, after a few days the mother starts to carry them around, and, as predicted, females are associated with males. A phylogeny-based test covering all primates provides further support for this hypothesis (van Schaik and Kappeler, unpubl. data).

*Varecia* is an apparent exception in that they are diurnal infant parkers with strong male-female bonds. Quite frequently, however, the male guards the parked infants (Pereira et al., 1987; White et al., 1993). This unique male behavior may be related to this species' diurnal life style, or to the large size of the nests and their conspicuousness. Since a male only guards the litter of one female at a time, and is unlikely to be selected to do so, unless he has a high probability of having fathered the offspring, a male-female pair bond is needed. Thus, this strange case in fact supports this hypothesis, and underscores that the primary function of the pair bond is to protect the infant(s), and not to provide other services to the female.

Where females are gregarious, individual pair-bonding is not needed because a single male could defend the infants of multiple females. Among at least some anthropoids (e.g., Janson, 1984; van Noordwijk and van Schaik, 1988), females converge in their mate choice on a single male, who is highly tolerant of, and spatially associated with, the infants. Indeed, pair-bonding is not pronounced in exclusively diurnal lemurs. In *L. catta*, one adult male has strong relationships with the group's females and is often their primary mating partner (Sauther, 1991; Kappeler, 1993b). This central male is known to defend the offspring against infanticide attempts by males known not to have sired the offspring (Pereira and Weiss, 1991). The proximity pattern in *Propithecus tattersalli* is also consistent with this idea.

## Pair-bonding, Mating and Reproductive Synchrony

The bonded pair is the basic social unit in *Varecia* and among the carrying species that are nocturnal or cathemeral. In wild *Varecia*, the pair is also the mating unit: virtually all matings are between female-male affiliates (Morland 1991b). Among *Eulemur* species, matings are not between exclusive male-female pairs (Colquhoun, 1987; Vick, 1977; Kappeler, unpubl. data). However, while some mechanism to ascertain paternal confidence, such as a first-male advantage, is required for this system to be evolutionarily stable, exclusivity is not required, as shown by the occurrence of friendships among promiscuous anthropoids (Smuts, 1987). Additional matings with other males are expected, because this mating strategy will assure high confidence of paternity for the primary social partner, but will cause other familiar males to refrain from attempting infanticide (Hrdy and Whitten, 1987).

The extent to which a female can express such preference for her male associate will depend on the extent to which females can eliminate male-male contest competition for mating access to females. The crucial determinant of male monopolization potential are the spatial distribution of females and the temporal distribution of their estruses. It is on this second aspect that we focus here. When estrus is brief and highly synchronized, males cannot monopolize access to more than one female. The highly seasonal breeding of most lemurs (Richard and Dewar, 1991) is a necessary condition for estrous synchrony because when females can breed throughout the year estrous synchrony is simply unattainable. Such strict breeding seasonality should be based on pronounced environmental seasonality, since adhering to a strict seasonal schedule may otherwise be costly due to potential loss of time. If some mishap, such as loss of infant or longer lactational amenorrhea than usual, throws a female out of phase with the environmental cycle, she will lose an entire year.

Strict breeding seasonality may be a necessary condition for estrous synchrony, it is not in itself sufficient. There must be a strong selective force actively maintaining estrous synchrony. Its main consequence is that it effectively eliminates any potential for males to monopolize breeding access to females. Thus, estrous synchrony enables females to enforce pair bonds rather than to let the system drift toward harem polygyny.

The hypothesis that infanticide avoidance was the selective force in the evolution of pair bonds and multiple pair groups makes specific predictions about the extent of estrous synchrony expected in different social systems (Table 3). Estrous synchrony should be most extreme in the cathemeral species that live in "multi-pair" groups, intermediate in pair-bonded species, and least extreme in the exclusively diurnal species or in the solitary nocturnal species. Females in pair-bonded species should be synchronous whenever population density is high and males could easily commute between females.

**Table 3.** Predicted patterns of estrous synchrony, sexual dimorphism and female dominance among lemur species, based on the infanticide prevention hypothesis.

| | **Solitary or Gregarious + Polygynous** | **Pair-living** | **Multiple Pairs** |
|---|---|---|---|
| **Examples:** | *Lemur catta*<br>*Propithecus*<br>*Cheirogaleids*<br>*Daubentonia* | *Phaner*<br>*Avahi*<br>*E. rubriventer* | *Eulemur*<br>*Varecia* |
| **Estrous Synchrony** | weak to none | some | strong |
| **Sexual Dimorphism** | present | none | none |
| **Female dominance** | absent | present | present |

For a proper test of this prediction, one would have to compare the degree of estrous synchrony among the females within a group or a local population. The available data are extremely limited, but are consistent with the hypothesis. In one captive *E.coronatus* group both resident females mated during the same 3-4 h period in two consecutive years (Kappeler, unpubl. data). *L.catta* is a diurnal species and should have abandoned this synchrony, yet it appears highly synchronous. However, unlike in *E.coronatus*, each female in the group is in estrus on a different day, each for about half a day, so one male can, in principle, fertilize them all (Pereira, 1991; Pereira and Weiss, 1991). Thus, while *L.catta* did not lose its synchrony entirely, the lengthening and staggering of estrous periods makes partial monopolization by a single male possible. Because lemurs can have multiple cycles within one season, their reproductive activity is not continuous. Instead there tend to be several clumped estrous events within groups or local populations.

A weaker test of the hypothesis is to compare birth seasonality across categories of species. Where the birth season is long, estrous synchrony is almost certainly absent; where it is short it is at least possible. These data, compiled from DUPC records (Table 1; Kappeler, unpubl. data), support the prediction. The solitary species have birth seasons lasting at least 3 months, sometimes much more. Multiple breeding events in a single season, as seen in *Microcebus* (Martin, 1972), will tend to break any precise synchrony. The highly seasonal *Cheirogaleus* species hibernate for at least 6 months each year, and therefore have only a limited time window for reproduction (Foerg, 1982; Petter et al., 1977). *Daubentonia* has been observed to be in estrus during a period of at least 5 months (Sterling, 1992). None of the multi-pair species, in contrast, have a birth season that lasts more than 3 months. The difference between the species living in multiple pairs and the solitary species is significant (Mann-Whitney, $n_1=6$, $n_2=4$, $U=2$, $P<0.05$). Although the strong observed seasonality among the diurnal *Propithecus verreauxi* goes against the hypothesis, the *post partum* estrus in captive *Propithecus* noted earlier indicates a flexibility in estrus hardly compatible with strict synchrony. Thus, while these results are very preliminary, they are not inconsistent with the hypothesis that estrous synchrony is found where male-female pairs occur.

At the same time, the data also indicate that many solitary lemurs (with the exception of *Mirza*) are more seasonal in their breeding than most lorisids, which can have very weak birth peaks in the wild (Charles-Dominique, 1977; but see Nash, 1983) and in captivity (Kappeler, unpubl. data). The western part of Madagascar, where many of the solitary lemurs occur, is a highly seasonal environment (Hladik, 1980). Thus, strong breeding seasonality, a precondition for extreme estrous synchrony among females, is present in the lemuriforms.

**Consequences for Morphology and Social Behavior**

The elimination of intrasexual competition for mates among males, due to estrous synchrony among females, should reduce selection toward larger male size and weaponry. Thus, one would predict a transition from sexual monomorphism to male-biased dimorphism in body size and weaponry from multi-pair societies, through pairs to gregarious diurnal species and nocturnal solitary species (see Table 3). However, this prediction is not supported by the distributional pattern of sexual dimorphism in body size and canine length (Kappeler, 1993d).

Indirectly, this argument makes predictions concerning the distribution of female dominance. Estrous synchrony and female dominance should be highly correlated across species. In the absence of sexual dimorphism, pay-off asymmetries can account for systematic differences in dominance between the sexes. Females usually gain more from access to food sources than males, so males are expected to defer. Consequently, females tend to be dominant in the monogamous anthropoids. This argument would predict a lack of female dominance in both parking species and consistently diurnal species. The evidence, however, is not consistent with this prediction. Female *Microcebus murinus* often attack and always win in contests with males (Pagès-Feuillade, 1988), although *Lepilemur leucopus* most likely does not (Russell, 1977). Among the diurnal *L. catta* and *P.verreauxi*, females consistently dominate males (Jolly 1966; Pereira et al. 1990; Richard and Heimbuch 1975). Thus, the female dominance puzzle remains as elusive as ever.

**The Power of the Infanticide-prevention Hypothesis**

In this chapter, we proposed the following two-step hypothesis. First, infanticide prevention has selected for the evolution of male-female associations among primates carrying infants. Second, the cathemeral activity period of lemurs necessitated the maintenance of pair bonds in the face of possible male monopolization attempts, and so explains the evolution of estrous synchrony, sexual monomorphism and female dominance - in short many of the features that set lemurs apart from other prosimians or anthropoids. Although it is too early to pass a verdict, this hypothesis provides a coherent explanation for the frequent occurrence of bonded monogamy, the solitary social structure of many other species, the absence of solitary species among anthropoids, the remarkable intraspecific variability from bonded pairs to large groups and the even adult sex ratios of many lemur groups, the pairwise bonding in larger groups of the cathemeral species and *Varecia*, and the stronger birth seasonality and estrous synchrony of the pair-bonded species.

At the same time, the hypothesis was inadequate in explaining the interspecific patterns among lemurs of other unusual lemur features: female dominance, relative testes size, and sexual dimorphism of body and canine size (Kappeler, 1993d). The lack of fit between these features and the various types of society can be interpreted in various ways. First, the hypothesis we examined here is wrong or incomplete. Second, all these features are subject to additional selective pressures which may vary in different ways across taxa.

There is also a third possibility, however. Some of the inconsistent phenomena may represent either non-equilibrium situations or local optima brought about by geologically recent changes in the life styles of these species. While this is of course difficult to test, the extinction of large numbers of especially large-sized lemurs about a millennium ago (see Tattersall, 1993) may have opened up niches for the remaining smaller species (that have now become the largest extant ones). All lemurs, including the diurnal ones have eyes that are basically adapted to nocturnal vision (Tattersall, 1982). Such a scenario may explain some of the inconsistent cases. A possible case of disequilibrium may be the bonding patterns in *Lemur catta* and *Propithecus diadema*. While they were consistent with the hypothesis that females of diurnal species should converge in their preference of male partners (just like anthropoids), the group compositions of these species still show an approximately equal adult sex ratio (Figure 1). One possible reason is that female dominance is retained, and that this allows the females to experience relatively little feeding competition from the males due to the peripheralization of all but one of the males: Jolly's (1966) drones' club in *L. catta*. Another example of a possible local optimum is the highly unusual estrous asynchrony among individuals within extreme estrous synchrony in *L. catta*.

Despite these shortcomings and uncertainties, however, we believe that this hypothesis is presently the only coherent idea to explain many typical lemur features. We therefore present the following minimum agenda for a more critical evaluation of the processes postulated by the infanticide-prevention/ cathemeral activity hypothesis:

(1) We predict that, among the cathemeral lemur species, careful descriptions of spatial patterning of animals and of their grooming and agonistic behavior should reveal that individual male-female pairs stand out as having strong affiliative and affinitive associations (bonds).

(2) We also predict that species living in multi-female, multi-male groups during the day should split up during periods of nocturnal acitivity (perhaps only on a seasonal or irregular basis); and that the smaller parties should consist mainly of male-female pairs, in fact the same pairs that were found to associate preferentially during the day.

(3) Observations on estrous synchrony in the wild or in captivity under natural light cycles should show the predicted maximum synchrony among females in multi-pair species living in the same group, and somewhat less, in the same population, while no synchrony beyond similar seasonality is expected in the solitary species.

(4) We predict that mating in the multi-pair species should not show the usual anthropoid pattern of rank-dependent mating success and paternity but rather that a significant proportion of the fertilizations are by male associates of the female.

**The Evolution of Primate Life Styles**

This discussion raises the question of the evolutionary origins of infant carrying and of diurnal activity among primates. The consensus among students of primate evolution is that the earliest primates were nocturnal and had nests, since both are ancestral mammalian traits (Martin, 1990). Provided that this point of departure is correct, we have to account for the evolution from a nocturnal nest-living primate with a litter of two or more infants to a diurnal one that carries its single infant around. Thus, there are several transitions: from nocturnality to diurnality, from larger litters to smaller ones, and from relatively helpless (altricial) young to relatively precocial young that can cling to the mother from birth onwards. How are these transitions related?

Although these are complex questions awaiting a more detailed treatment (van Schaik and Kappeler, unpubl. data), some general thoughts can be developed. Primates of all radiations are relatively precocial compared to other mammals (Eisenberg, 1981; Martin, 1990). Moreover, some nocturnal radiations, such as the lorisids, have surprisingly precocial young (Martin, 1990). These observations suggest that precociality preceded diurnality. Because precociality is generally correlated with reduced litter size (Derrickson, 1992), it can also account for the observed trend in litter size.

If precociality is the factor that made the other changes possible, two further questions arise. First, what caused the trend toward increased precociality? And second, why did these precocial species turn diurnal? Again, only highly speculative answers are available. As to the evolution of precociality, we suggest that the first step toward it was parking outside the nest rather than leaving the infants in the nest. After this, giving up the nest altogether is only a small step. Perhaps it is adaptive to hide infants, rather than leaving them in the nest in species where infants grow slowly, and thus have a longer period of vulnerability. Gradually, nests or holes would accumulate smells and waste products, thus attracting predators or hostile conspecifics (see also Tinbergen et al., 1967). The crucial selective factor favoring parking may thus have been a slowing of growth. Perhaps, this reduced growth in turn was partly brought about by an increase in body size. This scenario is compatible with the general correlation among mammals between precociality, slower growth and a slower life history track (cf. Harvey et al., 1989; Derrickson, 1992).

Correlates of diurnality are frugivory (colorful fruit is easier to find during the day), larger body size, and greater ease of movement through trees, which allows for far greater mobility and range covered during one day. Which of these correlates actually acted as selective agents is very difficult to reconstruct, but they are interrelated. Having precocial young allows the female to carry the infant around right from birth and thus makes it possible for her to wander freely without having to return to fixed sites. Frugivores tend to range more widely than folivores or insectivores (Clutton-Brock and Harvey, 1977). Larger body size also strongly constrains the feasibility of crypsis as a predation-avoidance strategy, and shows a strong correlation with diet (Kay, 1984). Thus, precociality may have facilitated the change to the more active life style required to maintain a diurnal-frugivore lifestyle. Whether or not an increase in body size underlies both the trend toward precociality and the need to become at least partly diurnal remains obscure.

## ACKNOWLEDGEMENTS

We thank many lemur specialists for patiently listening to and commenting on the ideas presented in this paper: Jörg Ganzhorn, David Haring, Alison Jolly, David Meyers, Hillary Morland, Deborah Overdorff, Elwyn Simons, Frances White, Pat Wright, and especially Michael Pereira. We thank D. Meyers and H. Morland for making available as yet unpublished data, and Robin Dunbar for comments. This is publication No. 545 of the Duke University Primate Center.

## REFERENCES

Arbelot-Tracqui, V., 1983, Etude étho-écologique de deux primates prosimiens: *Lemur coronatus* et *Lemur fulvus sanfordi*, Ph.D. thesis, Université de Rennes.

Aureli, F., van Schaik, C. P. and van Hooff, J. A. R. A. M., 1989, Functional aspects of reconciliation among captive long-tailed macaques (*Macaca fascicularis*), *Am. J. Primatol.* 19:39.

Bearder, S. K., 1987, Lorises, bushbabies, and tarsiers: diverse societies in solitary foragers, *in*: "Primate Societies," B. B. Smuts, D. L. Cheney, R. M. Seyfarth, R. W. Wrangham and T. T. Struhsaker, eds., University of Chicago Press, Chicago.

Budnitz, N., 1978, Feeding behavior of *Lemur catta* in different habitats, *in*: "Perspectives in Ethology, Vol. 3," P. Bateson and P. Klopfer, eds., Plenum Press, New York.

Charles-Dominique, P., 1977, "Ecology and Behaviour of Nocturnal Primates," Columbia University Press, New York.

Charles-Dominique, P., Cooper, H. M., Hladik, A., Hladik, C. M., Pages, E., Pariente, G. F., Petter-Rousseaux, A., Petter, J.-J. and Schilling, A., 1980, eds., "Nocturnal Malagasy Primates: Ecology, Physiology and Behavior," Academic Press, New York.

Clutton-Brock, T. H., 1989, Mammalian mating systems, *Proc. Roy. Soc. Lond. B* 236:339.

Clutton-Brock, T. H. and Harvey, P. H., 1977, Primate ecology and social organization, *J. Zool., Lond.* 183:1.

Colquhoun, I. C., 1987, Dominance and "fall fever": the reproductive behavior of male brown lemurs (*Lemur fulvus*), *Can. Rev. Phys. Anthropol.* 6:10.

Colquhoun, I. C., 1993, The socioecology of *Eulemur macaco*: a preliminary report, *in*: "Lemur Social Systems And Their Ecological Basis," P. M. Kappeler and J. U. Ganzhorn, eds., Plenum Press, New York.

Dague, C., 1986, Contribution a l'étude du *Lemur rubriventer* dans son milieu nature, *Mammalia* 50:561.
Derrickson, E. M., 1992, Comparative reproductive strategies of altricial and precocial eutherian mammals, *Funct. Ecol.* 6:57.
Eisenberg, J. F., 1981, "The Mammalian Radiations," University of Chicago Press, Chicago.
Engqvist, A. and Richard, A., 1991, Diet as a possible determinant of cathemeral activity patterns in primates, *Fol. Primatol.* 57:169.
Foerg, R., 1982, Reproduction in *Cheirogaleus medius*, *Fol. Primatol.* 39:49.
Ganzhorn, J. U., Abraham, J. P. and Razanahoera-Rakotomalala, M., 1985, Some aspects of the natural history and food selection of *Avahi laniger*, *Primates* 26:452.
Godfrey, L. R., Lyon, S. K. and Sutherland, M. R., 1993, Sexual dimorphism in large-bodied primates: the case of the subfossil lemurs, *Am. J. Phys. Anthropol.* 90:315.
Harcourt, A. H., 1979, Social relationships between adult male and female mountain gorillas, *Anim. Behav.* 27:325.
Harcourt, C. S., 1991, Diet and behaviour of a nocturnal lemur, *Avahi laniger*, in the wild, *J. Zool. Lond.* 223:667.
Harrington, J. E., 1975, Field observations of social behavior of *Lemur fulvus fulvus*., *in*: "Lemur Biology," I. Tattersall and R. W. Sussman, eds., Plenum Press, New York.
Harrington, J. E., 1978, Diurnal behavior of *Lemur mongoz* at Ampijoroa, *Fol. Primatol.* 29:291.
Harvey, P. H., Promislow, D. E. L. and Read, A. F., 1989, Causes and correlates of life history differences among mammals, *in*: "Comparative Socioecology," V. Standen and R. A. Foley, eds., Blackwell, Oxford.
Hausfater, G. and Hrdy, S. B., 1984, "Infanticide: Comparative and Evolutionary Perspectives," Aldine, Hawthorne, New York.
Hiraiwa-Hasegawa, M., 1988, Adaptive significance of infanticide in primates, *TREE* 3:102.
Hladik, A., 1980, The dry forest of the west coast of Madagascar: climate, phenology, and food available for prosimians, *in*: "Nocturnal Malagasy Primates: Ecology, Physiology, and Behavior," P. Charles-Dominique, H. M. Cooper, A. Hladik, C. M. Hladik, E. Pages, G. F. Pariente, A. Petter-Rousseaux, J.-J. Petter and A. Schilling, eds., Academic Press, New York.
Hrdy, S. B., 1977, "The Langurs of Abu," Harvard University Press, Cambridge.
Hrdy, S. B. and Whitten, P. L., 1987, Patterning of sexual activity, *in*: "Primate Societies," B. B. Smuts, D. L. Cheney, R. M. Seyfarth, R. W. Wrangham and T. T. Struhsaker, eds., University of Chicago Press, Chicago.
Janson, C. H., 1984, Female choice and mating system of the brown capuchin monkey *Cebus apella* (Primates: Cebidae), *Z. Tierpsychol.* 65:177.
Jenkins, P. and Albrecht, G., 1991, Sexual dimorphism and sex ratios in Madagascan prosimians, *Am. J. Primatol.* 24:1.
Jolly, A., 1966, "Lemur Behavior," University of Chicago Press., Chicago.
Jolly, A., 1984, The puzzle of female feeding priority, *in*: "Female Primates: Studies by Woman Primatologists.," M. F. Small, ed., A.R. Liss, New York.
Jolly, A., Gustafson, H., Oliver, W. L. R. and O'Connor, S. M., 1982, *Propithecus verreauxi* population and ranging at Berenty, Madagascar, 1975 and 1980, *Folia Primatol.* 39:124.
Jolly, A., Rasamimanana, H. R., Kinnaird, M. F., O'Brian, T. G., Crowley, H. M., Harcourt, C. S., Gardner, S. and Davidson, J. M., Territorialtity in *Lemur catta* groups during the birth season at Berenty, Madagascar, *in*: "Lemur Social Systems And Their Ecological Basis," P. M. Kappeler and J. U. Ganzhorn, eds., Plenum Press, New York.
Jones, K. C., 1983, Inter-troop transfer of *Lemur catta* males at Berenty, Madagascar, *Fol. Primatol.* 40:145.
Kappeler, P. M., 1990a, Female dominance in *Lemur catta*: more than just female feeding priority?, *Fol. Primatol.* 55:92.
Kappeler, P. M., 1990b, The evolution of sexual size dimorphism in prosimian primates, *Am. J. Primatol.* 21:201.
Kappeler, P. M., 1991, Patterns of sexual dimorphism in body weight among prosimian primates, *Fol. Primatol.* 57:132.
Kappeler, P. M., 1993a, Female dominance in primates and other mammals, *in*: "Perspectives in Ethology, Vol. X," P. Bateson, N. Thompson and P. Klopfer, eds., Plenum Press, New York.
Kappeler, P. M., 1993b, Variation in social structure: the effects of sex and kinship on social interactions in three lemur species, *Ethology* 93:125.
Kappeler, P. M., 1993c, Reconciliation and post-conflict behaviour in ringtailed (*Lemur catta*) and redfronted (*Eulemur fulvus rufus*) lemurs, *Anim. Behav.*, in press.

Kappeler, P. M., 1993d, Sexual selection and lemur social systems, *in*: "Lemur Social Systems And Their Ecological Basis," P. M. Kappeler and J. U. Ganzhorn, eds., Plenum Press, New York.
Kaufman, R., 1991, Female dominance in semifree-ranging black-and-white ruffed lemurs, *Varecia variegata*, *Fol. Primatol.* 57:39.
Kay, R. F., 1984, On the use of anatomical features to infer foraging behavior in extinct primates, *in*: "Adaptations for Foraging in Non-Human Primates," P. S. Rodman and J. G. H. Cant, eds., Columbia University Press, New York.
Klopfer, P. H. and Boskoff, K. L., 1979, Maternal behavior in prosimians, *in*: "The Study of Prosimian Behavior," G. A. Doyle and R. D. Martin, eds., Academic Press, New York.
Koyama, N., 1988, Mating behavior of ring-tailed lemurs (*Lemur catta*) at Berenty, Madagascar, *Primates* 29:163.
Martin, R. D., 1972, A preliminary field-study of the lesser mouse lemur (*Microcebus murinus* J.F. Miller 1777), *Z. Tierpsychol. Suppl.* 9:43.
Martin, R. D., 1990, "Primate Origins and Evolution," Chapman and Hall, London.
McGlynn, C. A., 1992, Special relationships: the redfronted lemur's (*Eulemur fulvus rufus*) alternative to female dominance?, Master's thesis, Duke University, Durham.
Meier, B. and Albignac, R., 1991, Rediscovery of *Allocebus trichotis* Gunther 1875 (Primates) in North East Madagascar, *Fol. Primatol.* 56:57.
Meier, B., Albignac, R., Peyrieras, A., Rumpler, Y. and Wright, P., 1987, A new species of *Hapalemur* (Primates) from South East Madagascar, *Fol. Primatol.* 48:211.
Mertl-Millhollen, A. S., Gustafson, H. L., Budnitz, N., Dainis, K. and Jolly, A., 1979, Population and territory stability of the *Lemur catta* at Berenty, Madagascar, *Fol. Primatol.* 31:106.
Meyers, D. M. and Wright, P. C., 1993, Resource tracking: food availability and *Propithecus* seasonal reproduction, *in*: "Lemur Social Systems And Their Ecological Basis," P. M. Kappeler and J. U. Ganzhorn, eds., Plenum Press, New York.
Morland, H. S., 1990, Parental behavior and infant development in ruffed lemurs (*Varecia variegata*) in a northeast Madagascar rainforest, *Am. J. Primatol.* 20:253.
Morland, H. S., 1991a, Preliminary report on the social organization of ruffed lemurs (*Varecia variegata variegata*) in a northeast Madagascar rain forest, *Fol. Primatol.* 56:157.
Morland, H. S., 1991b, Social organization and ecology of black and white ruffed lemurs (*Varecia variegata variegata*) in a lowland rainforest, Nosy Mangabe, Madagascar., Ph.D. thesis, Yale University, New Haven.
Nash, L. T., 1983, Reproductive patterns in galagos (*Galago zanzibaricus* and *Galago garnettii*) in relation to climatic variability, *Am. J. Primatol.* 5:181.
Newton, P. N., 1986, Infanticide in an undisturbed forest population of Hanuman langurs, *Presbytis entellus*, *Anim. Behav.* 34:785.
Overdorff, D., 1991, Ecological correlates of social structure in two prosimian primates: *Eulemur fulvus rufus* and *Eulemur rubriventer* in Madagascar, Ph.D. thesis, Duke University, Durham.
Pagès-Feuillade, E., 1988, Modalités de l'occupation de l'espace et relations interindividuelles chez un prosimien nocturne malgache (*Microcebus murinus*), *Fol. Primatol.* 50:204.
Pereira, M. E., 1991, Asynchrony within estrous synchrony among ringtailed lemurs (Primates: Lemuridae), *Phys. Behav.* 49:47.
Pereira, M. E., 1992, Agonistic interaction, dominance relations, and ontogenetic trajectories in ringtailed lemurs, *in*: "Juvenile Primates: Life History, Development, and Behavior," M. E. Pereira and L. A. Fairbanks, eds., Oxford University Press, New York.
Pereira, M. E., Kaufman, R., Kappeler, P. M. and Overdorff, D. J., 1990, Female dominance does not characterize all of the Lemuridae, *Fol. Primatol.* 55:96.
Pereira, M. E., Klepper, A. and Simons, E. L., 1987, Tactics of care for young infants by forest-living ruffed lemurs (*Varecia variegata variegata*): ground nests, parking, and biparental guarding, *Am. J. Primatol.* 13:129.
Pereira, M. E. and Weiss, M. L., 1991, Female mate choice, male migration, and the threat of infanticide in ringtailed lemurs, *Behav. Ecol. Sociobiol.* 28:141.
Petter, J. J., Albignac, R. and Rumpler, Y., 1977, "Faune de Madagascar 44: Mammifères Lémuriens (Primates Prosimien)," ORSTOM and CNRS, Paris.
Petter, J. J. and Peyrieras, A., 1970, New contribution to the study of a Malagasy lemur, the aye-aye (*Daubentonia madagascariensis* E. Geoffroy), *Mammalia* 34:167.
Petter-Rousseaux, A., 1962, Récherche sur la biologie de la réproduction des primates inferieurs, *Mammalia* 26 Suppl. 1:1.
Pollock, J. I., 1979, Female dominance in *Indri indri*, *Fol. Primatol.* 31:143.

Rasmussen, D. T., 1985, A comparative study of breeding seasonality and litter size in eleven taxa of captive lemurs (*Lemur* and *Varecia*), *Int. J. Primatol.* 6:501.
Richard, A., 1974, Intra-specific variation in the social organization and ecology of *Propithecus verreauxi*, *Fol. Primatol.* 22:178.
Richard, A. F., 1985, Social boundaries in a Malagasy prosimian, the sifaka (*Propithecus verreauxi*), *Int. J. Primatol.* 6:553.
Richard, A. F., 1987, Malagasy prosimians: female dominance, *in*: "Primate Societies," B.B. Smuts, D. L. Cheney, R. M. Seyfarth, R. W. Wrangham and T. T. Struhsaker, eds., University of Chicago Press, Chicago.
Richard, A. F. and Dewar, R. E., 1991, Lemur ecology, *Ann. Rev. Ecol. Syst.* 22:145.
Richard, A. F. and Heimbuch, R., 1975, An analysis of the social behavior of three groups of *Propithecus verreauxi*, *in*: "Lemur Biology," I. Tattersall and R. W. Sussman, eds., Academic Press, New York.
Richard, A. F., Rakotomanga, P. and Schwartz, M., 1991, Demography of *Propithecus verreauxi* at Beza Mahafali, Madagascar: Sex ratio, survival, and fertility, *Am. J. Phys. Anthropol.* 84:307.
Rigamonti, M.M., 1993, Home range and diet in red ruffed lemurs (*Varecia variegata rubra*) on the Masoala Peninsula, Madagascar, *in*: "Lemur Social Systems And Their Ecological Basis," P. M. Kappeler and J. U. Ganzhorn, eds., Plenum Press, New York.
Russell, R. J., 1977, The behavior, ecology, and environmental physiology of a nocturnal primate, *Lepilemur mustelinus*, Ph. D. thesis, Duke University, Durham.
Sauther, M. L., 1991, Reproductive behavior of free-ranging *Lemur catta* at Beza Mahafaly Special Reserve, Madagascar, *Am. J. Phys. Anthropol.* 84:463.
Sauther, M. L. and Sussman, R. W., 1993, A new interpretation of the social organization and mating system of the ringtailed lemur (*Lemur catta*), *in*: "Lemur Social Systems And Their Ecological Basis," P. M. Kappeler and J. U. Ganzhorn, eds., Plenum Press, New York.
Schilling, A., 1979, Olfactory communication in prosimians, *in*: "The Study of Prosimian Behavior," G. A. Doyle and R. D. Martin, eds., Academic Press, New York.
Smuts, B. B., 1985, "Sex and Friendship in Baboons," Aldine, Hawthorne.
Smuts, B. B., 1987, Sexual competition and mate choice, *in*: "Primate Societies.," B. B. Smuts, D. L. Cheney, R. M. Seyfarth, R. W. Wrangham and T. T. Struhsaker, eds., University of Chicago Press, Chicago.
Sommer, V., 1987, Infanticide among free-ranging langurs (*Presbytis entellus*) at Jodhpur (Rajasthan/India): recent observations and a reconsideration of hypotheses, *Primates* 28:163.
Sterling, E., 1992, Timing of reproduction in aye-ayes in Madagascar (*Daubentonia madagascariensis*), *Am. J. Primatol.* 27:59.
Sussman, R. W., 1977, Feeding behavior of *Lemur catta* and *Lemur fulvus*, *in*: "Primate Ecology," T. H. Clutton-Brock, ed., Academic Press, London.
Sussman, R. W., 1991, Demography and social organization of free-ranging *Lemur catta* in the Beza Mahafaly Reserve, Madagascar, *Am. J. Phys. Anthropol.* 84:43.
Sussman, R. W. and Tattersall, I., 1976, Cycles of activity, group composition, and diet of *Lemur mongoz* in Madagascar, *Fol. Primatol.* 26:270.
Tattersall, I., 1977, Ecology and behavior of *Lemur fulvus mayottensis* (Primates, Lemuriformes), *Anthropol. Pap. Am. Mus. Nat. Hist.* 54:425.
Tattersall, I., 1978, Behavioral variation in *Lemur mongoz*, *in*: "Recent Advances in Primatology III," D. Chivers and K. Joysey, eds., Academic Press, London.
Tattersall, I., 1982, "The Primates of Madagascar," Columbia University Press, New York.
Tattersall, I., 1987, Cathemeral activity in primates: a definition, *Folia Primatol.* 49:200.
Tattersall, I., 1993, Madagascar's lemurs, *Sci. Amer.* 1/93:90.
Tinbergen, N., Impekoven, M. and Franck, D., 1967, An experiment on spacing out as defense against predators, *Behaviour* 28:307.
van Hooff, J. A. R. A. M. and van Schaik, C. P., 1992, Cooperation in competition: the ecology of primate bonds, *in*: "Coalitions and Alliances in Humans and other Animals," A. H. Harcourt and F. B. M. de Waal, eds., Oxford University Press, Oxford.
van Noordwijk, M. A. and van Schaik, C. P., 1988, Male careers in Sumatran long-tailed macaques (*Macaca fascicularis*), *Behaviour* 107:24.
van Schaik, C. P., 1989, The ecology of social relationships amongst female primates, *in*: "Comparative Socioecology.," V. Standen and R. A. Foley, eds., Blackwell, Oxford.
van Schaik, C. P. and Dunbar, R. I. M., 1990, The evolution of monogamy in large primates: a new hypothesis and some crucial tests, *Behaviour* 115:30.
van Schaik, C. P. and van Hooff, J. A. R. A. M., 1983, On the ultimate causes of primate social systems, *Behaviour* 85:91.

Vick, L. G., 1977, The role of inter-individual relationships in two troops of captive *Lemur fulvus*, Ph.D. thesis, University of North Carolina, Chapel Hill.

Vick, L. G. and Pereira, M. E., 1989, Episodic targeting aggression and the histories of Lemur social groups, *Behav. Ecol. Sociobiol.* 25:3.

Watts, D. P., 1990, Ecology of gorillas and its relation to female transfer in mountain gorillas, *Int. J. Primatol.* 11:21.

White, F. J., 1991, Social organization, feeding ecology, and reproductive strategy of ruffed lemurs, *Varecia variegata*, *in*: "Primatology Today: Proceedings of the XIII Congress of the International Primatological Society, Nagoya and Kyoto 18-24 July 1990," A. Ehara, T. Kimura, O. Takenaka and M. Iwamoto, eds., Elsevier Science Publishers, Amsterdam.

White, F. J., Balko, E. A., Fox, E. A., 1993, Male transfer in captive ruffed lemurs, *Varecia variegata variegata*, *in*: "Lemur Social Systems And Their Ecological Basis," P. M. Kappeler and J. U. Ganzhorn, eds., Plenum Press, New York.

Wilson, J. M., Stewart, P. D., Ramangason, G. S., Denning, A. M. and Hutchings, M. S., 1989, Ecology and conservation of the crowned lemur, *Lemur coronatus*, at Ankarana, N. Madagascar, *Fol. Primatol.* 52:1-26.

Wrangham, R. W., 1980, An ecological model of female-bonded primate groups, *Behaviour* 75:262-300.

Wrangham, R. W., 1987, Evolution of social structure, *in*: "Primate Societies.," B. B. Smuts, D. L. Cheney, R. M. Seyfarth, R. W. Wrangham and T. T. Struhsaker, eds., University of Chicago Press, Chicago.

Wright, P. C., 1988, Social behavior of *Propithecus diadema edwardsi* in Madagascar, *Am. J. Phys. Anthropol.* 75:289.

Wright, P. C., 1989a, Comparative ecology of three sympatric bamboo lemurs in Madagascar, *Am. J. Phys. Anthropol.* 78:327.

Wright, P. C., 1989b, The nocturnal primate niche in the New World, *J. Hum. Evol.* 18:635.

Wright, P. C., Daniels, P. S., Meyers, D. M., Overdorff, D. J. and Rabesoa, J., 1986, A census and study of *Hapalemur* and *Propithecus* in southeastern Madagascar, *Primate Conserv.* 8:84.

# RESUMES

(Translated by Anne Arsène)

## MODELES D'UTILISATION DU DOMAINE VITAL ET ORGANISATION SOCIAL DU AYE-AYE (*DAUBENTONIA MADAGASCARIENSIS*) AU NOSY MANGABE

Eleanor J. Sterling

Lors d'une étude de terrain réalisée pendant deux ans sur l'environnement et le comportement du aye-aye (*Daubentonia madagascariensis*) sur l'île de Nosy Mangabe, j'ai récolleté des données sur l'accouplement, la proximité et le comportement social de six mâles et deux femelles. Sur Nosy Mangabe, le *Daubentonia* mâle n'utilise pas l'espace comme les autres primates nocturnes étudiés jusqu'à ce jour. Les domaines vitaux des mâles se chevauchent considérablement, contrairement à ceux des femelles. Ceux-ci étant très étendus, il est difficile pour les mâles de défendre une femelle isolée et son territoire. Ce facteur, ainsi que les résultats préliminaires des saisons asynchrones de reproduction, pourrait expliquer le fait que les mâles tendent à la polygynie et non à la monogamie, et qu'ils ne défendent pas les ressources à long terme. Cependant, certaines particularités de Nosy Mangabe, telles que sa surface restreinte, le manque de menaces prédatrices, le manque d'opportunités d'immigration et d'émigration, affectent vraisemblablement les espèces vivant sur cette île. Les résultats de cette étude montrent que les systèmes de répartition spatiale ne représentent pas de façon adéquate l'éventail des diversités dans les systèmes sociaux et les modes d'accouplement des différentes espèces; ils montrent également que ces critères ne devraient être considérés que comme des variables d'un premier ordre, déterminant les catégories de l'organisation sociale des primates nocturnes.

## SOCIO-ECOLOGIE DE L'*EULEMUR MACACO*: RESULTATS PRELIMINAIRES

Ian C. Colquhoun

Ce rapport décrit les résultats préliminaires d'une étude de terrain socio-écologique réalisée sur 4 groupes d'*Eulemur macaco macaco* vivant dans différents domaines du Massif d'Ambato. Durant cette étude, les 4 groupes étaient composés de 5 à 13 animaux; la taille moyenne du groupe était de 9,75. La répartition des sexes chez les adultes était de 1:1. Leurs domaines vitaux s'étendaient de 3,5 à 7,0ha, pour une surface moyenne d'environ 5,0-5,5ha. La densité de population des maki noirs du Massif d'Ambato était d'environ 200 animaux/$km^2$. Sur le plan social, les grands groupes étaient moins cohésifs que les petits et ils se sous-divisaient fréquemment. Le grand groupe que j'ai étudié, vivant dans une région forestière relativement perturbée, se divisa continuellement durant la période sèche: ce milieu écologique requérait une telle adaptation, que ce changement dans l'organisation sociale était prévisible. Nous avons pu également constater de nettes variations saisonnières dans le rythme d'activité et le régime alimentaire. Cependant, ces variations furent constatées chez tous les groupes étudiés, bien qu'elles ne soient pas liées à un type d'habitat particulier, comme cela était par contre le cas pour la taille des groupes et leur organisation sociale. *E. macaco* pourrait donc être considéré comme une espèce "généraliste" avec une niche écologique relativement étendue.

## DOMAINE VITAL ET REGIME ALIMENTAIRE CHEZ LE MAKI VARIE (*VARECIA VARIEGATA RUBRA*) SUR LA PENINSULE DE MASAOLA, MADAGASCAR

Marco M. Rigamonti

Pendant 11 mois, de décembre 1990 à novembre 1991, deux groupes de makis varies (*Varecia variegata rubra*) ont été observés dans les forêts primaires de la péninsule de Masaola. Des données sur le comportement par rapport au domaine vital, sur le régime alimentaire, l'activité, la composition et la dispersion du groupe, ont été enregistrées pour 8 animaux principaux. Les deux groupes, composés respectivement de 5 et 6 animaux, utilisaient des domaines vitaux de 23,3 et 25,8ha. La valeur moyenne du parcours journalier pour le groupe principalement étudié était de 436m. Pendant la saison fraîche et humide, les deux groupes se sub-divisèrent, se centralisant dans différentes endroits du domaine vital. Les endroits du domaine vital principalement utilisés changeaient de mois en mois. Les makis étaient plutôt frugivores (73,9%) et complétaient leur régime par des feuillages (20,9%) et des fleurs (5,3%). Pendant la période d'observation, les animaux principalement étudiés se nourrirent de 42 sortes d'arbres, dont ils préférèrent 7 variétés (72,5% des variétés recensées). Le domaine vital et son utilisation dépendaient de la disponibilité en grands arbres fruitiers.

## TRANSFEREMENT DES MALES CHEZ LE MAKI VARIE, *VARECIA VARIEGATA VARIEGATA*, EN CAPTIVITE

Frances J. White, Elisabeth A. Balko et ElisaBeth A. Fox

Depuis 1985, nous avons observé des groupes de makis noirs et blancs (*Varecia v. variegata*) en semi-liberté au Duke University Primate Center (DUPC). D'autres ont été observés en liberté au Parc National de Ranomafana dans le sud-est de Madagascar depuis 1988. En 1990, des clôtures séparant deux groupes de *V. variegata* établis au DUPC ont été déplacées. Dans chaque groupe, les mâles des rangs les plus inférieurs commencèrent par visiter le groupe voisin, puis ils émigrèrent de leur groupe natal vers l'autre groupe. Un des mâles émigra vers un groupe ne comprenant qu'un mâle et qu'une femelle solitaire et ne tarda pas à dominer l'ancien mâle résident. Le mâle qui émigra dans le groupe plus grand (6-8 animaux) passa progressivement à des rangs supérieurs et il s'accoupla avec les deux femelles résidentes adultes (mère et fille). Chez le *Varecia variegata* l'accouplement de la femelle avec différents mâles pourraît être une adaptation devant réduire les risques d'infanticide, tandis que les accouplements avec un mâle unique semblent être nécessaires à l'obtention de la protection paternelle (garde). Lors de certaines études de terrain, des mâles visitant d'autres groupes ont également été observés.

## PREDATION SUR LES LEMURIENS: IMPLICATIONS POUR L'EVOLUTION DU COMPORTEMENT DES PETITS LEMURIENS NOCTURNES

Steven M. Goodman, Sheila O'Connor et Olivier Langrand

Il est admis comme principe que la prédation est un facteur primordial dans le façonnage des comportements sociaux des mammifères. Cependant, les observations de prédateurs prélevant des primates sont rares et la plupart des cas rapportés sont relatifs à des primates diurnes de tailles relativement grandes. En ce qui concerne les lémuriens de Madagascar, peu d'informations quantitatives étaient disponibles et il a été considéré jusqu'à présent que les actes de prédation perpétrés par des carnivores ou des rapaces étaient rares. Contrairement à toute attente il s'avère qu'un grand nombre d'informations est disponible sur ce sujet, informations originaires de différentes sources et présenteées ci-après. Les éléments les plus précis sur la prédation se rapportent à *Microcebus murinus*. La dynamique de la population de cette espèce est étudiée en prenant en considération la pression de prédation importante exercée par deux espèces de rapaces nocturnes (*Tyto alba* et *Asio madagascariensis*), et particulièrement les implications de la prédation sur le comportement social et sur les aspects de l'histoire naturelle de cette petite espèce de primate.

## ADAPTATION ET CONTRAINTES PHYLOGENETIQUES DU COMPORTEMENT ANTIPREDATEUR DES MAKI CATTA ET VARIES

Joseph M. Macedonia

Au Duke University Primate Center (DUPC, Durham, NC, USA) j'ai enregistré pendant trois ans les réponses que les maki catta (*Lemur catta*) et variés (*Varecia variegata*) envoient à leurs prédateurs. Ces données comprennent les réponses vocales et non-vocales à des prédateurs naturels ou simulés; leurs fonctions ont été analysées d'après la taille physique, les habitudes écologiques et la biologie du système reproducteur des différentes espèces. Ce sont les réponses non-vocales qui différaient le plus chez ces deux espèces de lémuriens: tandis qu'en général le catta fuyait ses prédateurs, le maki varié, lui, semblait les affronter, voire même les attaquer. En ce qui concerne les réponses vocales, les variations interspécifiques reposaient sur la diversité des cris, la spécificité des stimuli et les fonctions attribuées aux réponses. Le comportement antiprédateur du maki catta (y compris la grande taille du groupe) est considéré comme une adaptation aux pressions, non-négligeables, des prédateurs, adaptation qui serait due à son assez petite taille, et au fait que c'est un primate semi-terrestre qui vit en terrain découvert. Contrairement au maki catta, le comportement antiprédateur agressif du maki varié serait dû, d'une part, à son assez grande taille, et d'autre part, à ses jeunes "sédentaires", c. à d. ne s'agrippant pas à leurs mères. Par conséquent, contrairement aux catta, les maki variés ne peuvent fuir leurs prédateurs sans mettre en danger leur succès reproducteur.

## TERRITORIALITE CHEZ LES GROUPES DE *LEMUR CATTA* A LA PERIODE DES NAISSANCES A BERENTY, MADAGASCAR

Alison Jolly, Hantanirina R. Rasamimanana, Margaret F. Kinnaird, Timothy G. O'Brian, Helen M. Crowley, Caroline S. Harcourt, Shea Gardner et Jennifer M. Davidson

Sept groupes voisins du *Lemur catta* ont été étudiés sur trois périodes de cinq jours: deux en 1990 et une en 1991 à la période des naissances, entre septembre et octobre. Les femelles menaçaient celles d'autres groupes en les fixant, en se ruant en avant et, occasionnellement, en les agressant physiquement. Ces confrontations semblent être une défense contre les intrus sur le territoire: en effet, les troupes se retiraient ensuite au centre de leur territoire. La répartition journalière moyenne (1377m, N=101) par rapport au diamètre du domaine vital (319m pour les polygones concaves) fait partie des plus élevées chez les primates, ce qui indique que les ressources qu'offre ce domaine méritent d'être défendues. Les répartitions sur 15 jours sont aussi élevées que sur 1 an. Les relations entre les groupes se divisent en deux catégories: les confrontations fréquentes (HC) (4 des couples voisins possibles, avec 1,0 confrontations/jour) et les confrontations rares (LC) (8 couples, avec 0,3 confrontations/jour). Ce sont les couples HC qui ont le plus de points communs: plus d'approches à courte distance, et un plus grand nombre d'approches aboutissant à des conflits. Les couples LC ont moins de zones de chevauchement: moins d'approches à courte distance, et moins d'approches aboutissant à des confrontations. Dans les deux cas, HC comme LC, les couples s'évitent. Les confrontations étaient précédées et suivies d'une grande vélocité des troupes qui se déplaçaient en angles, indiquant ainsi l'approche et la retraite. Les hurlements des mâles n'avaient pas d'effet direct sur les distances effectuées par les troupes. En 1990, les touristes donnaient des bananes aux lémuriens; les ressources alimentaires ont diminué considérablement en 1991, l'approvisionnement en nourriture s'étant effectué lors d'une année de sècheresse. En 1991, les couples HC et LC se révèlèrent être les mêmes que l'année précédente: l'approvisionnement en nourriture n'influença aucun des antagonismes. Nous pensons donc que chez le *L. catta*, la territorialité est indispensable aux femelles pour l'accès aux ressources nutritives (du moins pour les populations denses), que les grandes troupes ne gagnent pas nécessairement du terrain lors des confrontations, que l'agression ciblée des femelles est liée aux contraintes du système territorial, et que le mâle ne joue qu'un rôle bénin sur la distance entre les troupes durant la période des naissances.

## NOUVELLE INTERPRETATION DE L'ORGANISATION SOCIALE ET DU MODE D'ACCOUPLEMENT DES MAKI CATTA (*LEMUR CATTA*)

Michelle L. Sauther et Robert W. Sussman

Des études approfondies, réalisées à long terme sur des individus identifiés, nous ont permis de réexaminer la structure et l'organisation sociales du *Lemur catta*. Les makis catta ont des domaines vitaux qui se chevauchent considérablement, dont la taille varie selon les saisons et les régions, mais qui ne doivent pas être considérés comme strictement territoriaux. Les animaux de cette espèce vivent en groupes multi-mâles avec des femelles résidentes et une femelle dominante qui semble être le point d'attraction de tous les autres membres du groupe. La taille moyenne du groupe est de 13 individus (limites: 5-27) avec une répartition égale des sexes. Approchant de l'âge adulte, tous les mâles émigrent de leur groupe natal, et tous les mâles plus agés changent de groupe sur une période moyenne de 3 à 5 ans (bien que quelques mâles soient déjà restés dans le même groupe au moins 6 ans). Les mâles d'un même groupe se différencient par leur natalité, leur statut social et leurs relations avec les femelles. Parmis les groupes de notre étude approfondie, un seul mâle "central" immigré a été repèré. De tels mâles ont plus d'interactions avec les femelles, et ils sont les premiers à s'accoupler. Les femelles effectuent des choix dans leurs accouplements et refusent les mâles trop apparentés; elles s'accouplent avec les mâles n'appartenant pas à leur troupe, ce qui mène à de nombreuses stratégies d'accouplement chez les mâles.

## COMPORTEMENT NUTRITIF SELON L' ETAT PHYSIOLOGIQUE DES FEMELLES *LEMUR CATTA*

Hantanirina R. Rasamimanana et Elie Rafidinarivo

A la réserve de Berenty, dans le sud de Madagascar, la période des naissances correspond au début de la saison humide, celle de l'allaitement correspond à la pleine saison humide, celle de l'oestrus au début de la saison sèche, et celle de la gestation à la pleine saison sèche. Durant ces différents états physiologiques des femelles, la nourriture disponible varie considérablement. Nous avons comparé pendant 110 jours et sur une période de 20 mois le comportement nutritif, la nourriture et les saveurs à la disponibilité de deux groupes de *L. catta* vivant dans différentes parties de la réserve. Sur son territoire, la troupe H avait accès à de nombreuses plantations d'arbres variés ainsi qu'aux bananes offertes par les touristes. La troupe F vivait dans des espaces forestiers naturels et éloignés des touristes. La troupe H eut une nutrition plus variée, grâce à une plus grande diversification des ressources. Durant toute l'année, la troupe fut principalement folivore et montra moins de changements dans son régime alimentaire et son activité saisonnière. La troupe F s'alimenta moins et sa nutrition fut moins variée, et elle réduisit très nettement non seulement son activité, mais aussi la diversification de son régime alimentaire à la saison sèche; enfin, selon la saison, elle était plus frugivore que folivore. L'activité, la sélection alimentaire, ainsi que les "petits casse-croûtes" durant la saison des naissances pourrait avoir comme but d'optimiser le gain d'énergie. Il existe quelques signes indiquant un besoin d'équilibrer les composants alimentaires, mais il semblerait que, chez ces espèces s'adaptant facilement, ce soit la quantité qui prime. Cependant, le régime alimentaire plus limité de la troupe F ne se traduisit pas par un taux de reproduction moindre, puisqu'au contraire il augmenta de 50%, tandis que celui de la troupe H augmenta de 17%.

## COMPETITION POUR LES RESSOURCES CHEZ LES POPULATIONS DE MAKIS (*LEMUR CATTA*): CONSEQUENCES SUR LA DOMINANCE DES FEMELLES

Michelle L. Sauther

Le stress ressenti par la femelle chez le maki lors de l'adaptation de son cycle de reproduction à la disponibilité saisonnière des ressources peut être amplifié par son grand investissement maternel lors de sa pré- et post-maternité, ainsi que par un mode de vie sociale grégaire. Dans ce contexte, on peut considérer que, dans un environnement trés saisonnier, la dominance de la femelle est un comportement critique permettant aux femelles de co-exister avec plus d'un mâle non-natif du groupe. Les femelles peuvent tolérer l'appartenance des mâles au groupe au long de l'année étant donné que les mâles sont des sentinelles moins exigentes permettant de détecter les prédateurs et de défendre le groupe contre ceux-ci. De plus, accepter que les mâles apartiennent au groupe est peut-être une tactique de reproduction viable, tant pour la femelle que pour le mâle, d'autant plus que les femelles s'accouplent d'abord avec les mâles appartenant au groupe. C'est sans doute la préférence des femelles pour des mâles petits, ainsi que les effets écologiques et reproducteurs sur la taille du mâle qui ont conduit chez les makis au système actuel de dominance de la femelle.

## FLEXIBILITE ET CONTRAINTES ECOLOGIQUES CHEZ LE LEPILEMUR

Jörg U. Ganzhorn

En me basant sur le *Lepilemur mustelinus* comme représentant de la branche folivore chez les lémuriens, je cherche à savoir à quel point la compétition interspécifique potentielle avec l'*Avahi laniger* et l'adaptation à un habitat de plus en plus restreint affectent respectivement la sélection du feuillage et l'utilisation de l'habitat. La qualité des feuilles dont se nourrissent les *Lepilemur* a été étudiée dans quatre régions différentes de Madagascar. La qualité du feuillage était définie par le rapport entre sa teneur en protéines et celle en fibres. Dans les forêts où les *Avahi* ne vivaient pas, les *Lepilemur* se nourrissaient de feuillages de haute qualité; et inversement, dans celles que les *Avahi* habitaient, les *Lepilemur* consommaient des feuillages de basse qualité et de qualité inférieure à ceux consommés par les *Avahi*. A part la présence des *Avahi*, aucune autre cause n'a pu être constatée en ce qui concerne ce choix de feuillage. Ceci peut être considéré comme une perturbation caractérielle. Les bosquets d'arbres denses, requis par leur locomotion, font la particularité des microhabitats des *L. m. ruficaudatus*. Le *Lepilemur* tolère jusqu'à un certain point une certaine dispersion des arbres entre les roches et utilise des microhabitats attestant une plus grande distance entre les arbres. Cependant, dans les microhabitats des *Lepilemur*, la distance entre les arbres reste constante quand la distance moyenne entre les arbres augmente à cause des roches. Ceci indique donc une limite due aux dépenses d'énergie que les *Lepilemur* doivent fournir pour couvrir la distances entre les arbres dont ils ont besoin pour se déplacer. Le regroupement des arbres nourrissiers s'accentue selon la taille des groupes: des *Lepilemur* solitaires, aux *Avahi* vivant en couple, jusqu'aux *Propithecus verreauxi* grégaires. Les modèles établis à l'aide de la théorie d'optimisation avaient prévu ces trois résultats obtenus. Ceci laisse supposer que la qualité du feuillage et la structure de l'habitat limitent les ressources du *Lepilemur* et que la compétition interspécifique sur la qualité du feuillage exerce une pression sur la branche folivore des lémuriens malgaches.

## CORRELATION ENTRE PARAMETRES ECOLOGIQUES ET DE REPRODUCTION AVEC L'UTILISATION DES DOMAINES VITAUX DES MAKIS A VENTRE ROUX (*EULEMUR RUBRIVENTER*) ET LES MAKIS BRUNS (*EULEMUR FULVUS RUFUS*)

Deborah J. Overdorff

Dans cette étude, je décris l'influence de variables écologiques (régime, disponibilité en nourriture, caractéristiques des plants) et de variables indépendantes de la disponibilité en nourriture, telles que la reproduction et la territorialité, sur le domaine vital journalier et à long terme de deux espèces de primates du sud-est de Madagascar: les makis bruns (*E. fulvus rufus*) et les makis à ventre roux (*E. rubriventer*). Nous avons procédé à une étude comparative de la distance journalière parcourue (DPL), la surface du domaine vital et la distance parcourue entre les plants de végétation chez ces deux espèces de lémuriens. Journellement ou à long terme, les makis bruns se déplaçaient plus que ceux à ventre roux. Les domaines des différents groupes se chevauchaient considérablement et les makis bruns ne défendaient pas les frontières de leur territoire. Par contre, les makis à ventre roux utilisaient exclusivement leur domaine vital et défendaient activement leurs frontières. De plus, les makis bruns préféraient les plants disséminés et se déplaçaient donc plus entre les plants que les makis à ventre roux. Chez ces deux espèces, l'utilisation du domaine vital ne correspondait pas nécessairement à la nourriture disponible, elle était plutôt influencée par les trois variations saisonnières dans leur régime nutritif: le nombre de plants visités, la diversité du régime et le nombre de repas. C'est lorsque ces trois variables étaient le plus élevées et qu'elles coïncidaient avec la lactation des femelles et que les makis bruns se déplaçaient le plus. Je n'ai pu observer une telle pointe saisonnière dans l'utilisation du domaine vital chez les makis à ventre roux; cependant, ils se déplaçaient plus loin lorsque leur régime était plus diversifié. Par conséquent, je suppose que, chez le maki brun, la présence de quelques femelles en phase de reproduction influence le choix de nourriture, et par conséquent le choix du domaine vital. Par contre, il semblerait que chez le maki à ventre roux, ce soit la territorialité qui limite les mouvements et les contraintes journalières et saisonnières du groupe.

## DEPISTAGE DES RESSOURCES: NOURRITURE DISPONIBLE ET SAISONALITE DE LA REPRODUCTION CHEZ LES *PROPITHECUS*

David M. Meyers et Patricia C. Wright

Afin d'examiner les causes ultimes de la saisonalité strictement reproductive que l'on trouve chez le primate malgache, nous avons comparé l'environnement alimentaire et le rythme de reproduction des deux espèces de *Propithecus* dans la forêt sèche et humide. Nous avons analysé, sur un cycle d'un an, les relations entre la phénologie, le comportement nutritif et la recherche de nourriture chez le propithèque à tête jaune de Tattersall (*Propithecus tattersalli*) et le propithèque diadème de Milne-Edwards (*P. diadema edwardsi*). Nous avons plus particulièrement décrit le dépistage et le cycle de reproduction de *P. tattersalli* dans les forêts sèches du nord, puis vérifié si le *P. diadema edwardsi*, qui vit dans les forêts humides du sud-est, se comportaient de façon similaire. Les *P. tattersalli* recherchent tellement les feuillages jeunes, que, lorsque ceux-ci apparaissent, ils se concentrent sur ce type d'alimentation sans même faire cas de l'abondance en denrées premières telles que les graines et les feuillages arrivés à maturité. Nous avons analysé les relations entre la nourriture disponible, le régime alimentaire, et la recherche de nourriture à l'aide des matrixes de corrélation. Chez *P. tattersalli*, la consommation de feuillages jeunes co-variait de façon positive avec leur disponibilité et le parcours journalier (DPL). C'est entre la disponibilité en feuillages jeunes et le parcours journalier que la corrélation était le plus positif, et entre l'absorption de feuillages arrivés à maturité et le parcours journalier qu'elle était le plus négatif. Les feuillages jeunes abondaient au début de la saison humide, et c'est à cette époque que les *Propithecus* les consommaient le plus. Chez les *Propithecus diadema edwardsi*, la fréquence de consommation de feuillages jeunes ainsi que le temps d'absorption augmentaient également durant les périodes d'abondance de ces feuillages. La période de consommation de feuillages jeunes et de leur abondance était similaire dans les deux régions. Chez les deux espèces de *Propithecus*, les enfants naissaient entre juin et juillet et étaient sevrés en novembre et décembre. Chez les mammifères, la période de lactation exige une nutrition plus importante. Etant donné que les propithèques sont des folivores de par leur anatomie, il se pourrait que les femelles en lactation, ayant besoin d'une nutrition plus importante, recherchent une nourriture plus riche et plus abondante. Chez ces deux espèces, la période de consommation de feuillages jeunes la plus importante correspondait à la fin de la lactation et du sevrage. De légères différences entre les deux espèces dans la saisonalité de leur comportement nous amènent à supposer que, même si la période de reproduction est programmée de sorte que la lactation et le sevrage coïncident avec la période d'abondance en feuillages jeuneses, la saisonalité des ressources n'est pas la seule cause ultime de la saisonalité de la reproduction chez le propithèque.

## RELATION ENTRE LES CHANGEMENTS COMPORTEMENTAUX SAISONNIERS ET LA THERMOREGULATION CHEZ LE MAKI VARIE (*VARECIA VARIEGATA VARIEGATA*)

Hilary Simons Morland

Les variations saisonnières du comportement ont été analysées chez toutes les espèces de lémuriens de Madagascar et attribuées à la saisonalité de l'environnement, très prononcée sur toute l'île. Bien qu'il existe des différences de précipitation selon les régions, la température semble varier de façon assez fiable selon les saisons sur tout Madagascar. Les données obtenues lors d'une étude de terrain réalisée sur les makis variés noirs et blancs (*V. v. variegata*) dans le nord-est de Madagascar ont montré que ces makis réduisent leur activité et leurs parcours, et augmentent leur nutrition, leur temps d'ensoleillement et de repos sur les branches durant les mois les plus frais de l'année. Il semble que les lémuriens réagissent plus aux cycles de température annuels par des mécanismes comportementaux et sociaux que purement physiologiques. La température ambiante est vraisemblablement une déterminante majeure dans les changements de comportement saisonniers chez les lémuriens. Ces hypothèses portent à conséquence dans l'évolution de l'organisation sociale des lémuriens. Le but de cet article est d'attirer l'attention sur la relation entre la thermorégulation et le comportement des lémuriens et d'émettre des hypothèses qui mériteraient d'être étudiées avec précision.

## ADAPTATION SAISONNIERE DE LA VITESSE DE CROISSANCE ET POIDS ADULTE CHEZ LE MAKI CATTA

Michael E. Pereira

Des signes photopériodiques fournissent aux animaux des régions non-équatoriales des repères qui leur permettent d'adapter leur métabolisme, leur taux de développement, leur reproduction et leur comportement. Cette étude a révèlé des adaptations de la vitesse de croissance et du poids adulte chez le maki (*Lemur catta*) vivant en semi-liberté, qui correspondaient à la vitesse de croissance des poils, au comportement thermorégulateur et social, et ce, indépendamment des changements de disponibilité en nourriture. Le poids des jeunes augmentait de 3 à 12 g/jr, accélérant ainsi leur vitesse de croissance annuellement après les solstices d'été. Les mâles et les femelles adultes prenaient du poids un ou deux mois après eux, juste avant l'arrêt de croissance à long terme des jeunes. Ce sont les corrélations observées lors de ces résultats, ainsi que le plan des précipitations à Madagascar qui m'ont amené à établir un nouveau modèle de stratégie démographique chez les lémuriens de grande taille. La saisonabilité, pénible, mais prévisible des ressources alimentaires semble avoir amené ces primates à évoluer en fonction de l'emmagazinement d'énergie, tout en supprimant le développement du métabolisme pendant la plus grande partie de la saison sèche à Madagascar. Cette tactique serait analogue aux divers états de torpeur qui engendrent la survie et la reproduction des mammifères non-primates vivant dans des milieux saisonnaux rigoureux. Cependant, dans le cas d'une stratégie démographique générale chez les primates grégaires, les taux de croissance des enfants devraient être exceptionnellement élevés par rapport à la taille des mères. Comme beaucoup d'autres parents, chez les lémuriens, les mères doivent investir énormément pour aider leur progéniture à atteindre un certain seuil de croissance: les jeunes lémuriens doivent grandir beaucoup et sans doute emmagaziner de l'énergie avant la saison sèche, afin d'être en mesure de réduire sans risque leur métabolisme et de grandir lentement pendant la saison sèche. Les exigences liées à la lactation sont susceptibles d'avoir contribué à l'évolution générale de la dominance de la femelle, ainsi que de l'infanticide des mâle chez les lemuridae et les indridae. D'autres recherches sur les stratégies d'évolution basée sur l'environnement seront nécessaires à une compréhension plus approfondie de l'histoire naturelle des primates.

## SELECTION SEXUELLE ET SYSTEMES SOCIAUX DES LEMURIENS

Peter M. Kappeler

Chez les lémuriens de Madagascar, le manque de dimorphisme sexuel de taille affecte les relations agonistiques entre mâles et femelles et contribue à cette caractéristique remarquable du système social des lémuriens: la dominance de la femelle sur le mâle. Ce chapitre est destiné à l'analyse de la nature et des mécanismes de la sélection intrasexuelle chez les lémuriens, et il tentera d'éclaicir le rapport entre la sélection sexuelle et les différences des sexes dans la relation de dominance. Je commence par vérifier l'hypothèse selon laquelle le manque de dimorphisme sexuel de taille des lémuriens polygynes résulterait du coût élevé de viabilité. Cette hypothèse suppose que les mâles des lémuriens polygynes sont pourvus de canines beaucoup plus grandes que celles des femelles dans le cas où le mâle doit, par ses combats, jouer un rôle dans la diversification de la reproduction. Cependant, ce n'est que chez quatre espèces polygynes que l'on trouve des canines biaisées, indiquant un dimorphisme sexuel chez les mâles. L'inverse est vrai chez le *Propithecus diadema*, et, d'après la taille des canines, il n'existe pas de différences sexuelles significatives chez les huit autres espèces. De plus, d'après les canines, le degré moyen du dimorphisme sexuel ne montre aucune hétérogénéité chez les lémuriens ayant des modes d'accouplement différents. Par conséquent, le manque général de dimorphisme sexuel de taille chez les lémuriens polygynes n'est pas dicté par leur viabilité. Il reflète même plutôt une faible proportion des critères liés au combat du mâle. L'habituelle absence de dimorphisme sexuel d'après la physionomie et les canines pourrait également être conciliable avec la théorie sur la sélection sexuelle, si la compétition mâle chez les lémuriens polygynes était post-copulatoire, c'est à dire si elle prenait la forme d'une compétition de sperme. Cependant, contrairement à cette hypothèse, j'ai constaté que la taille moyenne des testicules des lémuriens solitaires, vivant en couples ou en groupes ne variait pas pendant la période de reproduction. Enfin, ces analyses comparatives laissent à supposer que l'intensité de la sélection intrasexuelle est similaire chez les lémuriens monogames et polygynes. La faible variance dans le succès reproducteur du mâle chez les espèces non-monogames pourrait être due à la pénible compétition de polygynie chez quelques espèces solitaires, ainsi qu'à l'existence de couples "mâle-femelle" chez certaines espèces grégaires. La dominance des femelles apparaît chez les espèces où les mâles sont mieux armés, mais ceci n'est pas une caractéristique communes à tous les lémuriens polygynes. Ainsi, il est possible que la faible sélection intrasexuelle ait facilité la dominance de la femelle, cependant, cette hypothèse ne suffit pas à expliquer ce phénomène.

## CRITERES REPRODUCTEURS ET DEMOGRAPHIQUES, PERIODE D'ACTIVITE ET SYSTEMES SOCIAUX DES LEMURIENS

Carel P. van Schaik et Peter M. Kappeler

Chez les primates lémuriformes et anthropoïdes encore vivants, les comportements sociaux, les critères de regroupement et de liaison, les structures sociales et les dimorphismes sexuels sont extrêmement différents. Nous avons procédé à une étude comparative des critères communs aux lémuriens, et nous suggérons des critères suceptibles d'expliquer ces multiples différences dans leur organisation sociale. Les lémuriens ont trois types de structure en commun: les individus solitaires, ceux vivant en couples et ceux vivant en groupes. Chez quelques espèces, il arrive que des transitions aient lieu entre les couples et les groupes, même parmis les populations cathémérales d'*Eulemur* et de *Varecia*. Chez les plus grands groupes de ces espèces, les sexes sont répartis de façon égale chez les adultes, et les regroupements ainsi que le comportement social des individus laissent supposer que ces lémuriens vivent en couples "mâle-femelle" stables. On observe également un nombre égal d'adultes de sexe différent chez les espèces complètement diurnes *Lemur* et *Propithecus*, mais aucun signe de vie en couple. Chez les lémuriens, ces différences sont probablement dues aux différentes périodes d'activité des espèces cathémérales dont les groupes sont supposés se déplacer par couple durant l'activité nocturne. Le couple stable est donc l'unité de base dans beaucoup de groupes de lémuriens. Ceci est un cas particulier parmis les associations permanentes entre mâle et femelle rencontrées chez la majorité des primates. Nous supposons que cette évolution vers des associations permanentes s'est faite dans le but de prévenir l'infanticide: le mâle peut en effet ainsi protéger les femelles contre les tentatives d'infanticides des autres mâles. Dans le cas où les femelles vivent, en permanence ou temporairement, seules (espèces cathémérales), cette association prend la forme du couple stable. Lorsqu'elles vivent en groupe, elle prend la forme de groupes constamment bisexuels. Ceci nous amène à supposer que l'association mâle-femelle n'existe pas chez les espèces où les soins maternels n'exigent pas d'absences temporaires. Cette supposition est en effet confirmée: la variation interspécifique en ce qui concerne les soins maternels (porter/cacher l'enfant) co-varie parfaitement avec les critères communs, tant chez les prosimiens que chez les autres espèces de primates. Nous en concluons que la période d'activité, de même que les soins maternels sont déterminants pour les critères de regroupement et d'accouplement, tant pour les sociétés de prosimiens que pour celles des anthropoïdes. Il est probable que ce soit une précocité croissante qui ait rendu possible les activités cathémérales et le port des enfants chez les primates.

## INDEX

MIX
Papier aus verantwortungsvollen Quellen
Paper from responsible sources
FSC® C105338

If you have any concerns about our products,
you can contact us on
**ProductSafety@springernature.com**

In case Publisher is established outside the EU,
the EU authorized representative is:
**Springer Nature Customer Service Center GmbH**
**Europaplatz 3, 69115 Heidelberg, Germany**

Printed by Libri Plureos GmbH
in Hamburg, Germany